THE CENOZOIC ERA
Tertiary and Quaternary

ELLIS HORWOOD SERIES IN GEOLOGY

Editor: D. T. DONOVAN, Professor of Geology, University College London

This series aims to build up a library of books on geology which will include student texts and also more advanced works of interest to professional geologists and to industry. The series will include translations of important books recently published in Europe, and also books specially commissioned.

FAULT AND FOLD TECTONICS
W. JAROSZEWSKI, Department of Geology, University of Warsaw

A GUIDE TO CLASSIFICATION IN GEOLOGY
J. W. MURRAY, Professor of Geology, University of Exeter

THE CENOZOIC ERA
C. POMEROL, Professor, University of Paris VI.
Translated by D. W. HUMPHRIES, Department of Geology, University of Sheffield, and E. E. HUMPHRIES. Edited by Professors D. CURRY and D. T. DONOVAN, University College London

INTRODUCTION TO PALAEOBIOLOGY: GENERAL PALAEONTOLOGY
B. ZIEGLER, Professor of Geology and Palaeontology, University of Stuttgart, and Director of the State Museum for Natural Science, Stuttgart

BRITISH MICROPALAEONTOLOGICAL SOCIETY SERIES

This series, published for the British Micropalaeontological Society, will gather together knowledge of a particular faunal group for specialist and non-specialist geologists alike. The scope of the series has been broadened to include the common elements of the fauna, whether index or long-ranging species, and to convey a broad impression of the fauna and allow the reader to identify common species as well as those of restricted stratigraphical range.
The synthesis of knowledge presented in the series will reveal its strengths and prove its usefulness to the practising micropalaeontologist, and those teaching and learning the subject. By identifying some of the gaps in the knowledge, the series will, it is believed, promote and stimulate further active research and investigation.

STRATIGRAPHICAL ATLAS OF FOSSIL FORAMINIFERA
Editors: G. JENKINS, The Open University, and J. W. MURRAY, Professor of Geology, University of Exeter

MICROFOSSILS FROM RECENT AND FOSSIL SHELF SEAS
Editors: J. W. NEALE, Professor of Micropalaeontology, University of Hull, and M. D. BRASIER, Lecturer in Geology, University of Hull

FOSSIL AND RECENT OSTRACODS
Editors: R. H. BATE, British Museum of Natural History, London, E. ROBINSON, Department of Geology, University College London, and L. SHEPPARD, British Museum of Natural History, London

A STRATIGRAPHICAL INDEX OF CALCAREOUS NANNOFOSSILS
Editor: A. R. LORD, Department of Geology, University College London

THE CENOZOIC ERA
Tertiary and Quaternary

Charles Pomerol
Professor, University of Paris VI

Translated by:
Dr. DEREK W. HUMPHRIES
Department of Geology
University of Sheffield
and EVELYN E. HUMPHRIES

Edited by:
Professors D. CURRY and D. T. DONOVAN
University College London

ELLIS HORWOOD LIMITED
Publishers · Chichester

Halsted Press: a division of
JOHN WILEY & SONS
New York · Brisbane · Chichester · Toronto

published in 1982 by
ELLIS HORWOOD LIMITED
Market Cross House, Cooper Street, Chichester,
West Sussex, PO19 1EB, England
*The publisher's colophon is reproduced from
James Gillison's drawing of the ancient Market
Cross, Chichester.*

Distributors:
Australia, New Zealand, South-east Asia:
Jacaranda-Wiley Ltd., Jacaranda Press,
JOHN WILEY & SONS INC.,
G.P.O Box 859, Brisbane, Queensland 40001,
Australia
Canada:
JOHN WILEY & SONS CANADA LIMITED
22 Worcester Road, Rexdale, Ontario, Canada
Europe, Africa:
JOHN WILEY & SONS LIMITED
Baffins Lane, Chichester, West Sussex, England.
North and South America and the rest of the world:
Halsted Press: a division of
JOHN WILEY & SONS
605 Third Avenue, New York, N.Y. 10016, U.S.A.

This English edition is translated from the author's
original French edition Ère Cénozoique published in 1973
by DOIN Éditeurs©, Paris, the copyright holders.

© Ellis Horwood Limited Publishers 1982
British Library Cataloguing in Publication Data
Pomerol, Charles
The Cenozoic era. –   (Ellis Horwood series in geology)
1. Geology, Stratigraphic – Cenozoic
I. Title   II. Curry, D.
III. Donovan, D. T.   IV. Ere cenozoique.
English
551.7'8   QE690   80–42073

ISBN 0-85312-256-3 (Ellis Horwood Ltd., Publishers)
ISBN 0-470-27140-X (Halsted Press)

Printed in Great Britain by R. J. Acford, Chichester.

Author's Preface

As a result of intensive research by both academic and commercial groups there has, in recent years, been a great increase in the understanding of the stratigraphy, correlation and palaeogeographical relations of rock sequences throughout the world. The new information so acquired, which has been disseminated during the proceedings of many International Colloquia and documented in large and growing numbers of specialist papers, has raised the need for a compact but comprehensive overview of present-day knowledge and theory which is at the same time comprehensible by and accessible to students, teachers, research workers and any others having a special interest in Earth History. This book has been written to fill that need. Fundamental principles are examined in the light of the most recent data, and the book includes a large series of illustrations and tables, most of which are new, accompanied by commentaries by specialists; the whole providing valuable source-material for teachers from high-school to university levels. The present volume is translated from the third of a series in the French language, the others being "Precambrian and Palaeozoic (1977)" and "Mesozoic (1975)". A fourth volume, also in French, deals with the principles and methods of Stratigraphy and Palaeogeography.

Inevitably the main emphasis has been on Western Europe, but the remainder of Europe and North Africa are covered in some detail and summary accounts deal with events in the rest of the world. The scope of stratigraphy and palaeogeography are today so wide, however, that even within this relatively restricted brief I have found it necessary to ask for help from many co-workers, both French and foreign. They have willingly provided me with information and figures, and have read and corrected drafts. There are too many of these collaborators to be listed here, but their names appear in the body of the work where appropriate. I am very grateful for all their assistance, without which this book would never have seen the light of day. I would also like to thank my immediate collaborators, Mme. CAMUS, who checked the manuscript and prepared the indexes, M. PERREAU, who chose the molluscs, and M. FAY, who photographed them.

The need to contain this book within a reasonable length has made it impossible to provide an exhaustive bibliography but nevertheless it contains several hundred references, including a list of important reference works. I hope that my co-workers will forgive any omissions, whether deliberate or accidental. Many of the ideas set out here were originated by them. However I must accept responsibility for the opinions expressed when dealing with controversial topics. In such cases I have endeavoured to summarise opposing arguments in the text, but the need to prepare definitive tables, for instance, has meant that I have finally had to take a stand in favour of one point of view, rather than another.

Following upon the studies of classical macro- and micropalaeontology, the new fields opened up by the electron microscope and the mass spectrometer have provided additional means of dating and have called into question a number of correlations which were thought to be fully established. For the good scientist, the recognition of such errors, even when they have been demonstrated by others, is by no means a threat to his self-esteem but rather a challenge to fresh research. For the knowledge of the precise age of a formation and its rigorous correlation with sequences deposited at the same period in other parts of the world form the foundation on which reconstructions of palaeogeography and structure are based, and the touchstone for modern hypotheses of global tectonics.

Table of Contents

Late Eocene planktonic foraminiferid: *Cribrohantkenina inflata* of the Yazoo formation, Mississippi, x 1000. See also the photograph of the complete foraminiferid in Figure 2.2, p. 48 (photo: Bolli).

Introduction: The Cenozoic Era | 1

The Cenozoic era (from Greek *kainos*; recent) encompasses the history of the earth from the end of the Cretaceous to the present day, about 65 million years (Fig. 1.1).

It is the shortest of the geological eras. It has not lasted as long as some periods of other eras (Cambrian or Cretaceous – 70Ma); it is equal in length to the Carboniferous (65Ma) and it is only a little longer than the Ordovician (60Ma). Its duration is only one fifth of that of the Palaeozoic era and rather more than a third of that of the Mesozoic era.

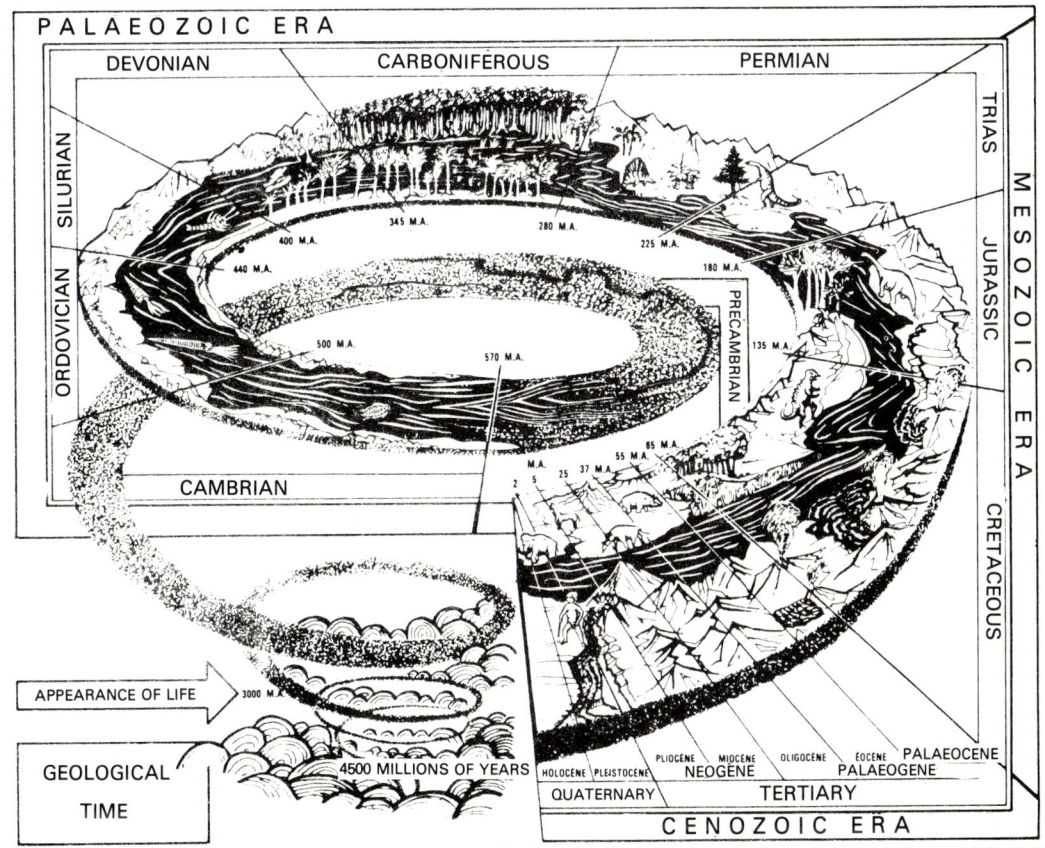

1.1 **Geological time: the succession of eras and periods illustrated by some outstanding events in World history.** The time scale is not uniform but has been exaggerated for the most recent periods. Some dates are still very approximate. (Drawn by M. Petzold from an illustration by the U.S. Geological Survey, Washington).

The maximum thickness of Cenozoic deposits is about 30km, a little less than those of the Mesozoic (35km). The Tertiary era witnessed the uplift of the Alpine chains and also the beginnings of their destruction, with the formation of abundant clastics and, in stable areas, important carbonate deposits. The palaeogeography during the Cenozoic gradually approached the present-day geography; the opening of the North Atlantic was completed while the South Atlantic continued to widen. But it was only in the Pliocene that the Mediterranean took on its familiar outline.

The animals and plants also foreshadowed those of the present day. The only great explosion was that of the monocotyledons in the vast, rich grasslands where the rapidly evolving mammals browsed. This evolution is comparable in many ways with that of the reptiles in the Mesozoic. But the comparison stops there for one of the evolutionary branches of the mammals has led to the appearance of Man and if one day – which will mark the end of the Cenozoic – the mammals must disappear from the face of the earth, they will owe this to mental disequilibrium rather than to physical overdevelopment.

I. – SUBDIVISIONS AND LIMITS

The Cenozoic era is generally divided into three periods: **Palaeogene** or **Nummulitic, Neogene** and **Quaternary** (Table I). The Quaternary, because of its relatively short duration (some 2Ma), is hardly separable from the Tertiary, which comprises the Palaeogene and Neogene periods.

The terms *"Tertiary era"* (introduced by BRONGNIART in 1810 to describe the beds which followed the Chalk) and *"Quaternary era"* are still in common use, though they have been condemned by International Geological Congresses. In fact, the Quaternary era, characterised by the development of Man and a succession of glaciations is, palaeogeographically speaking, no more than an extension of the Neogene into the present day.

The two periods of the Tertiary are divided into epochs. Initially, Charles LYELL in 1833 subdivided the Tertiary era into three parts, the **Eocene,** the **Miocene** and the **Pliocene,** though he still called the Quaternary by the older name *"Diluvium"*. In 1854, BEYRICH introduced the term **Oligocene** for the upper part of the Eocene s.s. and SCHIMPER in 1874 renamed the lower part of the Eocene, the **Palaeocene.** At about the same time HOERNES (1853) grouped the Miocene and Pliocene into the Neogene, while NAUMANN (1866) grouped the three earlier

epochs into the Palaeogene. This latter unit was, later, called the Nummulitic by RENEVIER in 1873.

The Quaternary (DESNOYERS 1829) is subdivided into two epochs. LYELL (1839) called the major part of the period the Pleistocene and GERVAIS (1867) named the period of about 10,000 years, embracing late and post-glacial times, the **Holocene** (Table I).

The upper limit of the Cenozoic is the most precise of all stratigraphic boundaries for it corresponds to the present instant. But the "present" of today is 86,400 seconds after the "present" of yesterday; in other words, the present is always moving. By convention, the "present" began in 1950 and it is from this date that we indicate time with the initials B.P. (before the present).

We cannot, of course, predict the foundations on which a new era may start, but let us hope that it may not be, once more, "azoic".

By contrast, the lower limit is a fruitful source of discussion, of a kind which is dear to stratigraphers and to some extent their "raison d'être". Geologists have long believed in the existence of **a major break between the Mesozoic and Cenozoic.** At this moment the giant reptiles, the ammonites, the belemnites and the rudists all disappeared and there is an

Table I
Principal subdivisions of the Cenozoic Era
The age of the limits of the Oligocene is uncertain
The names for tectonic phases are taken from the Mediterranean region
except for Laramide (from Laramie, Wyoming in the Rocky Mountains)
and Pasadenean (from Pasadena, California)

Systems or Periods		Series or Epochs	Duration in millions of years	Age of commencement of unit (Ma)	Tectonic phases (Alpine orogeny)
QUATERNARY		Holocene (*holos:* wholly *kainos:* recent)	0.01 (10000 yr)		
		Pleistocene (*pleistos:* most)	2		
TERTIARY	NEOGENE (*neos:* new *genos:* birth)	Pliocene (*pleios:* more)	3	— 2 —	*Valache = Valaque* or *Pasadenean*
		Miocene (*meios:* less)	18	— 5 —	*Rhodanian* *Attic* *Styric*
	PALAEOGENE or NUMMULITIC	Oligocene (*oligos:* few)	13	— 23 —	*Savic* *Helvetic*
		Eocene (*eos:* dawn of)	19	— 36 —	*Pyreneo-Provencal* *Pre-pyrenean* or *Illyric*
		Palaeocene (*palaeos:* ancient)	10	— 55 —	
CRETACEOUS				— 65 —	*Laramide* or *Arvinchean*

abrupt change of facies with the widespread appearance of shelly neritic deposits. On the continental margins there were new transgressions following the major regression at the end of the Cretaceous. From a tectonic viewpoint the limit is marked by the end of the Laramide orogeny and the beginnings of the Alpine orogeny.

Of course, a discontinuity is not a boundary because the former necessarily corresponds to a certain period of time. And, in fact that controversial stage, the **Danian,** which was created in 1846 by DESOR in Denmark to designate the bryozoan limestones which rest on the Maastrichtian chalk at Faxöe (Fakse) and at Stevns Klint to the south of Copenhagen would appear to be continuous with the Cretaceous, despite a slight discordance (Fig. 2.1, p. 28). Nevertheless, ammonites, belemnites, inoceramids and *Globotruncana* are now absent, while echinoderms, bryozoans and brachiopods are more like those of the Montian (see below). In addition, *the first characteristic Cenozoic planktonic organisms* appear in the Danian. These include the foraminiferids *Globorotalia danica* and *G. pseudobulloides* and the coccolith *Markalius astroporus* (see Table II, p. 22). Furthermore, the Danian deposits are not followed by a regression but continue upwards into the overlying Tertiary marls (Selandian stage). **It seems preferable, therefore, to place the Danian at the base of the Cenozoic.** Its upper part is equivalent to the Montian of Belgium which is, indisputably, a Tertiary stage.

II. – PALAEONTOLOGY (Fig. 1.2)

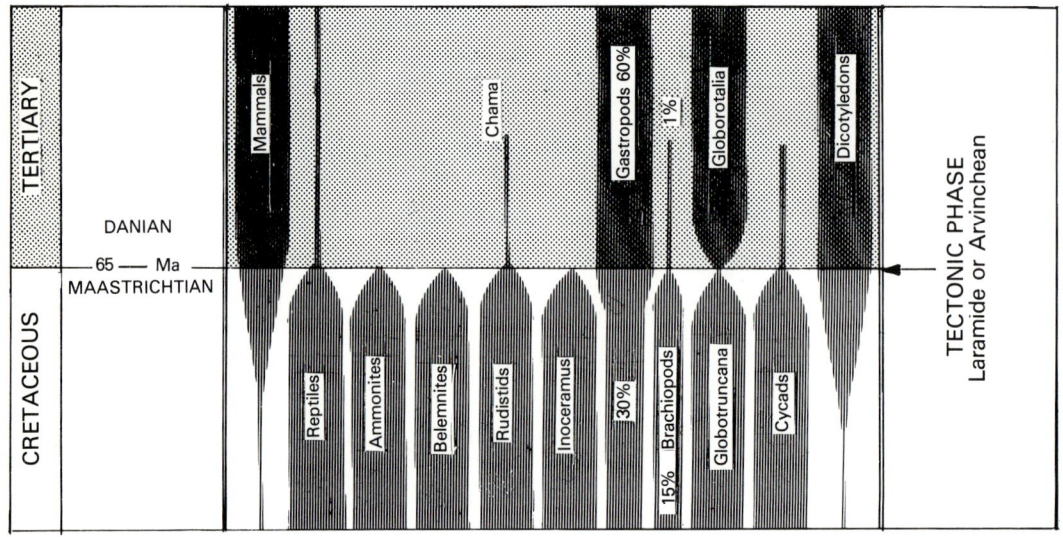

1.2 **Principal palaeontological events at the Cretaceous-Tertiary boundary.**

The disappearance of many groups of organisms during the late Cretaceous truly marked the end of an era. On land, the Cycads died out almost entirely, while the Dicotyledons flourished. Finally, the Monocotyledons, which had appeared at the end of the Jurassic but had remained insignificant, became abundant in the Neogene, when the grasses (Gramineae) proliferated.

The great reptiles which had dominated the world in Mesozoic times vanished abruptly and during a brief period of ten million years (the Palaeocene) the mammals began to establish themselves prior to their great diversification at the beginning of the Eocene. With their increase in size, especially in the Neogene, the mammals dominated the land surface, and it was only in the Quaternary that they were finally supplanted by the last evolutionary branch to which they had given birth, the hominids.

In the seas, the changes in the fauna were no less spectacular. While certain groups disappeared completely (belemnites, ammonites, inoceramids, rudists, globotruncanids, rugoglobigerinids) those which survived were profoundly changed. Among the coccoliths only five Cretaceous forms seem to have survived, while the discoasters made their first appearance. *Globorotalia, a planktonic foraminiferid appeared*, while the genus *Globigerina* was revitalized, though early Cenozoic forms are of small size and simple structure and bear little or no relation to those of the Upper Cretaceous.

This great biological upheaval, the most pronounced in the Earth's history, was probably a result of *a lowering of world temperature*, due either to a cosmic event or to a temporary connection between the North Atlantic and the Arctic Ocean, accompanied by *palaeogeographic changes*: the acceleration of sea-floor spreading, deepening of the Atlantic, and the separation of the Americas, allowing carbonate-poor Pacific waters to penetrate into the Atlantic Ocean. As well as the direct consequence of the lowering of temperature on the growth and reproduction of poikilothermic animals such as the reptiles, these phenomena profoundly disturbed the food chains. The reduced abundance of plankton resulted, for example, in the extinction of some groups of pelagic molluscs.

The **giant reptiles of the Cretaceous** show evidence which suggests that they **had attained at least partial homothermy** (constant body temperature): both anatomical (bone structure, pillar-like limbs supporting the body from beneath, like those of the elephant, rather than projecting to the side as in the crocodile) and physiological (living in groups and the need to produce large quantities of energy). It is believed that *the absence of hair or feathers* which would have protected them from the cold, and their *inability to hibernate* because of their great size, rendered them particularly vulnerable to a marked lowering of temperature.

The definitions of the Cenozoic stages by the older authors were based largely on the echinoderm, gastropod and bivalve faunas. The molluscs were extremely abundant; more than 1000 species have been recorded in the Lutetian at Grignon, near Paris.

By comparison with the Mesozoic, the proportion of gastropods increases from 30% to 60% of the shelly fauna, while the cephalopods fall from 20% to 1% and the brachiopods from 15% to 1%. Although the molluscan faunas, and especially the Miocene pectens, are still of great value, progress in Cenozoic stratigraphy must in future rest mainly on the study of the microfauna. In this context the word microfauna means not only the Protozoa and in particular, the foraminiferids, but also the ostracods (Crustacea), pollen, dinoflagellates, small mammals (the teeth of insectivores and rodents) and especially planktonic micro-organisms. The latter include foraminiferids such as the Globigerinidae and the Globorotaliidae (Fig. 2.22, p. 48) and the nannoplankton, a term which includes the coccoliths and some organisms which it is impossible to attach to a specific group (Fig. 2.24, p. 50). The scanning electron microscope with its high magnification and resolution has given, over the past decade, great impetus to the study of these organisms. Formerly, the larger foraminiferids, such as the *nummulites* (which first appeared in the late Palaeocene and continued to the end of the Oligocene) were used to subdivide the first period of the Cenozoic, for this reason called the Nummulitic.

Stratigraphers still rely on the larger foraminiferids, and have established evolutionary series not only in the nummulites (Fig. 2.14, p. 41) but also in other genera such as *Discocyclina*, and *Assilina* in the Eocene, *Lepidocyclina* in the Oligocene, *Miogypsina* in the Miocene and *Alveolina* throughout the Palaeocene and Eocene (Fig. 2.15, p. 42)

III. – PALAEOGEOGRAPHY

The widening of the oceans, and the Alpine and Laramide orogenies, profoundly affected the evolution of the Cenozoic, which thus witnessed the continuation or completion of events begun in the Mesozoic (Fig. 1.3).

During the *Maastrichtian (70Ma)*, just before the Mesozoic-Cenozoic boundary, *the North Atlantic continued to open*. In the Palaeocene, sea floor spreading occurred first between Canada and Greenland, and later in the Eocene, between Greenland and Scandinavia forming the Reykjanes-Iceland-Jan Mayen-Spitzbergen ridge, with the Faeroes and Rockall Plateau splitting off the European continent (Fig. 1.4, p. 17). This movement was accompanied by the sliding of Eurasia towards the south east which resulted in compression and shear in the Mediterranean region, the driving

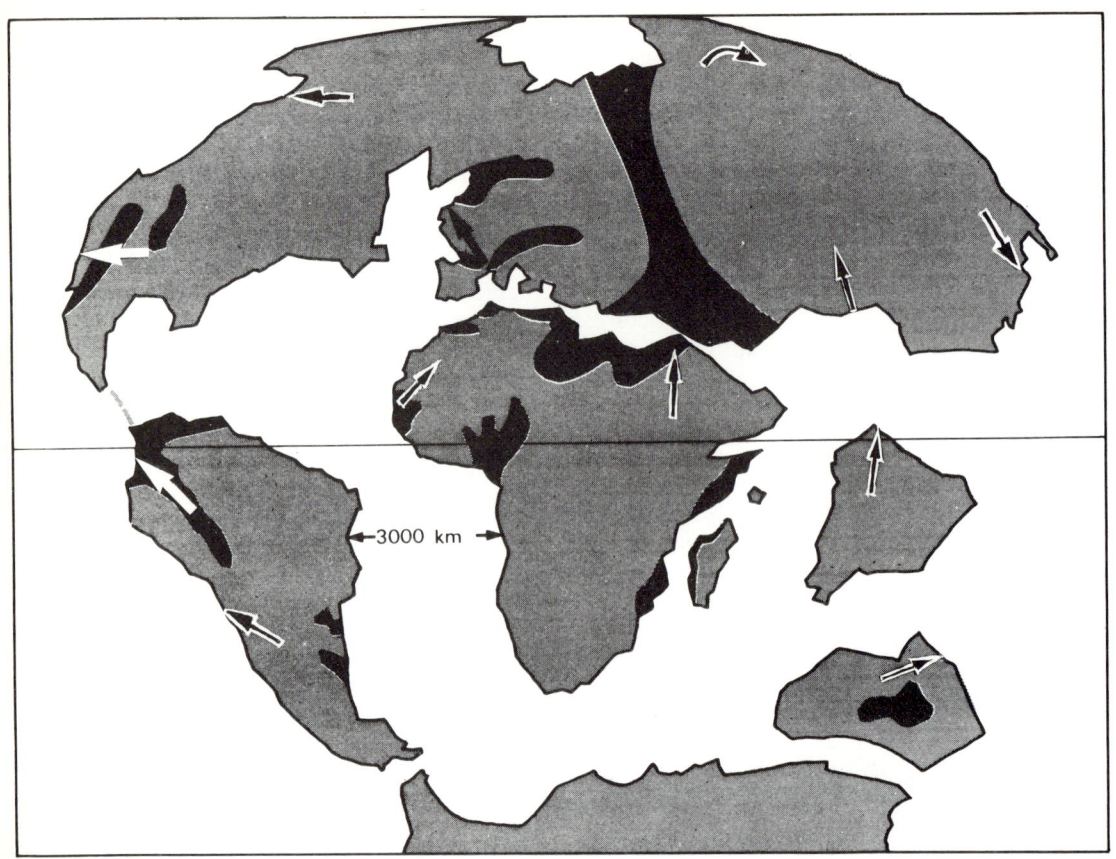

1.3 **The World at the beginning of the Cenozoic Era, about 65Ma ago.** The North Atlantic was open between Gibraltar and Newfoundland, but North America and Europe were still joined by Greenland, which was just beginning to separate. The Atlantic Ocean, which was initiated in the Jurassic, continued to widen during the Cenozoic. Madagascar had separated from Africa, but Arabia was still attached to the latter. India was moving northwards but had not yet joined up with Asia. A slender link had been established between North and South America at the beginning of the Palaeogene, but was followed by a separation which lasted until the Pliocene. Epi-continental seas are shown in black. The most important of these is the Ural Sea, which separated Europe from Asia until the end of the Oligocene. Gulfs almost bisected the African continent in Palaeocene times. The migration of mammals between America and Europe took place via Greenland in Palaeocene times, but after the opening of the North Atlantic had isolated Greenland from both America and Europe by the end of the Eocene, and following the closing of the Ural Sea in the Oligocene, migration occurred by way of the Bering Isthmus in Neogene times. The arrows indicate the direction of movement of the continental plates and clearly illustrate the break-up of the southern continent of Gondwanaland (modified, after Dietz and Holden, 1970).

forces of the Alpine orogeny.

This *compressive phase* reached its maximum at the Eocene-Oligocene boundary with the final uplift of the Pyrenees followed by the most intense Alpine movements and coincided with a plate collision which momentarily stopped the expansion of the North Atlantic. Nevertheless, by that time, the break between America and Eurasia was complete. The mammals, which during the Palaeocene and early Eocene had migrated from North America towards Europe across the northern continent, could now only cross by way of the Bering isthmus, impermanent because of its low elevation. Conveniently, the Ural Sea disappeared at the end of the Oligocene. This change in the migration routes resulted in a profound modification of the mammalian fauna of western Europe: this is Stehlin's "great break" (see p. 38).

Sea floor spreading speeded up during the late Miocene from two to four centimetres per

1.4 **Submarine relief of the North Atlantic** (after Heezen & Tharp, Lamont Geological Observatory). The mid-Atlantic ridge, emergent at the Azores, is clearly visible. Extending northwards to form the Reykjanes ridge, it is interrupted by Iceland and its satellite, Surtsey, which formed in 1963. It then continues to Jan Mayen Island on the edge of the Arctic ocean. Note the channel between the Faeroes and Scotland, the path of the first incursion of the waters of the Northern Ocean into the Atlantic, possibly from earliest Tertiary times and before the complete opening of the Atlantic in late Eocene-Oligocene times. Another link was established in the Palaeocene by way of the epicontinental seas of the North Sea and the English Channel. Note the shape of the Gulf of Gascony (south-east corner of Bay of Biscay), resulting from the rotation of the Iberian peninsula through 30° during the Jurassic and the Cretaceous. Rockall Bank, west of Scotland, is a submerged fragment of the European continent. (Depths in feet: numbers in brackets are heights above the abyssal plain which is about 16,000 feet below sea level). (Reproduced by permission of the National Geographic Society, Washington).

year, resulting today in a system of **six principal plates**: African, Eurasian, American, Pacific, Indian and Antarctic (Fig. 1.5). These are bounded either by mid-ocean ridges or by trenches at the margins of continents.

The **Alpine orogeny** led to the disappearance of the Mesozoic Tethyan Sea which, prior to the existence of the Atlantic Ocean, stretched from Gibraltar to south east Asia and beyond. According to FOUCAULT (1972), the circum-Pacific and Alpine chains formed a single orogenic belt (Fig. 1.6). The closure of the Tethyan part of this system during the Cenozoic implies a sinistral rotation of Africa with respect to Europe. The pivotal point appears to have been in the region of Gibraltar. The outcome of this movement was, by the end of the Miocene (about six million years ago), the *complete closure of the Mediterranean* to form an immense evaporite basin which only regained connection with the open ocean by the fracturing of the isthmus of Gibraltar during the Pliocene.

The Alpine orogeny was expressed in North America by the **contraction of the Caribbean area and by the uplift of the Sierra Nevada and the Coast Ranges** (an uplift which continues at the present day). In South America the Andes, which had begun to rise in the Mesozoic, were further elevated. It is worthy of note that on the borders of these active tectonic zones, but in the less folded areas, the *principal crude oil deposits of the world occur* (California, Texas, Venezuela, Libya, Galicia, Caucasus, Middle East, Iran . . .). The orogenic movements on the west of the Americas were accompanied by *major faulting and by intense volcanism*, both of which continue today.

The intense orogenic movements of the Palaeogene were followed in the Neogene and in the Quaternary by epeirogenic movements of great amplitude. Meanwhile, many thousands of metres of detrital material were deposited in the continental and marginal basins of southern Europe and the far west of North America. At the present time this tectonism is still active, not only in California where the horizontal movement of the San Andreas fault is about two centimetres per year (Fig. 1.7), and in the Afar region of Ethiopia where the vertical movement

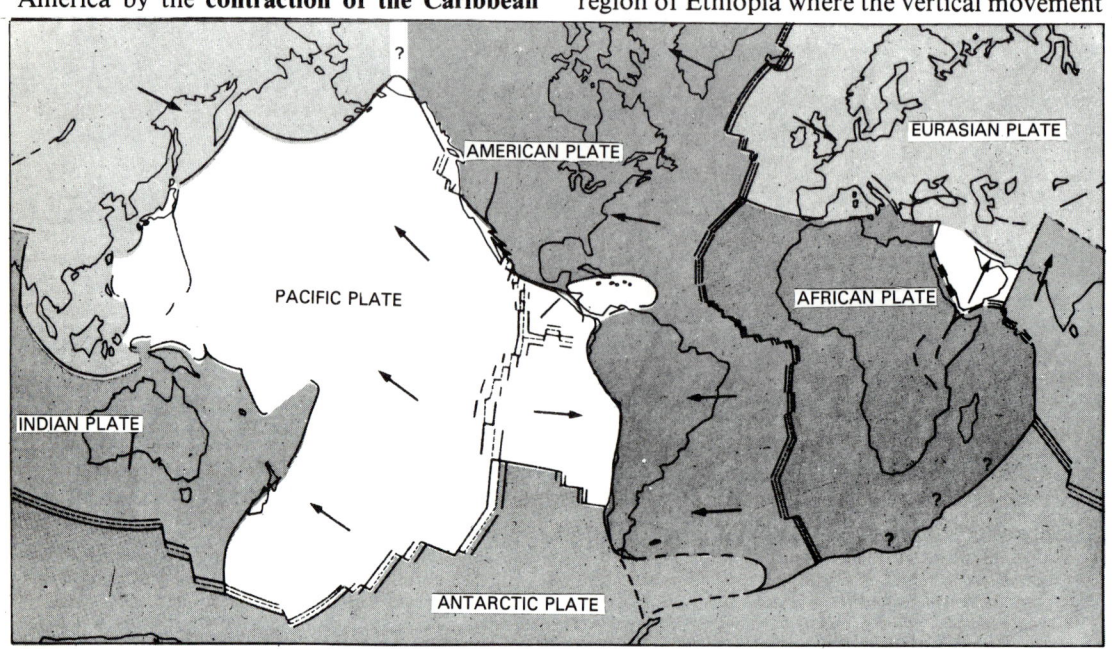

1.5 **The system of six principal plates resulting from the expansion of the ocean floors.** The plates are separated either by mid-ocean ridges (as between the American and African plates) or by trenches at the edges of continents (as between the Pacific and Eurasian plates). Their movement is indicated in relation to the African plate, which is considered fixed (after E. Bullard, *Scientific American*, 1969).

of some faults is of the same order, but also in many other regions such as the Paris basin, although the deformation there is very slight.

The movements which finally led, in the Pliocene, to the *joining up of North and South America* allowed a fauna of higher placental mammals to invade the South American continent where marsupials, ungulates and edentates had evolved in an environment isolated since the Mesozoic, except for a brief link at the end of the Cretaceous – beginning of the Tertiary. The only continent which was completely isolated during the Cenozoic was Australia.

The *disappearance of the epicontinental Ural Sea*, which allowed the reversal of the direction of migration of mammals between America and Europe at the end of the Oligocene, reminds us that other seas of the same type had inundated the continental margins in roughly the same areas as they had done in the Mesozoic, though they were now less extensive (Fig. 1.3, p. 16). They reached their maxima in the Eocene and the Oligocene, then retreated. Among the

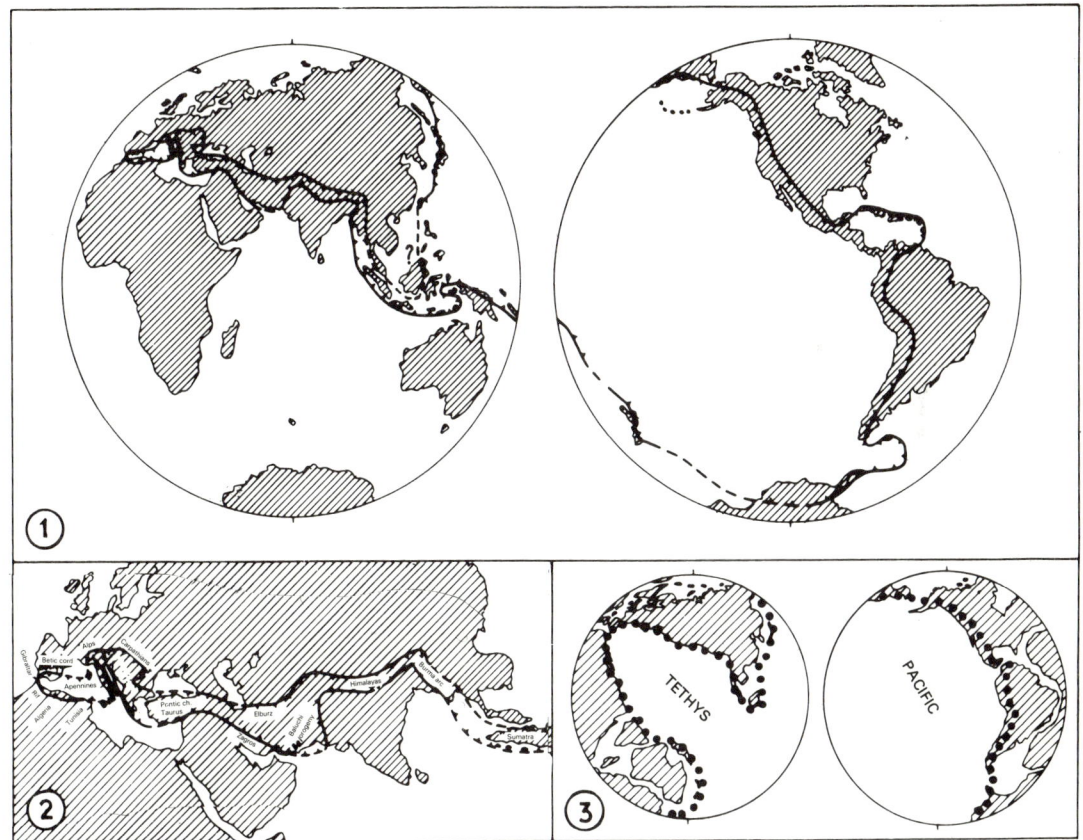

1.6 **The relationship between the Alpine and Circumpacific chains** (after A. Foucault, 1972).
1 – The present-day margins of the Alpine and Circumpacific chains (marked by triangles pointing to the interior of the chains).
2 – Detail of figure 1 in the Tethyan region.
3 – The supposed margins of the Alpine and Circumpacific chains (line of dots) at the beginning of their history (beginning of the Mesozoic). The position of the continental blocks (shaded) constituting the hinterland of the fold system is based on the work of S. W. Carey (*Continental Drift*, Hobart, Tasmania, 1958) and of E. Irving (Paleomagnetism, 1964). Eurasia is supposed fixed.

sediments formed in these basins, those of western Europe were especially fossiliferous and it is notable that most of classical stratotypes have been chosen from these deposits.

The Miocene period saw the beginnings of our present geography but it was in the Pliocene that the present distribution of land and sea was finally established, though continental shore-lines were considerably affected by the eustatic

1.7 **Horizontal movement of the San Andreas fault at Almaden, California,** about 200km south of San Francisco. The irresistible movement, which has broken the concrete walls of a channel constructed 10 years earlier, is recorded electrically at the University of Berkeley. The average movement is about 2cm per year and generally proceeds by imperceptible jumps (Photo: Brabb).

transgressions and regressions consequent upon the Quaternary glaciations.

In the Holocene, Man has added his own touches by severing the connections between continents at Suez and Panama, thus giving back to two great continents of the southern hemisphere the isolation which they had only recently lost in the Pliocene. But if Man believes he can master certain violent geological phenomena (floods, storm surges), control some (river courses and steam jets) and predict others (volcanic eruptions and earthquakes) he is totally powerless in the face of the great pulsations of the Earth which result in epeirogenic movements (as in the Mediterranean area), active faults (Pacific border of America, East African rift valley) and continuing orogenesis (Pacific borders of Asia and the Caribbean). These are the consequences of plate movements as crust is created along the ocean ridges and consumed along Benioff zones.

Great climatic changes have occurred during the Cenozoic as a result of the movements of the polar axis and the Plio-Quaternary glaciations. In the Palaeogene the climate of present temperate zones was tropical. For example, the present day Caribbean is a good model of carbonate sedimentation for the Paris Basin in the Eocene. At the end of that epoch there was a fall in temperature and an increase in aridity. The evaporite belt occupied central Europe and moved southwards in the Neogene to reach the Sahara in the Quaternary (Fig. 2.27, p. 56). The climate became uniformly warm in the Miocene, but more varied in the Pliocene with the first indications of the onset of the Quaternary glaciations.

General character of the Palaeogene | 2

The **Palaeogene** is the first period of the Cenozoic and it is also the longest (42Ma). It extends from the end of the late Cretaceous (the Maastrichtian ends about 65Ma) to the beginning of the Neogene (23Ma). The term Palaeogene was created by NAUMANN in 1866 to unite the Eocene and the Oligocene. The Montian and *a fortiori* the Danian were excluded from the Eocene, while the Chattian and the Aquitanian were included in the Oligocene. In Europe, though not in America where nummulites are rare, the synonymous term Nummulitic has been used. This latter term was introduced by d'ARCHIAC and HAIME in 1853, precisely defined in 1896 by RENEVIER, who included the beds from the Montian to the Stampian in the Nummulitic, and adopted by HAUG in 1907. Strictly speaking, the two terms Palaeogene and Nummulitic are not therefore synonymous, especially as RENEVIER had already modified the original concept of the Nummulitic by including the Montian. Today the limits and sub-divisions of the Palaeogene are still the subject of much discussion.

I. – LIMITS AND SUBDIVISIONS

The Palaeogene is sub-divided into three epochs: **Palaeocene** (SCHIMPER, 1874), **Eocene** (LYELL, 1833), **Oligocene** (BEYRICH, 1854). The fact that these three terms were defined by different authors, in different regions, based in part on marine and in part on continental formations, has led to arguments about their respective limits which have still not been resolved. The term "Palaeocene" has been disregarded by many French and Italian authors, who consider it to be merely the lowest part of the Eocene. Elsewhere, the Palaeocene is almost universally recognized. Here, the latter point of view will be adopted, so that the Eocene *s.s.* will be divided into three parts: the early, middle and late.

1. Palaeocene (Table II)

The lower limit – discussed earlier, p. 13 – coincides with that of the Cenozoic era. The upper limit was defined by SCHIMPER in 1874 (*Traité de Paléontologie végétale*) as follows [translation]:

"*Palaeocene period:* Bracheux sands, the older travertines of Sézanne, lignites and sandstones of the Soissonnais (*Suessonian*).

The Palaeocene flora is represented only by remains found in two restricted localities, of which one is in the caves near Sézanne (Champagne) and the other in the neighbourhood of Soissons (Bracheux sands, lignites and sandstones of the Soissonnais).

Eocene period: Monte-Bolca; '*Calcaire grossier*' of Paris and the marl beds of the Trocadero; gypsum and arkose of the Puy; London Clay and the contemporary deposits of the Isle of Sheppey; sandstones of the Sarthe and the neighbourhood of Angers; the Skopau lignite; the green clays of Montmartre; the gypsum of Aix-en-Provence."

From this first definition it follows that, for SCHIMPER, the Suessonian is included in the Palaeocene. D'ORBIGNY (1852), however, had

Table II
Correlation between the principal biozones and stages of the Palaeocene

Epochs	Western European stages	Planktonic foraminiferids	Nannoplankton	Continental stages in N. America (mammalian)	Marine stages (California)	Principal appearances and extinctions
LOWER EOCENE	Lower Ypresian / Sparnacian / Ilerdian			Wasatchian		Appearance of rodents, carnivores, artiodactyls, perissodactyls, proboscidians
		Globorotalia velascoensis	Marthasterites contortus			Appearance of nummulites and assilinids
55 MA	Selandian / Landenian / Thanetian		Discoaster multiradiatus	Clark-forkian	Ynezian	Extinction of condylarths
		Globorotalia pseudomenardii	Heliolithus riedeli			
			Discoaster gemmeus			
		Globorotalia pusilla pusilla	Heliolithus kleinpelli	Tiffanian		
PALAEOCENE		Globorotalia angulata	Fasciculithus tympaniformis			
	Montian / Danian / Dano-Montian	G. uncinata		Torre-jonian		
		Globorotalia trinidadensis	Cruciplacolithus tenuis			
		Globorotalia pseudobulloides and Globigerina daubjergensis	Markalius astroporus	Puercan Dragonian	Montian	
65 MA		Globigerina eugubina				Appearance of globorotalids
UPPER CRETA-CEOUS	Maastrichtian	Globotruncanella mayaroensis	Arkhangelskiella			Extinction of globotruncanids, ammonites, belemnites, rudistids
		Globotruncana gansseri				

included not only the Bracheux sands and the lignitic clays, but also the overlying Cuise sands, in the Suessonian stage. SCHIMPER was in no doubt that the "clays and sandstones of Soissonnais" should be placed in the Palaeocene, although the clays are Sparnacian and the sands are Cuisian. It can be deduced therefore, that for SCHIMPER, both the Sparnacian and the Cuisian were part of the Palaeocene, and not just the first of these stages as is often supposed.

SCHIMPER's definition of the Eocene period implies that the London Clay, now considered to be older than the sandstones of the Soissonnais, should be placed in the Eocene. Thus, there is a contradiction in SCHIMPER's proposals in that, on page 680, he placed the Sparnacian and the Cuisian in the Palaeocene, whereas on page 684, he considered that the London Clay should be included in the Eocene. However, he clearly had reservations since he wrote that "the Palaeocene flora is only represented by remains found in two restricted localities".

Unhappily, this original ambiguity has become, as so often happens in stratigraphy, the source of inexhaustible controversy. Thus has arisen the notion that the Sparnacian is part of the Palaeocene and that the Eocene begins with the Cuisian (zone of *Nummulites planulatus*). In terms of the general principles of stratigraphy, a boundary in this position is most unfortunate in that it places two important stages, the Ilerdian and the Ypresian, astride the boundary between two epochs.

On the other hand, in the Paris Basin *the most important break in the whole Palaeogene*, in the unanimous opinion of vertebrate palaeontologists, occurs between the Thanetian fauna of Cernay-les-Reims (which includes *Plesiadapis, Pleuraspidotherium, Arctocyon* and *Neoplagiaulax*) and the Sparnacian fauna containing artiodactyls, perissodactyls, rodents, carnivores, bats, flying lemurs and prosimians. **This break corresponds to the Thanetian-Sparnacian boundary** (Table II) and in the United States to the Clarkforkian-Wasatchian boundary. In Belgium this boundary would be placed between the Landenian and the Ypresian (see Table VIII, p. 114).

In Southern Europe the boundary is placed, in terms of stages, *below the Ilerdian* and, in terms of biozones, at or a little above the base of the zone of *Globorotalia velascoensis*. It lies approximately in the zone of *Discoaster multiradiatus*. The boundary is marked by *the appearance of the first nummulites and the first assilines*. On this basis, the Palaeocene would have lasted about 10Ma – from – 65Ma to – 55Ma.

The Palaeocene is divided into lower and upper parts, the lower corresponding to the Dano-Montian and the upper to the Thanetian.

a) Dano-Montian. This hybrid stage results from the combination of the terms Danian (DESOR, 1846) and Montian (DEWALQUE, 1868). It illustrates a state of affairs to which we shall frequently refer: the plethora of stages and consequent overlap of some of them. Thus the base of the Montian at Mons in Belgium corresponds to the upper part of the Danian at Fakse in Denmark. The hiatus between the Cretaceous and the Tertiary is therefore more important in Belgium than in Denmark (Fig. 2.1). The upper part of the Danian is overlain by a local stage, the *Selandian* (corresponding in part to the Montian) on which the Eocene *s.s.* is discordant (see p. 124).

b) Thanetian. The **Thanet Sands,** described by PRESTWICH (1852), were raised to the status of a stage by RENEVIER (1873). They outcrop in a restricted area in the London Basin (see p. 119) and correspond to the **Sables de Bracheux** of the Paris Basin. Alongside the Thanetian, the term *Landenian* (DUMONT, 1839) based on formations in Belgium, is also used. It began earlier than the Thanetian and finished a little later with a continental facies.

Before leaving the Palaeocene, a stage name rarely used today may be mentioned. This is the Suessonian (from Soissons: Suessonum, D'ORBIGNY, 1852), which united the Thanetian, Sparnacian and Cuisian stages.

Table III: The principal stages and biozones of the Eocene
(The correlations between the various biozonations are imprecise at some levels)

Epochs	Western European stages	Planktonic foraminiferids	Nannoplankton	Nummulites	Alveolinids	Continental (Mammalian) Stages in North America	Marine stages in California
OLIGOCENE	Early Stampian (Sannoisian)	Hastigerina micra / Cassigerinella chipolensis	Helicopontosphaera reticulata	N. intermedius		Chadronian	Refugian
36 MA / LATE EOCENE	Priabonian	Globigerina gortanii or equivalent	(Ericsonia subdisticha)	N. retiatus		Duchesnian	Narizian
		Globorotalia cerroazulensis	Isthmolithus recurvus	N. fabianii s. str. — N. fabianii s.l.	Neoalveolina		
	Bartonian	Globigerapsis semiinvoluta		N. aff. fabianii	A. elongata	Uintan	
MIDDLE EOCENE	Lutetian — late (or Biarritzian)	Truncorotaloides rohri	Discoaster tani nodifer	N. brongniarti / N. perforatus	A. prorrecta		Ulatisian
		Orbulinoides beckmanni		N. carpenteri / N. aturicus	A. munieri		
	Lutetian — middle	Globorotalia lehneri	Chiphragmalithus alatus	N. sordensis	A. stipes	Bridgerian	
	Lutetian — early	Globigerapsis kugleri	Discoaster sublodoensis	N. laevigatus	A. violae		
		Hantkenina aragonensis		N. manfredi	A. daniellii		
	Cuisian	Globorotalia palmerae	Discoaster lodoensis	N. praelaevigatus	A. oblonga		Penutian
EARLY EOCENE		Globorotalia aragonensis		N. planulatus	A. trempina	Wasatchian	
	Ypresian — Ilerdian	Globorotalia formosa formosa	Marthasterites tribrachiatus	N. involutus	A. corbarica		Bulitian
	Ypresian — Sparnacian	Globorotalia aequa / G. subbotinae = G. rex	Discoaster binodosus	N. exilis	A. moussoulensis		
		Globorotalia velascoensis	Marthasterites contortus	N. praecursor	A. ellipsoidalis / A. cucumiformis		
55 MA / PALAEOCENE	Thanetian	Globorotalia pseudomenardii	Discoaster multiradiatus	N. fraasi	A. levis / A. primaeva		Ynezian

2. Eocene (Table III)

This epoch, the commencement of which is placed at the junction between the Thanetian and the Sparnacian, consisted, according to its creator, Charles LYELL, of the following beds from bottom to top:
- plastic clay and lower sand (cf. Sparnacian and Cuisian)
- "calcaire grossier" (cf. Lutetian)
- siliceous limestone (cf. St. Ouen limestone)
- gypsum and marls (cf. Ludian)
- upper marine group (Fontainebleau sand, Stampian)
- third freshwater formation (cf. Étampes limestone and Montmorency sandstone, upper Stampian).

In 1859, LYELL removed the upper part (the Fontainebleau sands) from his Eocene and put them in the lower Miocene. In 1854, BEYRICH introduced a new period between the Eocene and the Miocene s.s. which he called the Oligocene, and this, of course, included the Fontainebleau Sands.

The removal of the upper part of LYELL's original Eocene to the Oligocene was soon followed by the removal of the lower beds by SCHIMPER, in 1874, when he defined the Palaeocene (see above, p. 21), to include the Bracheux sands, the lignites and the Soissonnais sandstones. At the present time, the Bracheux Sands (Thanetian) are considered to belong to the Palaeocene, while the Soissonnais sandstones (Cuisian) are placed in the lower Eocene.

The upper part of the Eocene poses problems still more complex than those of its lower limit. As with the Dano-Montian, there is a superabundance of stages defined by different authors in different regions, including some in lagoonal facies. The four controversial stages are *the Ludian and the Sannoisian* in the Paris Basin, *the Tongrian* in Belgium and *the Lattorfian* in Germany. Also a matter for dispute are the horizons of the Headon, Osborne and Bembridge beds in England.

There is, however, a clear correlation between the Lattorf sands of Germany, the Grimmertingen sands of Belgium (lower Tongrian), the *Pholadomya ludensis* clays of the Paris Basin (lower Ludian) and the sands and clays of the Brockenhurst Beds (the middle part of the Headon Beds in England). This correlation is based on molluscs, foraminiferids and nannoplankton, although the last suggest a slightly greater age for the Brockenhurst Beds.

This important correlation corresponds to a marine transgression across earlier lagoonal deposits. This situation is comparable to the "Biarritzian" episode in the late Lutetian (see p. 98) and it may be questioned, therefore, whether these four units belong to the upper Eocene or the lower Oligocene.

Historical criteria (BEYRICH, 1854) favour the second alternative, but in the author's opinion this cannot be accepted for two reasons. In the first place, earlier writers did not have access to the mass of information now available. For example BEYRICH included lignite beds, now known to be Eocene, at the base of his Oligocene. Secondly, MAYER-EYMAR, proposer of the Lattorfian stage (1893) placed this stage at the top of the "Lower Tertiary" and rejected the validity of BEYRICH's Oligocene, a unit which he considered to be "as useless and artificial as the Quaternary".

Since the geochronological dating of these beds is still imprecise (35-38Ma), it is necessary to consider further the palaeontological evidence.

In 1909, the celebrated mammalologist from Basle, STEHLIN placed his "great break" (see p. 38) between the fauna of Montmartre and that of Ronzon, i.e. between the Ludian and the Sannoisian. This view has been questioned by THALER (1966) who has relied more on historical considerations than on palaeontological argument. He admits, a priori, that the Lattorfian and the Brockenhurst Beds are Oligocene and places, ipso facto, the Montmartre fauna and the equivalent fauna at Bembridge in the Isle of Wight, in the Oligocene.

However, MARTINI (1970) placed the Brockenhurst Beds in the late Eocene on the evidence of the nannoplankton. The beds of Latdorf (=Lattorf) which, because of the presence of *Ellipsolithus subdistichus*, were placed in the lower Oligocene, are now of uncertain horizon as this species, now called *Ericsonia subdisticha*, also occurs in the upper part of the Priabonian of Italy. This is a fundamental discovery for, if one accepts that the Lattorfian forms the base of the Oligocene, it is necessary to reconsider the definition of the Priabonian; but if it is considered that the Priabonian as defined by HARDENBOL in 1968 is equivalent to the upper Eocene in southern Europe, then the Lattorfian must be removed from the Oligocene. On the other hand, *the molluscan fauna of the Latdorf beds is similar to that of Mandrikovka*, near Dnepropetrovsk, in the Ukraine (Fig. 4.17 and 4.27, p. 134), which is undoubtedly late Eocene and may even be slightly older according to the planktonic organisms. *Nummulites germanicus* which occurs at Latdorf is, according to BLONDEAU (1969) descended from the upper Lutetian *N. anomalus*. This is a Southern European Priabonian species, known also in the upper Eocene of the Ukraine, which reached Germany by way of Poland.

Thus on the one hand we have the historical arguments (which we reject) and the uncertainty associated with the position of the zone of *Ericsonia* (ex *Ellipsolithus*) *subdisticha*, and on the other, the weight of evidence of the mammals, the molluscs, *N. germanicus* and the planktonic organisms occurring both in the beds at Latdorf and Helmstedt and in those of the Priabonian.

In this situation we adopt the opinion of CAVELIER (1972), who placed the Lattorfian, the lower Tongrian, the Ludian, and the Headon, Osborne and Bembridge Beds in the upper Eocene. It should be noted that there is no place for the Lattorfian in the unbroken Italian succession, where there is a direct passage from the Priabonian to the Oligocene, the boundary being placed between the zones of *N. fabianii* and *N. intermedius*, and between the zones of *Globigerina gortanii* and *Cassigerinella chipolensis* – *Hastigerina micra* (Table III).

The Eocene thus defined lasted about 19Ma (55 to 36Ma). It is divided into three parts:

1) *Lower Eocene* (55-48Ma) comprising the Ypresian stage, which can be divided into two sub-stages in the Paris Basin, the Sparnacian (below) and the Cuisian (above), see p. 23. A nearly-equivalent term, the Londinian (London Clay, MAYER-EYMAR, 1857) had a shorter duration and has now fallen into disuse. In southern Europe, the Ilerdian (see p. 128) is a marine stage at the base of the Eocene, lying between the Thanetian and the Cuisian.

2) *Middle Eocene* (48-43Ma) comprising the Lutetian stage, which also has three divisions. In Aquitaine the upper part has been given the superfluous name of Biarritzian; this unit is in fact equivalent to the upper part of the Lutetian of the Paris Basin (see p. 66).

3) *Upper Eocene* (43-36Ma). This includes the Bartonian (*s.s.*) and the Priabonian. However, according to the accepted biozonal schemes (Table III, p. 24) the Bartonian may in part belong to the Middle Eocene. In the Paris Basin the term Bartonian is used in a broad sense and it is divided into three substages called, from bottom to top, the Auversian, the Marinesian and the Ludian, the last corresponding to the Priabonian.

3. Oligocene (Table IV)

This epoch was, as noted above, defined by BEYRICH in 1854. His object was to unite in a new period (in fact, in an epoch) the formations between the Upper Eocene and the Lower Miocene as defined by LYELL.

In his definition of 1854, BEYRICH considered that the Oligocene commenced with lignite beds in north Germany, and included the marine sands of Alzey, the brackish and freshwater beds of the Mainz Basin, the marine clays with septaria and the marine Cassel sands of central and northern Germany. In BEYRICH's opinion, the equivalents in the Belgium were the Tongrian and Rupelian "systems", and, in the Paris Basin, the beds above the Beauchamp sands, that is the "calcaire moyen de Paris", the Fontainebleau sands and the Beauce limestone.

It must be stressed that there are two important points of confusion in BEYRICH's definition. The lignite beds of north Germany which he placed at the base of the Oligocene are now regarded as Eocene and the "calcaire moyen de Paris", that is to say, the St. Ouen limestone (Marinesian substage), which he considered to be Oligocene, is now also regarded as of undoubted late Eocene age.

The *lower limit* of the Oligocene, as discussed above in connection with the Eocene, is placed,

Table IV
Correlation between the principal biozones and stages of the Oligocene

Epochs	Western European stages	Planktonic foraminiferids	Nannoplankton	Continental stages in N. America	Marine stages in California
MIOCENE		*Triquetrorhabdulus carinatus*	*Triquetrorhabdulus carinatus*	Arikareean	Saucesian
— 23 MA —		1st *Globigerinoides*			
OLIGOCENE — Chattian	Chattian	*Globorotalia kugleri*	*Sphenolithus ciperoensis*	Orellian	Zemorrian
		Globorotalia opima opima	*Sphenolithus distensus*		
	Stampian or Rupelian	*Globigerina ampliapertura*	*Sphenolithus predistensus*	Chadronian	Refugian
— 36 MA	(Sannoisian facies)	*Hastigerina micra Cassigerinella chipolensis*	*Helicopontosphaera reticulata*	Duchesnian	

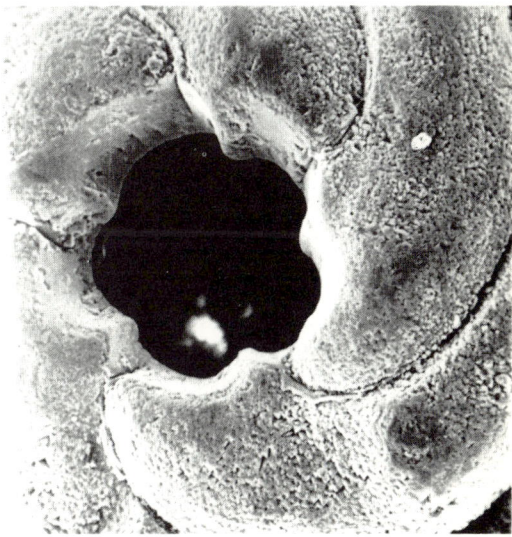

2.0 **Oogonium of a charophyte,** *Raskyella vadaszi*, Marinesian, St. Ouen limestone, Paris Basin. Left: side view x 300; right: apex, without apical cap x 1200 (S.E.M. photographs: J. Riveline).

in the Paris Basin, at the base of the green clay unit in the "Glaises à Cyrènes" (*Cyrena convexa* and *Psammobia plana*). In Belgium, it is placed at the base of the Hoogbutsel mammal horizon (middle Tongrian) and in Germany, above the Latdorf beds and the Silberberg beds of Helmstedt.

The upper limit of the Oligocene has often been placed by French geologists at the top of the Aquitanian, although in Aquitaine the Miocene transgression occurs at the base of that stage. This interpretation rests on the presence of "Oligocene" mammals in the Aquitanian Beds of the Bordeaux region. Micropalaeontological study and other recent research in Aquitaine have shown, however, that the

MAASTRICHTIAN ←→ DANIAN

2.1 **Danian-Maastrichtian contact at Stevns Klint, south of Copenhagen** (see map, Fig. 9.5, p. 205). *Maastrichtian:* 1 – White chalk with flints; 2 – Grey chalk separated from the white chalk by a hard-ground. *Danian:* 3 – Fish bed (very thin); 4 – *Cerithium* limestone; 5 – Bryozoan limestone (Photo: Håkansson).

Aquitanian belongs to the Miocene, as the Congress on the Mediterranean Neogene held in Vienna in 1959 had recommended.

The same Congress proposed that the Oligocene should end with the *Chattian*, but this has raised considerable problems. FUCHS, who had proposed the Chattian in 1894, considered it to be a northern equivalent of the Aquitanian. However, in 1911, HAUG used the term Chattian for the upper part of the Stampian as it had been defined by D'ORBIGNY. This consisted of the Ormoy beds and the overlying lacustrine limestone, although the correlation with the northern area had not been established.

The discovery of *Miogypsinoides* in the Chattian links it with the Aquitanian as defined by FUCHS. The marine molluscs are, however, inadequate to distinguish these stages, so that according to LORENZ (1964) the term Chattian is not usable in the Mediterranean area. Once more, therefore, we meet with a superfluous stage name (i.e., the Chattian) which cannot be used in Southern Europe. This stage is probably Aquitanian, i.e., at the base of the Miocene in Aquitaine, but it is still in an uncertain position in the northern basins where it has been dated at about 30Ma. The name "Chattian" is still often used but it must be remembered that its status is uncertain and confirmation of its validity depends on future research. Deprived of the Lattorfian at its base and the Chattian at its top, the Oligocene is thus reduced to *a single marine stage, the Stampian or Rupelian* (see p. 76).

Although there is some uncertainty in the correlation with the type areas, absolute dating of continental Oligocene deposits of the New World gives a duration of perhaps 13Ma (from 36Ma to 23Ma). The Lattorfian of Germany and the Grimmertingen sands of Belgium both yield ages of about 37Ma, but they are here regarded as upper Eocene. Several dates from the Chattian of North Germany indicate an age of about 30Ma, so that at 23Ma the base of the Miocene would seem to be too young.

It is possible that the Oligocene had a duration of no more than 8 to 10Ma, which is little more than the duration of each of the major cycles of the Eocene. There may thus be some truth in the comment made by MAYER-EYMAR nearly a century ago, that there was no justification for the existence of the Oligocene *epoch*.

The Oligocene ended with the disappearance of the nummulites and a large number of other genera of foraminiferids. At the same time *Globigerinoides* and *Miogypsinoides*, the latter ancestral to *Miogypsina* of the Miocene, made their appearance.

The place of the Stampian in the Oligocene

D'ORBIGNY (1852) defined the Stampian as follows: "In adopting this lowest division of the earliest beds that one used to regard as 'Miocene' we have named it the 'Stampian stage' from the best type section in France at Étampes (Stampae)".

The Stampian of D'ORBIGNY corresponds, in part, to the earlier definitions of GRAVES (1847) ("étage des sables et grès supérieurs") and BRONGNIART (1810) ("marnes gypseuses marines"), but includes the "beds from the green marine marls containing *Ostrea cyathula*, *O. longirostris*, etc." to the upper lacustrine limestones of Étampes inclusive.

In current terminology, this much disputed definition by D'ORBIGNY undoubtedly indicates that he intended that the Stampian should begin at the "glaises à Cyrènes" (base of the green clay of Romainville).

Following the creation by BEYRICH in 1854 of the Oligocene, which contemporary French geologists did not recognise, there was a period of intense confusion. This was due to the difficulty of correlating the type Oligocene with the sequence in the Paris Basin. According to MUNIER-CHALMAS and DE LAPPARENT (1893), the Oligocene consisted of two stages, the Tongrian (divided into the Sannoisian and the Stampian) and the Aquitanian, now regarded as Lower Miocene.

The *Sannoisian*, based on sections at Sannois and Orgement, near Argenteuil, consists of the beds between the main mass of gypsum and the marls with *Ostrea cyathula* and *O. longirostris*. These beds include the blue marls of Argenteuil, the white marls of Pantin, the green clay of Romainville and the "marine series of Sannois or Orgement" which MUNIER-CHALMAS (1891) correlated with the Brie limestone. However, CAVELIER (1964) claimed that the marine Sannoisian of the Paris Basin was, palaeontologically, indistinguishable from the overlying Stampian beds. Only their lithological character, due to conditions of deposition, enabled them to be differentiated. CAVELIER, therefore, thought it necessary to abolish the Sannoisian and to regard the Oligocene as a single stage (the Stampian). The Étampes limestone, which overlies the Fontainebleau sands, was formerly considered to be Aquitanian because of the presence of *Helix ramondi* in the marls of Trappes. However, the molluscs and the charophytes indicate a Stampian age. The placing of the

Étampes limestone in the Upper Stampian (DENIZOT, 1927) has recently been confirmed by the discovery at Étampes of mammals of Middle Stampian age in the Ormoy sands immediately beneath the Étampes limestone.

The term *Rupelian*, defined in Belgium, is often used synonymously with the Stampian (see p. 119). In our view the Rupelian covers a shorter period of time than the Stampian of France because it lies above the upper Tongrian (Hoogbutsel beds, Henis clay and Vieux-Joncs sand), which is correlated with our Lower Stampian (see Table VIII, p. 114).

II. – PALAEONTOLOGY

1. General

All the major present-day groups of organisms were well represented at the beginning of the Palaeogene with the exception of the monocotyledons, which were still rare and which did not become abundant until the Oligocene. During the Palaeocene, there is evidence of the first radiation of the mammals, followed by a second during the Eocene. They remained small in size, however, and it was not until the Neogene that the food chain, mono-cotyledons – herbivores – carnivores reached its zenith. The flora and fauna of the Cenozoic are not only of stratigraphical, but also of palaeo-ecological interest. The plants became increasingly responsive to the variations in climate and ecological studies permit the reconstruction of environments. The marine algae, together with the foraminiferids, bryozoans and ostracods provide evidence of the depth of water, its degree of agitation, its clarity, temperature and salinity.

The mammals, together with the micro-organisms are the most useful fossils for

Palaeogene biostratigraphy (Fig. 2.2). Before discussing these in detail, a brief survey of other groups will be given.

Historically, the early growth of the young science of stratigraphy was based largely on the study of molluscan faunas: the Tertiary is the era of shellfish. Though today their role in this respect is much reduced in comparison with micro-organisms, they have given great pleasure to collectors, and often permit even an unskilled observer to recognise the stratigraphic horizon which yields them. For example, a single shell of the bivalve, *Avicula defrancei* (Fig. 3.28) is sufficient to identify the Morte-fontaine horizon of the Marinesian. In the description of the formations which follow, some of the common species of molluscs will be mentioned. Among the gastropods, the *Cerithiidae* are so abundant and well-preserved that they are ideal for the illustration of the study of fossil populations (Fig. 2.3). In contrast, the cephalopods, having dominated the marine world of the Mesozoic, have become very rare and are mainly represented by *Hercoglossa*, a near relative of *Nautilus*, and

2.2 **The evolution of some characteristic groups of organisms from the end of the Mesozoic to the end of the Cenozoic.** The Cenozoic witnessed a great development in the Monocotyledons and in particular, the *Gramineae* (grasses). Although the mammals existed in the Mesozoic and began to diversify in the Palaeocene, explosive evolution of the main groups did not occur until the Eocene. Extinction of some groups took place towards the end of the Pleistocene and in the Holocene, at a time when the Hominids were developing rapidly.

Palaeontology

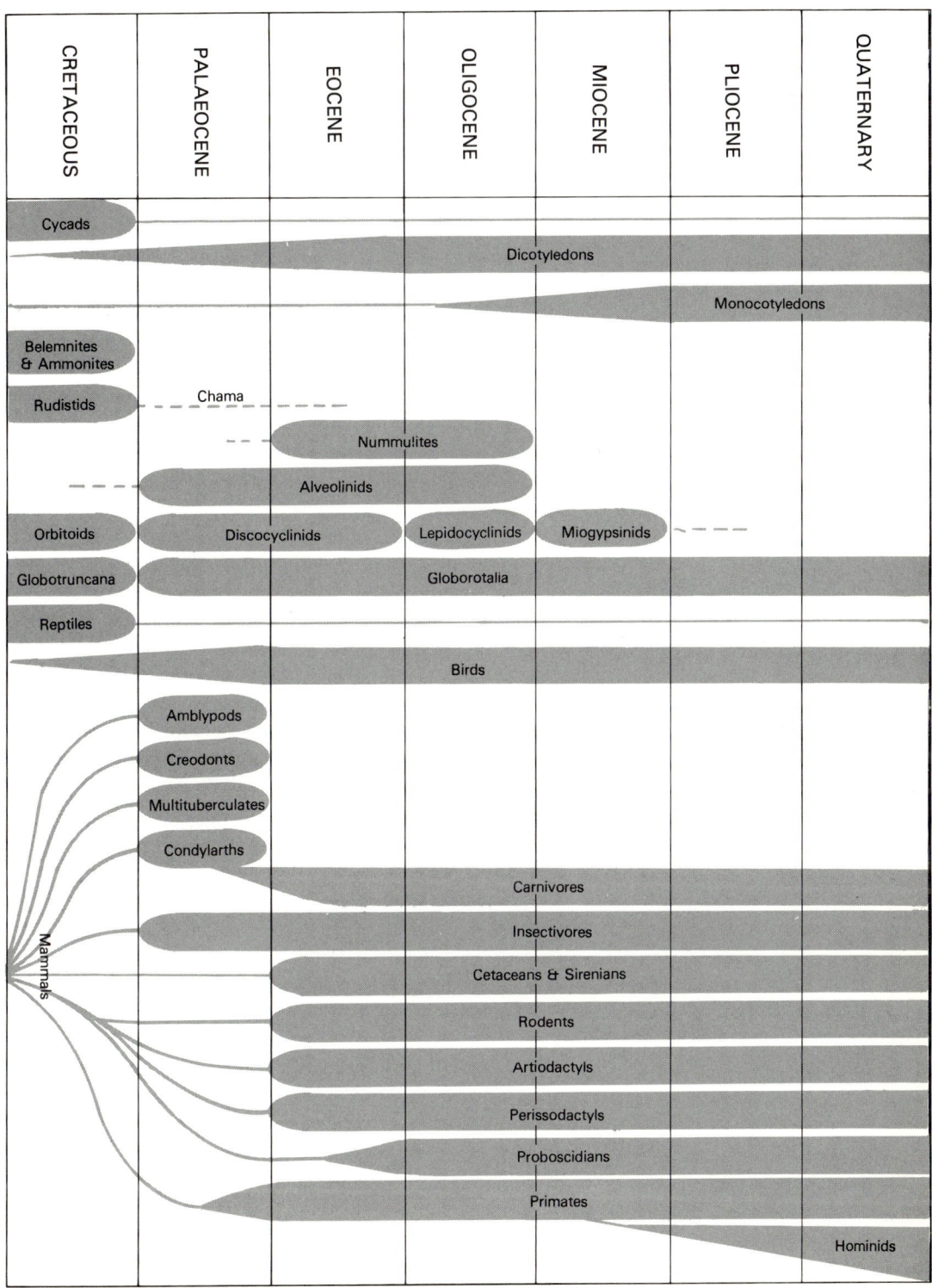

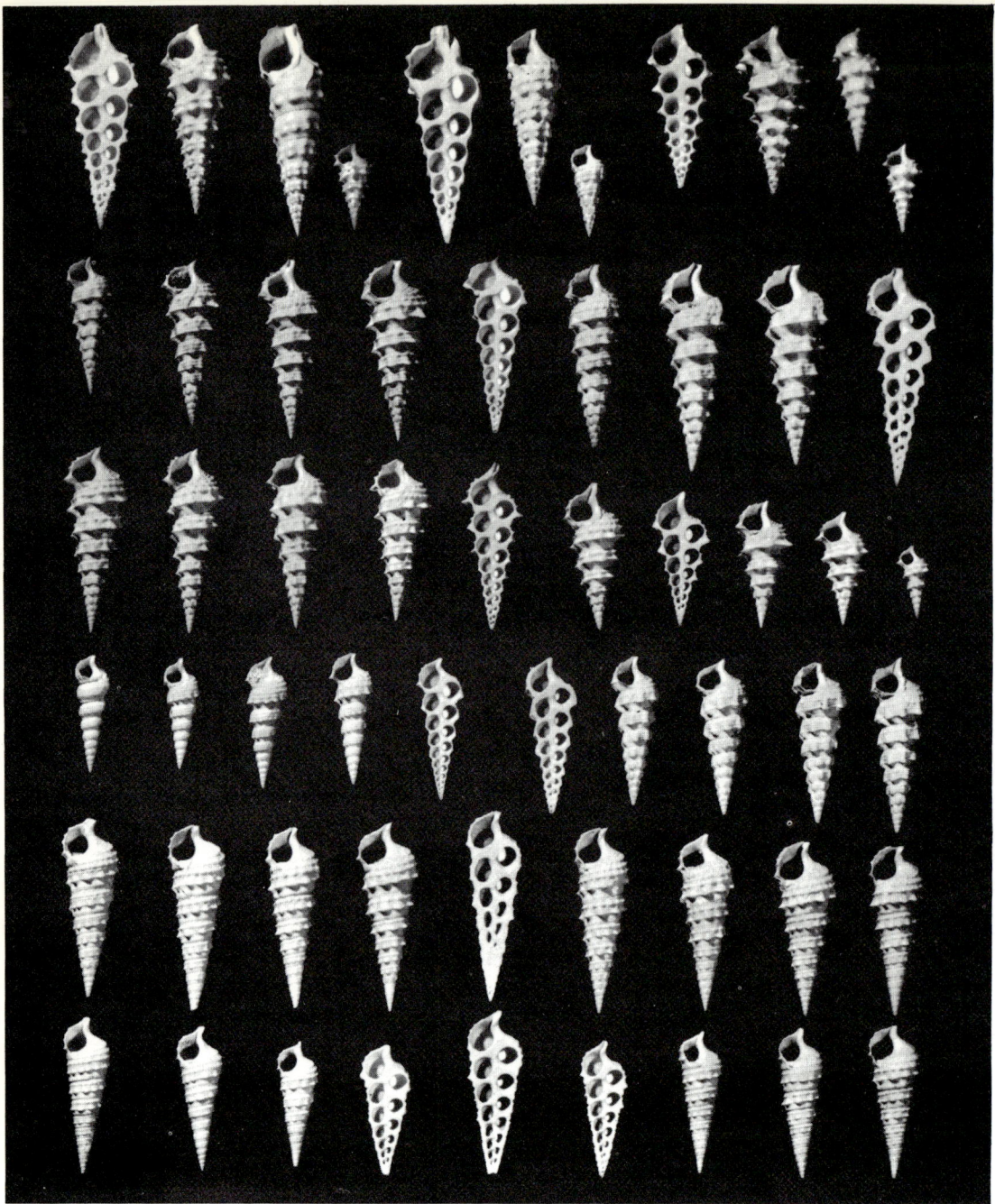

2.3 **Selected material for the study of the ontogeny of** *Cerithium* from the Lutetian and lower Bartonian (Auversian) of the Paris Basin. (The numbering of rows is from top to bottom and of individuals from left to right).

1 – (top row)   *Cerithium echinoides* (1-2-4)
                Cerithium pleurotomoides (3-5-6-7)
                Cerithium calcitrapoides (8-9-10-11)
2 –   *Cerithium bouei,* mixed forms
3 –   *Cerithium bouei,* carinate forms
4 –   *Cerithium bouei,* denticulate forms (ex. *C. coronatum*) (1-2-3-4-5)
                Cerithium bouei, costulate forms (6-7-8-9-10)
5 and 6 –   *Cerithium sowerbyi*   (after D. Camus, 1967)

the rostrum of the cuttlefish *Belosepia* (Fig. 3.28).

Often occurring in association with the molluscs in the neritic environment are the echinoderms (*Echinolampas*, Fig. 3.16), the hexacorals (*Eupsammia*, Fig. 3.16), the bryozoans (Fig. 2.4) and the calcareous algae (*Dactylopora*, Fig. 3.28).

Insects are, however, only preserved in exceptional circumstances. For example, they occur in spring-deposits, such as the travertine of Sézanne to the south of Épernay (Thanetian), the thinly-bedded argillaceous limestone of Monte-Postale, near Bolca, Trentino (Lutetian), the Quercy phosphorite (Eocene to Oligocene), the fine-grained limestones and marls of the lake of Aix-en-Provence (Eocene to Oligocene) and in Baltic amber (Oligocene). The most common groups are the Hymenoptera, the Diptera and the Lepidoptera.

In such insect-bearing beds plant debris (stems, leaves, flowers and fruits) and the remains of birds, reptiles and amphibians are often found. Fish remains are often abundant and the fish fauna of Monte-Bolca (Figs. 4.22 to 4.25) is found, like the molluscan fauna of Grignon, in collections throughout the world. Nevertheless, the teeth and otoliths of fishes and the scutes of tortoises and crocodiles, more resistant than their skeletons, are often found in neritic or lagoonal formations. Sharks' teeth, which reach a maximum length of 15cm in the Neogene genus, *Carcharodon* (Fig. 2.5), are common in the basal beds of transgressive sequences. Fish are important as stratigraphic markers in the Eocene of Belgium (CASIER, 1946). Although the reptiles which survived the Mesozoic hardly evolved during the Tertiary (the modern fauna has a decidedly archaic appearance), all the orders of birds are represented from Palaeocene times. Most of the

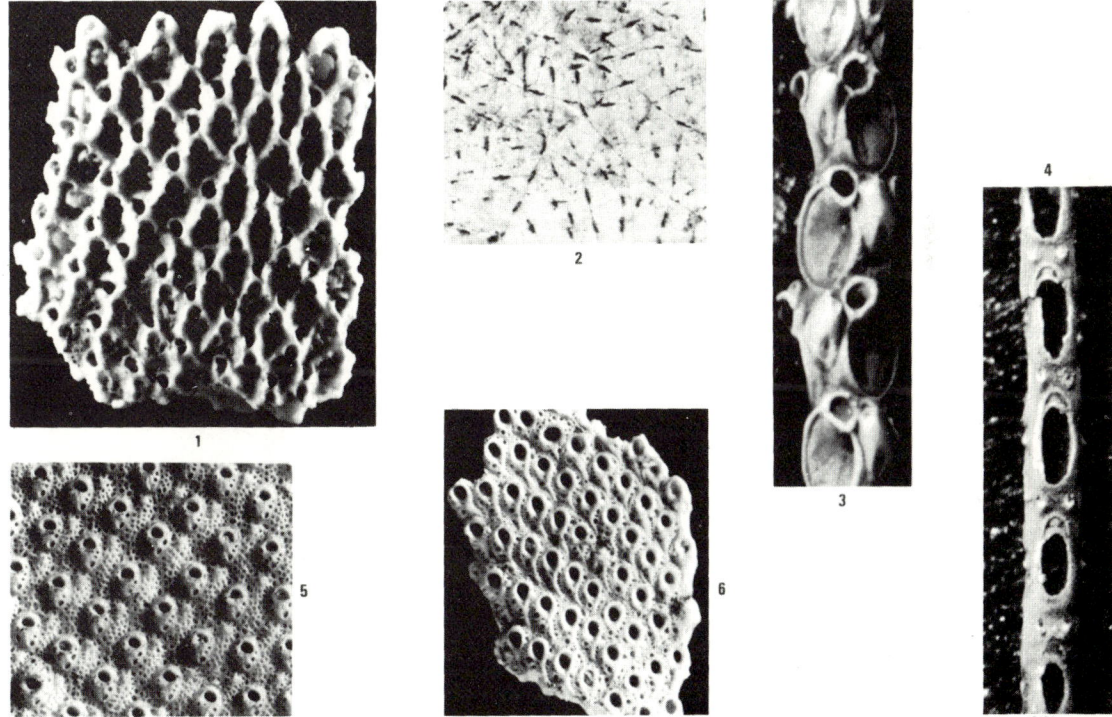

2.4 **Tertiary bryozoans of the Paris Basin** (*Buge Collection*. Photo: Serrette).
1 – *Cupuladria haidingeri* (Neogene) x 20; 2 – *Terebripora falunica* (Miocene) x 7 – a bryozoan which bored into mollusc shells; 3 – *Poricellaria alata* (Eocene) x 30; 4 – *Nellia bituberculata* (Eocene) x 30; 5 – *Tubucella mamillaris* (Eocene) x 15; 6 – *Schizostomella crassa* (Eocene) x 15.

2.5 **Jaws of Carcharodon** (Miocene), a 20m shark with teeth up to 15cm long (Photo: American Museum, New York).

living forms of Ratites and Carinates were already present in the Eocene, and their early forms are often of great size (e.g. *Gastornis* in the Paris Basin and *Diatryma* of Wyoming, Fig. 5.16, p. 151).

2. Mammals (Table V and Fig. 2.6)

Our knowledge of the mammals of the Palaeogene, on which Cuvier (1825) founded the sciences of vertebrate palaeontology and comparative anatomy, has in the last twenty years undergone great rejuvenation as a result of the systematic study of fragments of skeleton and in particular, of teeth from the earliest Tertiary mammals, such as the rodents and the insectivores. Their role is fundamental in the stratigraphy of continental formations, where the more primitive fauna of the Palaeocene can be clearly distinguished from that of the Eocene and Oligocene. Several attempts at zonation have been made, especially by Thaler (1966) on the basis of the rodents and by Ginsburg (1971)

using associations of genera.

In the Palaeocene, the principal mammal deposits are in North America (Torrejon, Tiffany, Clarkfork) and in Asia. In France, the *fauna of Cernay* (slightly later than that of Walbeck in Germany) is the most characteristic (Russell, 1964). None of the modern families are represented and all the mammals are of small size (less than 50cm). The Multituberculates, which appeared in the Jurassic, persisted with *Neoplagiaulax* but disappeared in the Eocene. Marsupials closely related to the opossum are present, and a wide variety of placental mammals, including small insectivores, which had first appeared in the Upper Cretaceous of Mongolia.

Arctocyon, related to the carnivores (and very common at Cernay) has an elongated body (45cm) supported by short, massive limbs and a long, thick tail. Despite its name, it was neither bear nor wolf, having a plantigrade gait. It was a good swimmer and was more omnivore than carnivore. It was a condylarth, distinguished from the true carnivores by its unspecialized shearing molars and by having a smaller, less convoluted brain.

The ancestors of the Ungulates were represented by other condylarths. *Phenacodus*, for example, was the size of a small wolf but had a large tail. Its dentition lacked a diastema and its molars were trituberculate. The herbivore *Pleuraspidotherium* browsed on the leaves of bushes some 30 million years before it was possible to graze on the prairie grasslands. The

condylarths shared their habitat on the river banks with a lemur, *Plesiadapis* (a small Prosimian ancestor of the Primates), with strong claws and teeth not unlike those of rodents. The large ungulates (*Palaeotherium*, *Coryphodon* and *Lophiodon*) characteristic of the Eocene, had not yet evolved.

With the extinction of the amblypods, multituberculates, creodonts (primitive carnivores) and the condylarths, and the appearance of the edentates, cetaceans (fresh water dolphins),

2.6 Locality map of the principal mammal deposits in France (see also Table V).

General Character of the Palaeogene
Table V
The stratigraphic distribution of the principal mammalian deposits of the Cenozoic in France and in other European localities
(see map of localities, Fig. 2.6 and Figs. 2.7 to 2.13, 5.6 to 5.12 and 12.11 to 12.13)

Epoch	Principal representatives	Localities
Early Pleistocene = late Villafranchian	*Elephas meridionalis, E. antiquus Equus stenonis, Rhinoceros mercki, Hippopotamus major, Rhino. etruscus* Extinction of *Mastodon* and *Dinotherium*	Saint-Prest, near Chartres (E. et L.) Val d'Arno (Tuscany) Saint Vallier (Drôme)
Late Pliocene = early Villafranchian	Appearance of *Bos, Equus, Elephas* Extinction of *Hipparion*	Villafranca d'Asti (Piedmont) Perrier, near Issoire (Puy-de-Dôme)
Early Pliocene	*Mastodon arvernensis Rhinoceros leptorhinus, Hipparion*	Montpellier (Hérault)
Late Miocene Pontian	*Dinotherium, Mastodon longirostris Machairodus* Appearance of *Hipparion*	Pikermi, near Athens (ex-Pikermian) Orignac-Luberon (Vaucluse) Teruel (Hautes-Pyrénées): TUROLIAN Valles-Penedes, near Sabadell: VALLESIAN
Basal late Miocene (early Tortonian)	*Mastodon angustidens Dinotherium Dryopithecus, Gazella*	La Grive Saint-Alban (Isère) Simorre (Gers) Saint-Gaudens (Haute-Garonne)
Middle Miocene Helvetian	*Pliopithecus Dinotherium Rhinoceros sansaniensis*	Sansan (Gers) Pontlevoy (Loir-et-Cher) Shell sands ("faluns") of Touraine
Early Miocene Burdigalian	Appearance of *Mastodon*, true Rhinoceros and *Anchitherium*	La Romieu (Gers) Orléans sands: Chitenay (Loir-et-Cher), Artenay, and Montabuzard (Loiret) Laugnac (Lot-et-Garonne)
Early Miocene Aquitanian	Impoverished Stampian fauna	Saint-Gérand-le-Puy (Allier) Paulhiac (Lot-et-Garonne) Beauce limestone: Selles-sur-Cher (Loir-et-Cher)
Middle Oligocene Stampian	*Anthracotherium* First Tapirs and deer (without antlers); last *Palaeotherium*	La Milloque (Lot-et-Garonne) Rabastens (Tarn), Etampes (Essonne) Villebramar (Lot-et-Garonne)
Early Oligocene Sannoisian	First *Anthracotherium* Rhinoceratoids and giant Artiodactyls Extinction of *Anoplotherium* and *Xiphodon*	Ronzon (Haute-Loire), Hoogbutsel (Belgium), Brie limestone
Late Eocene Bartonian	*Palaeotherium, Xiphodon Anoplotherium* Last *Lophiodon*	Montmartre (Paris) Euzet-les-Bains (Gard) Robiac (Gard) Le Guépelle (Val d'Oise)
Middle Eocene Lutetian	*Propalaeotherium Lophiodon Dichobune* Appearance of Proboscidians (*Moeritherium* of Fayum, Egypt)	Castres (Tarn) Issel (Aude), Bouxwiller (lower Rhine) Argenton-sur-Creuse (Indre)
Early Eocene (Cuisian and Sparnacian)	First rodents Artiodactyls, Perissodactyls Carnivores	Cuis, near Épernay (Marne) Mutigny (Marne) Meudon (Hauts-de-Seine)
Palaeocene	Multituberculate Creodonts Condylarths	Cernay-les-Reims (Marne) Walbeck (40km west of Magdeburg, Saxony)

2.7 **Dinoceras (= Uintatherium) from the Eocene of North America.** This massive mammal had short limbs and a small brain. The armament on the head (3 pairs of horns) is remarkable. The upper jaw has large canine teeth similar to those of *Machairodus* (Fig. 5.11, p. 148), but no incisors. Its thick-set frame resembled that of an elephant (1.5m to the shoulder) (after Augusta and Burian with the permission of the *Encyclopaedia Universalis*).

sirenians *(the manatee), rodents, artiodactyls, perissodactyls (a primitive four-fingered horse), proboscideans and cloven-footed carnivores in the Eocene, the present day mammalian fauna came into being.* Only the deer, giraffe, bovids and hominids were missing; these first appeared in the Neogene. The marsupials continued to be well represented. One of the best known is Cuvier's "opossum", *Peratherium cuvieri*, discovered in Montmartre, which was identified as a marsupial from the characters of its mandible. Later discovery of the pubic bones, which supported the ventral pouch, confirmed this hypothesis. This was a striking example of the principle of the correlation of characters.

Among the groups which became extinct in the Eocene, the heavy and thick set amblypods with nearly complete dentition and crested molars, were represented by *Coryphodon*, an animal intermediate in size between a tapir and a rhinoceros. Another amblypod, *Dinoceras*, had large "horns" as its name suggests, but these were only bony protuberances on the skull (Fig. 2.7).

The evolutionary leap occurred between the Thanetian and the Sparnacian (see p. 62), that is between the fauna of Cernay and that of Meudon. The perissodactyl *Hyracotherium* succeeded *Phenacodus* (Palaeocene). *Lophiodon*, a tapir-like animal, though without a trunk, appeared in the Cuisian and was followed in the Lutetian by *Propalaeotherium*.

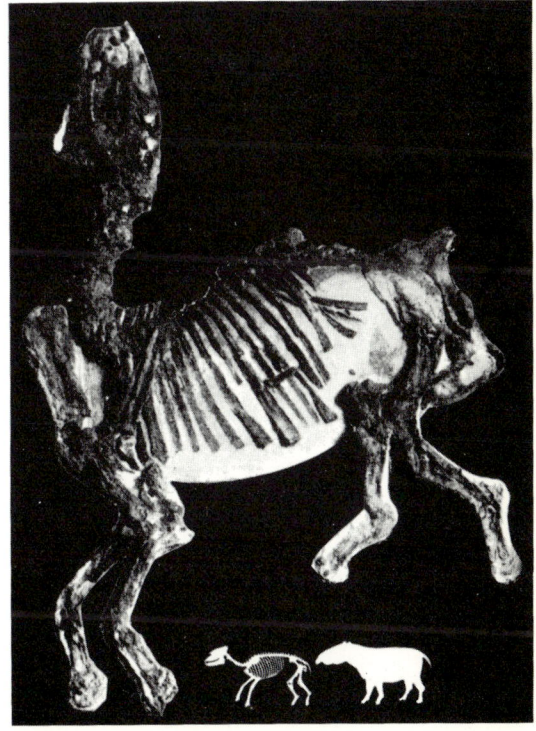

2.8 **Palaeotherium** (late Eocene and early Oligocene), height 1.2m to the shoulder. The body of the animal resembled that of a tapir and the head had a short trunk as shown in the small silhouette (bottom right) based on Cuvier's reconstruction. The attitude of the skeleton of *Palaeotherium magnum*, found in the gypsum beds (Ludian) of Vitry, near Paris, suggests that the animal made a supreme effort in trying to escape death by asphyxiation in the sulphate mud (photo: Serrette).

Study of the fauna of Egerkingen which occurs in fissures in the Middle Eocene oolitic ironstone (Siderolitic) of Switzerland reveals the appearance of several groups which were to expand in the Late Eocene: *Lophiodon, Propalaeotherium* and a great variety of artiodactyls (HARTENBERGER, 1970). The true *Palaeotherium* Upper Eocene, p. 37) had a full set of teeth, similar to that of a modern rhinoceros, the build of a tapir, five toes on the foreleg and three on the hind, indicating that it was in the line of evolution of the horses. These had made their first appearance in the Lower Eocene of North America with *Eohippus* which had four toes on the foreleg, three on the hind, and was the size of a fox (Fig. 2.9).

In the Ludian (latest Eocene) of Montmartre, as well as *Palaeotherium*, there are artiodactyls resembling the gazelle (*Xiphodon*), the hare (*Dichobune*) and a tapir with a flat tail like that of an otter (*Anoplotherium*). The graceful *Xiphodon* and the heavy *Anoplotherium* are today regarded as members of the Tylopoda, which were then highly diverse but had little resemblance to their modern counterparts, the camels and the dromedaries.

Before leaving the Eocene, we may recall *the appearance of the proboscidians*, ancestors of the modern elephants. The first member of this group was *Moeritherium*, which was about the size of a wild boar. It had an almost complete set of teeth of which the upper and lower

Eohippus, Eocene

Orohippus, Eocene

|—————|
10 cm

second incisors are specialized into small tusks. It also had bunodont molars and a very small trunk. In the Eocene of the Fayum in Egypt it is accompanied by the remains of an enormous and unusual hyracoid, *Arsinoitherium* (Fig. 2.10). According to HARTENBERGER (1970), the Eocene was a long period of endemism in Europe during which faunal exchanges with North America were totally interrupted by the opening of the North Atlantic. Exchanges with Asia seem to have been of little importance and (because of the presence of the Ural Sea) may not even have existed. This isolation favoured the adaptive radiation of those groups which were present at the beginning of this period. However, these groups ultimately became extinct in the Oligocene in the face of the arrival from Asia of newcomers which reached Europe as a result of the draining of an arm of the Ural Sea.

A new evolutionary jump, the **"great break"** of STEHLIN (1909), occurs between *the Eocene fauna of Montmartre and that of Ronzon* (Upper Loire), of early Oligocene age. The Eocene perissodactyl, *Palaeotherium*, died out, as did the prosimians. In their place appeared the rhinoceros (*Aceratherium*), the tapir and *Anthracotherium*, an artiodactyl, similar to the wild boar, with strong incisors and large canine teeth. A new group of perissodactyls, the Titanotheres (Fig. 2.13) attained their climax and then rapidly died out by the end of the Oligocene. The Equidae (*Mesohippus*, Fig. 2.11 and *Miohippus*) continued to increase in size (65cm to the withers) and at the same time became three-toed. The proboscidians (*Palaeomastodon* of the Fayum, see Fig. 5.8, p. 145) developed four tusks and a trunk. *Hyaenodon* was also present, as has been shown by the presence of good specimens in the

2.9 **Eohippus (or Hyracotherium),** early Eocene of Europe and North America, is considered to be the ancestor of the modern horse. It was no larger than a fox. Its premolars were sharp. The front feet had four functional digits, while the hind feet had three functional digits flanked by two rudimentary ones. It was probably digitigrade, i.e. it walked on its toes. (See also Fig. 5.9, p. 146.)
Orohippus, from the middle Eocene, was slightly larger in size. The median toes in the front feet were longer than those of *Eohippus* and the rudimentary toes on the hind leg of *Eohippus* had disappeared (after Augusta and Burian, by permission of the *Encyclopaedia Universalis*).

2.10 **Arsinoitherium:** Eocene, from the Fayum, Egypt. This hyracoid was as large as a rhinoceros and was the largest animal of its time. It had two enormous horns on the nasal bone and two smaller ones on the frontal bone of the skull (after Augusta and Burian in the *Encyclopaedia Universalis*).

2.11 **Mesohippus:** Oligocene of North America. This was about the size of a sheep and had teeth with well-developed cusps. It had only three functional toes on the front feet. (See also Fig. 5.9, p. 146) (after Scott).

2.12 **Hyaenodon:** Oligocene (Quercy phosphorites) was a carnivore with well-developed canine teeth and sharp molars, similar to those of a modern cat (after Scott).

2.13 **Brontotherium = Titanotherium:** Oligocene of North America. This perissodactyl was 2m at the shoulder. The concave skull had large horn-like proturberances above the eye-sockets. *Titanotherium* disappeared abruptly at the end of the Oligocene shortly after attaining its climax (after Augusta and Burian, *Encyclopaedia Universalis*).

Oligocene phosphorite deposits at Quercy in France (Fig. 2.12). Among the primates, one of the early simians, *Propliopithecus* of Egypt was the first member of the lineage leading up to the present-day gibbon.

Because the Aquitanian mammals were closely related to their predecessors, this stage is placed in the Oligocene by vertebrate palaeontologists, whereas the marine fauna indicates that its age is early Miocene (see p. 78).

The uniformity of the mammalian fauna presupposes communication between the continents. This is true for North America, Asia and Africa during at least part of the Tertiary period. By contrast, South America and Australia, because of their isolation, each had a very different mammalian fauna which had evolved in its own closed environment (p. 54 and 219).

3. Micro-organisms (Figs. 2.14 to 2.26, 3.31 to 3.33, 3.36 and 4.11).

The need to determine the stratigraphic horizon of formations sampled by drilling, which provides only small samples, has led petroleum geologists to study in detail the remains of micro-organisms contained in sedimentary rocks. These include the *foraminiferids, nannoplankton, radiolarians, ostracods, pollen, phytoplankton, charophytes* and, in the older rocks, *conodonts*. The science of micropalaeontology has developed very widely since the end of the Second World War.

In the last few years, the use of the scanning electron microscope has greatly added to our knowledge of micro-organisms and has led to the establishment of biozones which their authors believe to be of world-wide application.

In establishing these world-wide biozonations the study of the large foraminiferids (nummulites, discocyclines, and alveolines) has been largely neglected. (The use of the term ''large'' in relation to unicellular organisms may seen incongruous, but it is, in fact, justified since some nummulites such as *N. millecaput* can attain 10cm in diameter.) The study of other benthonic foraminiferids for stratigraphic purposes has also been largely abandoned, in both cases because they are believed to be greatly influenced by their environment. Effort has been largely concentrated on *the planktonic foraminiferids and the nannoplankton* since their distribution is believed to be world-wide. But after spectacular progress, there has been some disenchantment. It has been found that these micro-organisms also are sensitive to climatic variations and that equatorial globigerinids are different from those of polar seas. By contrast, Caribbean forms are similar to those of the Mediterranean.

From another point of view, if one considers not merely isolated benthonic foraminiferids, but associations of species in formations of comparable facies, the results can be of great stratigraphic importance. Their study should not, therefore, be neglected. Because of their

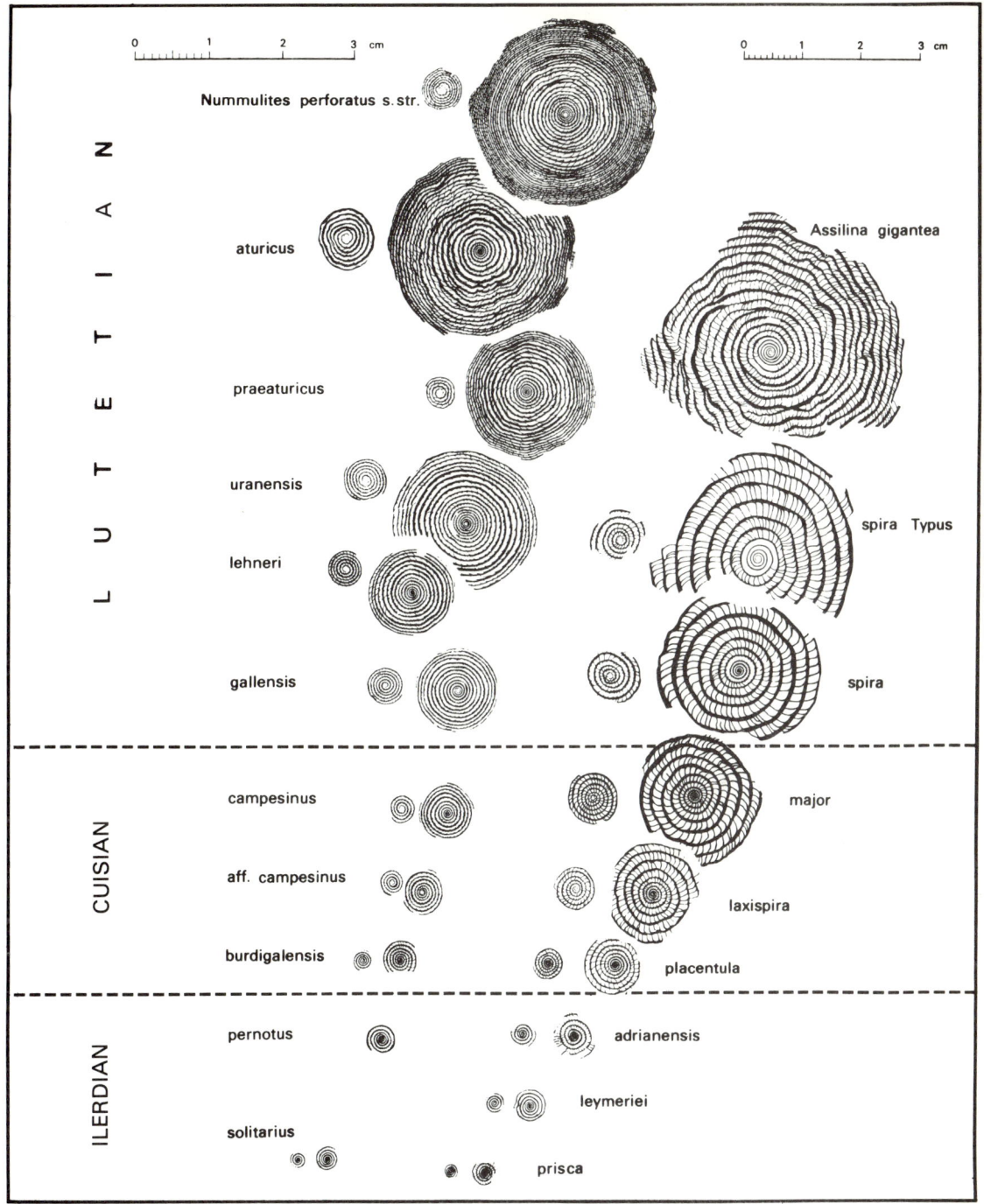

2.14 **Evolutionary lineages of Nummulites perforatus** (left) and **Assilina spira** (right) during the Eocene (after Schaub, 1963). For each species, form A (on the left) is the *megalospheric* form, having a larger proloculus (or initial chamber of the test) than form B (on the right) which is known as the *microspheric* form. Note however, that the adult of the microspheric form is larger than that of the megalospheric one.

The nummulites show an alternation of generations. Form A are the gamonts and produce the gametes which, by fusion, give rise to a schizont (or agamont) generation – form B. These in turn, give rise to embryos with 2n chromosomes which are gamonts.

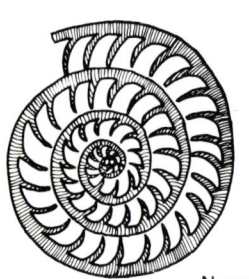

Nummulites

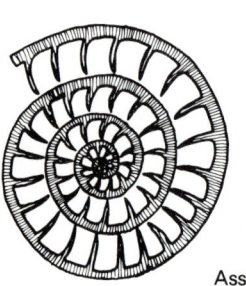

Assilina

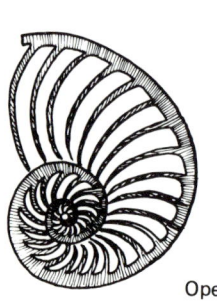

Operculina

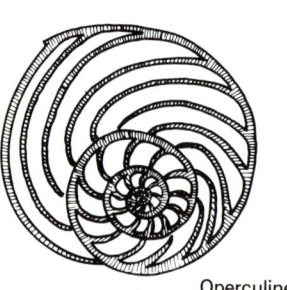

Operculinella

2.15 **The structure of Nummulites, Assilina, Operculina and Operculinella** (after Nemkov, 1967).

Members of this group have discoidal tests which may be flattened, lenticular or globular. The test is made up of two parts:

1 – A framework of lamellar calcite in the form of a plane spiral, the dorsal cord, the rate of increase of which is specific. The septa are extensions of the dorsal cord and the trabeculae extend outwards from the septa. The septa abut against the previous whorl. A basal opening or foramen in each septum allows the inter-connection of protoplasm between chambers. The whole is traversed by a number of canals (longitudinal, lateral or transverse) which connect with the exterior by pores.

2 – The dorsal cord and septa support the outer wall, or spiral lamina, which is arched in cross-section, and is pierced by a large number of small tubes, which are manifested on its surfaces by pores. The sub-family *Nummulitinae* is characterised by the presence of a large number of simple chambers. The following genera can be distinguished:

1 – Spiral lamina and chamber cavities *in*volute:
   a) spiral increasing rapidly: septa, long and sickle-shaped, tangential to the whorl at their base: **Operculinella** (Recent).
   b) spiral increasing slowly; septa short, curved and oblique to the radius of the whorl: **Nummulites.**
2 – Spiral lamina *in*volute, chamber cavities *e*volute: **Assilina.**
3 – Spiral Lamina and chamber cavities *e*volute: spiral increasing rapidly in both generations: **Operculina.**

Nummulites are marine benthonic organisms. They are both facies fossils and stratigraphic fossils. *Nummulites* (sensu stricto) lived in warm (25°C) shallow waters of normal salinity from the early Eocene (Ilerdian) until the end of the middle Oligocene. They often accumulated in banks, as in the nummulitic limestone (with abundant *N. laevigatus*) in the lower Lutetian in the Paris Basin. They are often found associated with *Discocyclina* in algal reefs.

The large number of species and the variety of associations and microfacies suggests that the various species of *Nummulites* had different ecological requirements.

1 – *N. planulatus*, Cuisian (Fig. 7. 14. p. 178) is often associated with *Alveolina* and miliolids in a sandy facies.

2 – *N. gizehensis* (Fig. 3.16, p. 70) is most often found in the sub-reef limestones of the upper Lutetian of the southern part of the Mediterranean region.

3 – *N. variolarius* (Fig. 3.28, p. 79) occurs in calm, shallow water. It is often found in association with miliolids in calcareous sands (middle and late Lutetian) or in quartz sands (Auversian, Marinesian).

Interest in *Nummulites* is mainly stratigraphical. Modern techniques of measurement make it possible to define specific lineages which have a distribution related to the climate and hence to the latitude. For example, in the late Lutetian, large nummulites with granulations associated with their septal filaments, such as *N. perforatus* and *N. aturicus*, are restricted to the northern part of the Tethyan area (from the Himalayas to Aquitaine, by way of Iran, Armenia, Turkey, the Carpathians, Switzerland, the French Alps and Catalonia). At the same time its southern part was occupied by a different species, *N. gizehensis*.

Three ancestral species appeared in the late Palaeocene: *N. fraasi*, *N. deserti*, *N. solitarius*. The first-named of these was widespread in the north of the Tethys and gave rise to the characteristic species; *N. planulatus* (Cuisian), *N. laevigatus* (early Lutetian), *N. millecaput* (middle Lutetian), *N. brongniarti* (late Lutetian = Biarritzian). The line became extinct at the end of the middle Eocene with other forms of large size.

1. See "Les Nummulites" by A. Blondeau, Ed. Vuibert, Paris.

importance, they are included in the illustrated tables of the distribution of the principal groups of micro-organisms in the Palaeogene. These groups include the ostracods (Fig. 3.31), nummulites (Fig. 2.14), benthonic and planktonic foraminiferids (Fig. 3.36 and 2.21), nannoplankton (Fig. 2.23), microflora (Fig. 2.25), phytoplankton (Fig. 2.26) and charophytes (Fig. 3.33). The lists are printed for reference, since some of the names will occur repeatedly in the text. The reader will find it useful to refer to these tables in order to place

the fossils referred to later in their proper stratigraphical context. The figures will also assist in the recognition of these organisms in thin sections or in preparations made from non-consolidated sediments.

Finally, it should be noted that it is not necessary to look for strict coincidence of the

form   A

form   B

2.16 **Assilina exponens** from the upper Lutetian of Cap Mortola (Italian Maritime Alps), form A (megalospheric) and form B (microspheric) x 4 (photo: Blondeau).
Species of *Assilina* are planispiral, with an *in*volute spiral wall but *e*volute chamber cavities, which distinguish this genus from *Nummulites*. The septa are straight and slope slightly backwards. *Assilina* appeared at the beginning of the early Eocene (Ilerdian) and died out at the end of the middle Eocene (Biarritzian biozone). The genus is often associated with *Nummulites*, sometimes forming banks above those of the latter. It lived in a marine off-shore environment. *Assilina exponens* first appeared in the early Lutetian of the northern shores of the Tethys and evolved during the Lutetian. The megalospheric and microspheric forms shown above are the latest forms of the species, collected from the Biarritzian biozone of the Palaeogene succession of the Italian Maritime Alps (Cap Mortola).

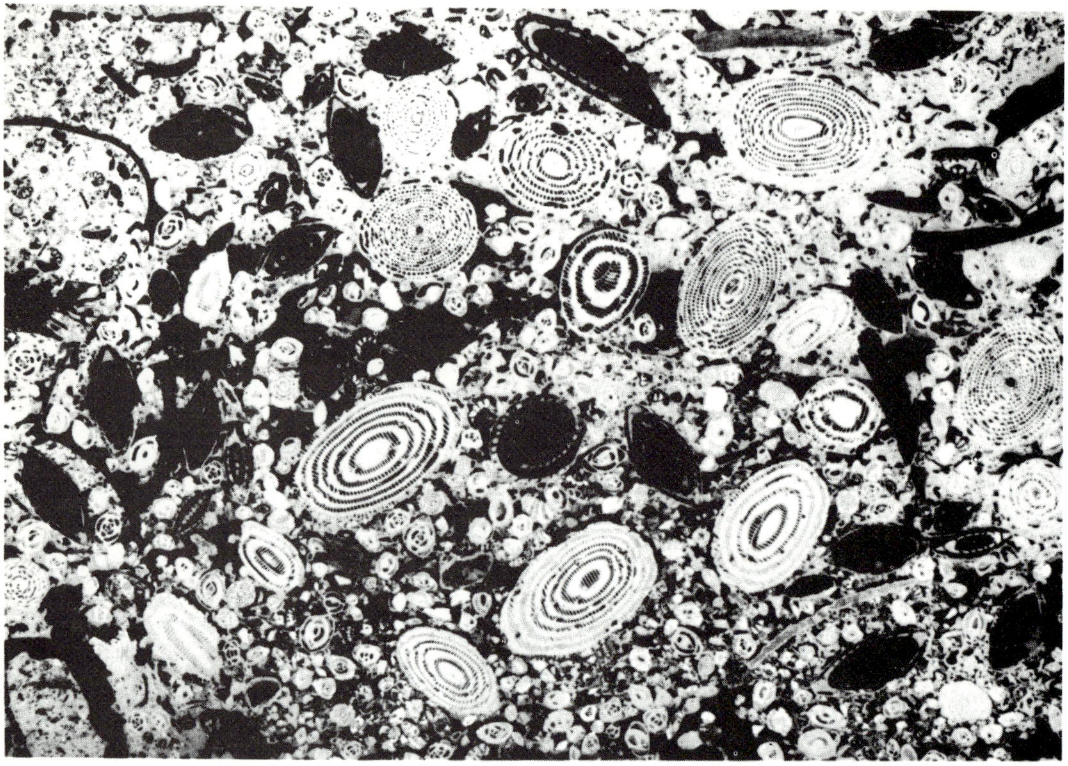

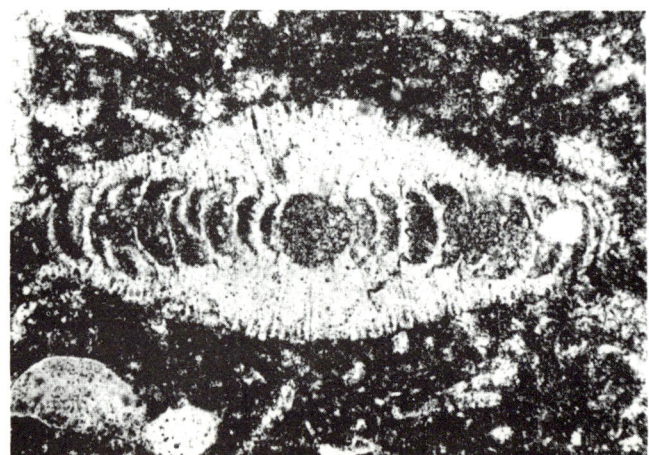

2.17 **Alveolina limestone** from the middle Ilerdian of the Aude (south east of Toulouse) x 10. Alveolinids are axially coiled and have a test which is elongated, fusiform or subglobular. The test is strengthened by partitions which may have minute blind cavities or "alveoli". In some tests the wall is greatly thickened in the axial region ("flosculinisation").

Note the presence of numerous miliolids and *Nummulites globulus* (the black patches) in the figure.

The alveolinids appeared in the mid-Cretaceous in warm shallow seas and died out at the end of the middle Eocene. Three species penetrated into the Paris Basin: *Alveolina oblonga* in the Cuisian, *A. bosci* in the middle Lutetian, *A. elongata* in the late Lutetian.

2.18 **Linderina paronai.** Species of the genus *Linderina* are foraminiferids with a flat discoidal test, which may be inflated or thickened at the centre, made up of chamberlets arranged planispirally x 40 (Photo: Blondeau).

The genus lived in company with miliolids in the middle Eocene. However, the species *Linderina brugesi* appeared in the Paris Basin only at the end of the Lutetian (Biarritzian biozone) and persisted into the upper Eocene (Auversian, Marinesian). *Linderina paronai* lived in the warm seas of India, Qatar, Arabia, Somalia and Madagascar in the Lutetian. It is very similar to *L. brugesi* (a Mediterranean species), but is more squat.

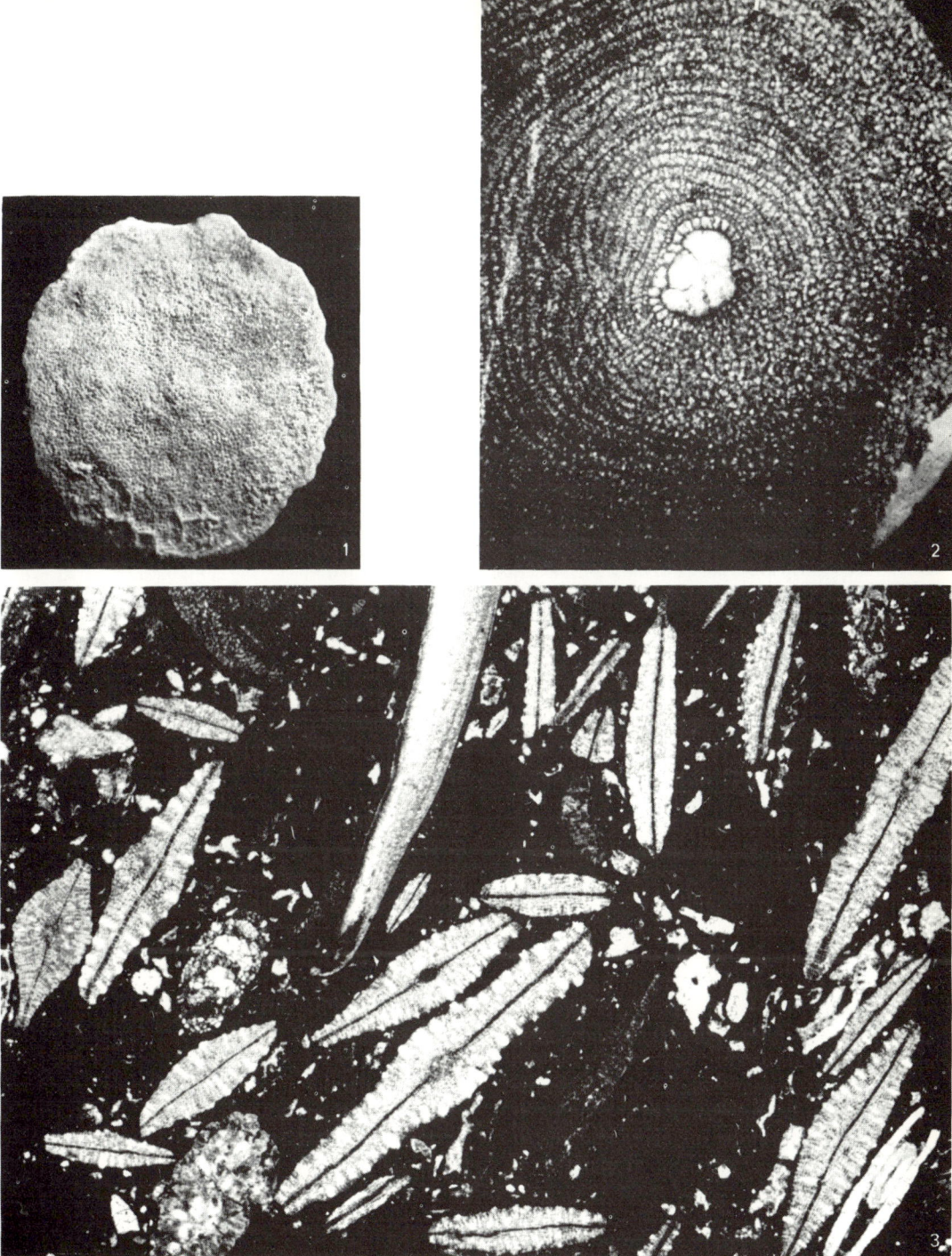

2.20 **Discocyclinids.** These are discoidal, biconvex foraminiferids in which the test is covered with granules of varying size. The test is 1mm to 40mm in diameter. Internally, an equatorial layer of rectangular chambers surrounds the micro- or megalospheric embryo (Fig. 2) and is covered on both sides by layers of lateral chambers, which give the test a more or less swollen appearance. The discocyclinids belong to the superfamily *Orbitoidacea* and, like the lepidocyclinids, evolved from the Cretaceous "Orbitoids". The discocyclinids died out at the end of the Eocene. Together with the nummulites and alveolinids, they are important in the investigation of the stratigraphy of the neritic formations of this epoch (Photo: Neumann).

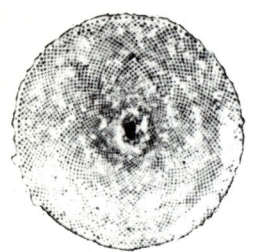

2.19 **Orbitolites complanatus** (x 2): A foraminiferid with a thin discoidal test (Ilerdian to middle Eocene). *Orbitolites* lived in warm, calm seas amongst sea grasses and seaweeds.

limits of horizons in various tables of biozones. It would indeed be surprising if the appearances or disappearances of organisms belonging to different evolutionary lines were strictly isochronous. As a result it may be difficult to agree on the precise position of the limits of different stages which, in essence, consist of several biozones. Discussion of this problem is outside the scope of this book, however.

2.21 **Planktonic foraminiferids of the Cenozoic Era**[1] (x 50): The planktonic foraminiferids are extremely sensitive to changes in their environment and are most abundant in an open sea of normal salinity. A fall in sea temperature at the end of the Cretaceous might thus explain the sudden world-wide extinction of many advanced forms, such as *Globotruncana*, *Rugoglobigerina* and many of the *Heterohelicidae* at the end of the Maastrichtian.

A fauna of small globigerinids with simple structure (e.g. *Globigerina eugubina* – see table opposite) with apparently little or no relationship to the faunas of the late Cretaceous, began to evolve at the beginning of the Palaeocene. As had happened in the Cretaceous, rapid development and diversification occurred in the Palaeocene beginning with species of *Globigerina* and inornate species of *Globorotalia* (*G. trinidadensis*). These were followed in the middle and late Palaeocene by carinate (keeled) *Globorotalia* (*G. acuta, G. velascoensis*) and forms with well-marked spines on the surface of the test (*Globigerinita echinata* (Fig. 2.22). This proliferation of types continued in the early Eocene (*Globorotalia aragonensis*) and accelerated in the middle and late Eocene, where several new genera appear, notably *Hantkenina* with long tubular spines, *Cribrohantkenina* with long spines and multiple apertures, *Globigerinatheka* characterised by several apertures in the last chamber and the presence of *bullae* (blister-like structures) (*G. semiinvoluta, G. barri*) and *Orbulinoides*, a large spherical form, which had only a short existence and was derived from *Globigerinatheka* (see Table opposite).

The further evolution of this characteristic fauna was abruptly cut short at the end of the Eocene. However, this interruption in the development of the planktonic foraminiferids was not as great as that which had occurred at the end of the Cretaceous. The forms which suffered most were the highly evolved ones of short duration, such as *Hantkenina* and *Globorotalia cerroazulensis* s.l., whereas several species of *Globigerina* persisted into the Oligocene. It appears likely that a sudden change in ecological conditions, probably a fall in temperature, caused the disappearance of the highly specialised species.

Beginning in the early Oligocene there was, for the third time (as had occurred in the early Cretaceous and the early Palaeocene), a rapid evolution of planktonic foraminiferids from simple globigerinid types, first of unkeeled forms of *Globorotalia* and, later, of keeled forms (*G. fohsi lobata, G. exilis, G. truncatulinoides*). Many more specialised forms appeared during the Neogene, for example, *Globigerinatella, Orbulina, Sphaeroidinella* and *Bolliella*. This tendency for diversification in the foraminiferids has continued to the present day.

Worldwide, it can be shown in the Palaeocene, as in the Cretaceous, that the distribution of foraminiferids was very uniform in latitudes up to about 50°. This allows their use for stratigraphical correlation on almost a worldwide scale. However, climatic changes have influenced the geographical distribution of the planktonic foraminiferids, as can be seen at the end of the Palaeogene and more markedly in the Neogene when climatic zones (tropical-subtropical-temperate-cold) were developed. As a direct consequence, distinct associations of foraminiferids occupied different temperature zones, though the boundaries of the zones overlapped. Correlation on a world wide, or even an inter-regional scale, therefore becomes more difficult. This is especially true in the Miocene.

Many Cretaceous and Tertiary planktonic foraminiferids show rapid evolution, reflected in the morphological modification of their tests. This has allowed the identification of a very large number of genera and species of short stratigraphic range which can serve as excellent "markers". The short range of certain species is illustrated in the table opposite.

The Cenozoic is now divided into 40 zones (Bolli, 1966) on the basis of the vertical range of planktonic foraminiferids, each zone occupying a mean of 1.6 million years. Other planktonic micro-organisms, such as the calcareous nanno-plankton occupying similar ecological niches, show a similar number of zones. The Tertiary can now be subdivided very precisely, using these zonations.

SOME CENOZOIC PLANKTONIC FORAMINIFERIDS

PALAEOCENE	EOCENE	OLIGOCENE	MIOCENE	PLIOCENE	PLEISTOCENE	Figure No.	Species
■						1	Globigerina eugubina
■						2	Globorotalia trinidadensis
■						3	Globorotalia angulata
■	■					4	Globorotalia velascoensis
	■					5	Globorotalia aragonensis
	■					6	Hantkenina aragonensis
	■					7	Globigerinatheka mexicana barri
	■					8	Orbulinoides beckmanni
	■					9	Globigerinatheka semiinvoluta
	■					10	Cribrohantkenina inflata
		■				11	Globigerinita dissimilis
		■				12	Globorotalia opima opima
		■	■			13	Globigerina ciperoensis angulisuturalis
			■			14	Globorotalia kugleri
			■			15	Globigerinatella insueta
			■			16	Orbulina universa
			■			17	Globorotalia fohsi lobata
			■	■		18	Globigerina nepenthes
				■		19	Globorotalia margaritae
				■	■	20	Sphaeroidinella dehiscens
				■		21	Globorotalia exilis
					■	22	Globorotalia truncatulinoides truncatulinoides
					■	23	Bolliella adamsi

Planktonic foraminiferids have a very wide distribution in the open sea. Because of their buoyancy they are often transported by ocean currents into zones where they do not normally live and are often deposited there. Their wide distribution and their rapid evolution make them an important group, not only for detailed stratigraphic subdivision, but also for correlation over long distances.

1. Text and plate by H. M. Bolli (Eidg. Technische Hochschule; geol. Inst. Zurich)

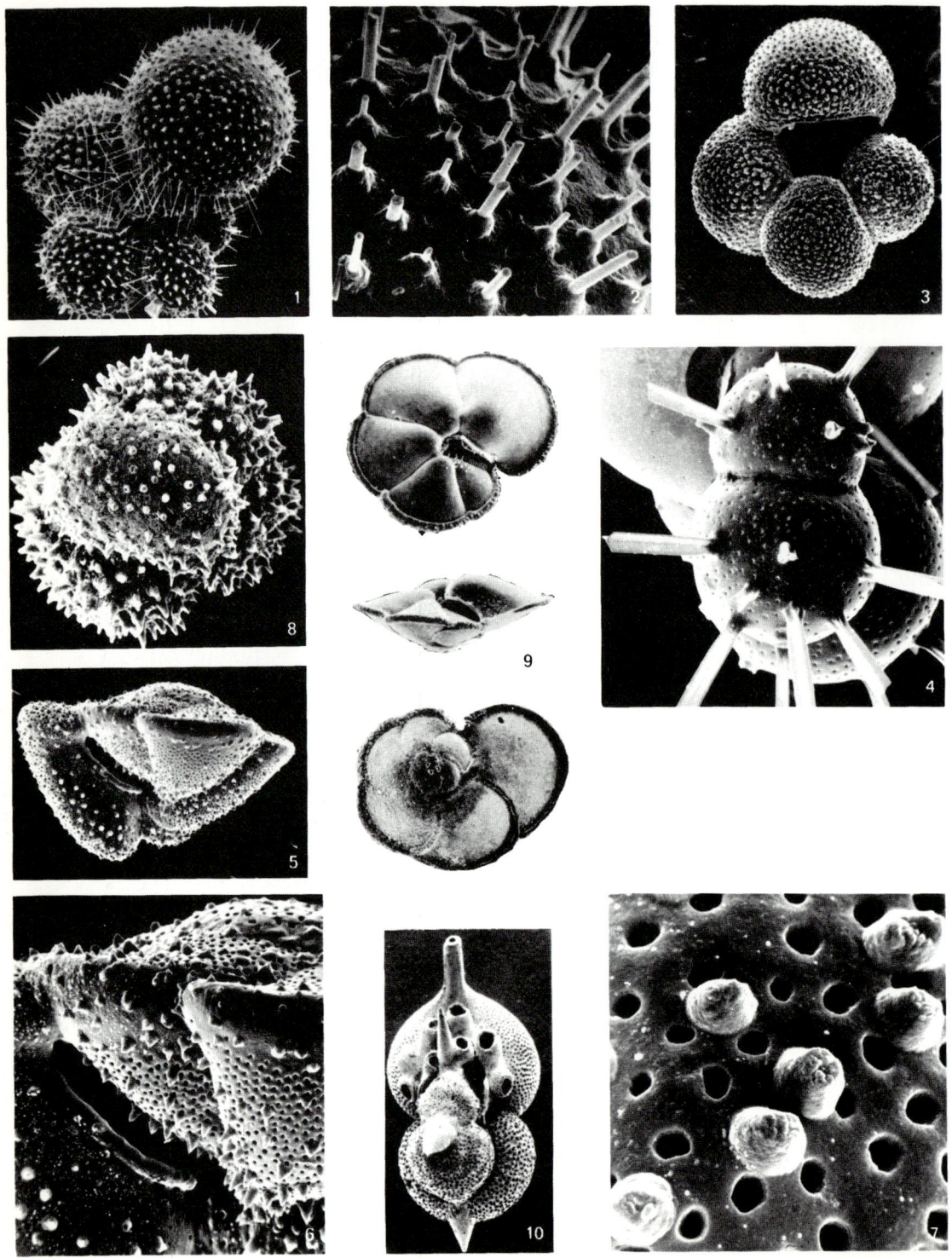

2.22 **Planktonic foraminiferids;** as seen under the scanning electron microscope (photo. Bolli).
Holocene: 1 – *Globigerina calida* (x 150) young, umbilical view. 2 – the same specimen, detail of wall of last chamber (x 1000). 3 – *Globigerina bulloides* (x 100) umbilical view. 4 – *Hastigerina pelagica* (x 250) detail of last whorl. 5 – *Globorotalia crassaformis* (x 100) lateral view. 6 – the same specimen, showing aperture and imperforate keel (x 250). 7 – another specimen (x 1000), umbilical view of the penultimate chamber.
Pleistocene: 9 – *Globorotalia menardii cultrata* (x 30) spiral, lateral and umbilical views.
Middle Eocene of Trinidad: 8 – *Globigerinita echinata* (x 200).
Upper Eocene, Yazoo formation, Mississippi: 10 – *Cribrohantkenina inflata* (x 100).

see opposite p. **48.**

see next page

2.23 **Nannoplankton of the Cenozoic**[1] (x 1000).
Coccoliths are composite calcite discs which envelop marine unicellular algae (coccolithophorids). The morphology of the discs preserved in marine sediments is very variable and it is on this basis that the fossil coccoliths are classified, although the complete envelope, or coccosphere, is more rarely preserved in sediments. In some recent and fossil coccolithophorids, coccoliths of different morphology may be found on the same coccosphere. Coccolithophorids are known to occur in all oceans at the present time. The diversity of species generally increases from the cold waters of high latitude to the warmer waters of the equatorial regions. Certain species show both "cold" and "warm" forms. The small size (1-20µ) of coccoliths is both an advantage and a disadvantage in biostratigraphy. Their minuteness means that the age of a very small sample (from a borehole chip for example) can often be determined. On the other hand, they can easily be reworked into newer deposits or carried long distances and may not, therefore, always indicate the true age or environment of the rock.
The coccoliths, discoasters and other calcareous organisms of similar dimensions have only recently been recognised as useful "markers" in the zonation of the Tertiary. Martini (1971) has proposed 25 zones in the Palaeogene and 21 in the Neogene. Several characteristic forms and their stratigraphical distribution are shown in the figures overleaf.

The Cretaceous – Tertiary boundary at the end of the Maastrichtian is marked by the extinction of the majority of Cretaceous species. Only five forms seem to have survived, but new forms slowly appeared. At the base of the Palaeocene, *Biantholithus sparsus* (see Table) is the only new species to occur. It is the first discoaster, a group which developed rapidly during the Palaeocene. The early discoasters were mostly discs, followed later by massive star-shaped forms. These were followed by slender star-shaped forms which finally became extinct at the end of the Pliocene
The coccoliths of the family *Coccolithaceae* developed in the lower Palaeocene and were present throughout the Tertiary sometimes becoming predominant. Their stratigraphical use is at present somewhat limited and is likely to remain so, until the details of their morphological variations in relation to time are better known. The evolution of *Sphenolithus* since the Palaeocene and of *Helicopontosphaera* since the Eocene are sufficiently well-known to allow these forms to be used stratigraphically. Apart from these, forms with a short range, such as *Fasciculithus* and *Heliolithus*, which are restricted to the Palaeocene, *Marthasterites* in the lower Eocene, *Nannotetrina* in the middle Eocene, *Isthmolithus* at the Eocene – Oligocene boundary, *Triquetrorhabdulus* at the Oligocene – Miocene boundary and *Catinaster* in the Miocene, have also been used as stratigraphic "markers".
The number of species of nannoplankton increased rapidly from the beginning of the Tertiary until the early Eocene, then slowly diminished until the early Miocene and has since remained fairly constant. However, the number of living species of nannoplankton is three times that known from recent sediments, which indicates the considerable loss of organisms which occurs during fossilization.

1. *Text and Table by K. Perch-Nielsen, Institute for Palaeontology, Copenhagen.*

PALAEOCENE	EOCENE	OLIGOCENE	MIOCENE	PLIOCENE	QUATERNARY	Figure No.	CENOZOIC NANNOPLANKTON	
						1	Biantholithus sparsus	
						2	Chiasmolithus danicus	
						3	Ellipsolithus macellus	
						4	Fasciculithus janii	
						5	Heliolithus kleinpelli	
						6	Discoaster multiradiatus	
						7	Discoaster binodosus	
						8	Marthasterites tribrachiatus	
						9	Neococcolithes dubius	
						10	Discoaster lodoensis	
						11	Nannotetrina fulgens	
						12	Rhabdolithus gladius	
						13	Sphenolithus furcatolithoides	
						14	Isthmolithus recurvus	
						15	Helicopontosphaera reticulata	
						16	Sphenolithus distentus	
						17	Sphenolithus ciperoensis	
						18	Triquetrorhabdulus carinatus	
						19	Sphenolithus belemnos	
						20	Discoaster exilis	
						21	Catinaster coalitus	
						22	Discoaster hamatus	
						23	Discoaster quinqueramus	
						24	Ceratolithus rugosus	
						25	Cyclococcolithus leptoporus	
						26	Discoaster surculus	
						27	Discoaster pentaradiatus	
						28	Pseudoemiliania lacunosa	

Palaeontology

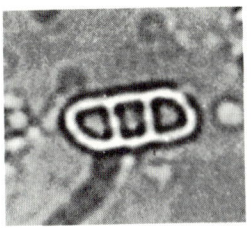

1

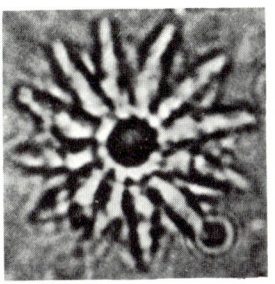

2

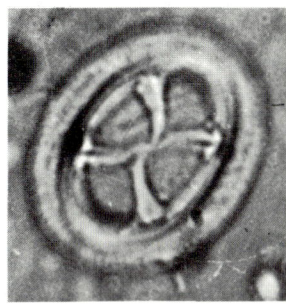

3a

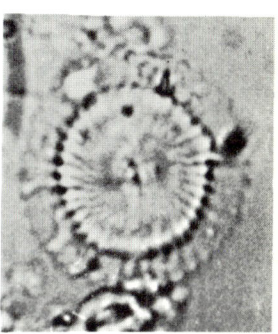

4a

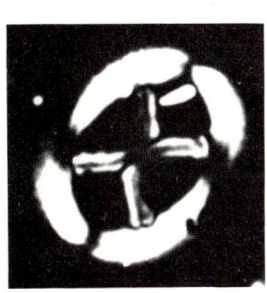

5a

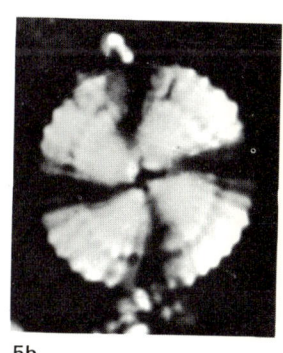

3b   4b   5b

2.24 **Nannoplankton:** seen by the light microscope (photos: Lezaud).
1 – *Isthmolithus recurvus*, Bartonian (x 2000). 2 – *Discoaster boulangeri*, early Lutetian (x 2000). 3 – *Chiasmolithus expansus*, early Lutetian: a) ordinary light; b) crossed nicols (x 2000). 4 – *Discoaster lodoensis*, early Lutetian: a) focussed on median level; b) focussed on high level (x 2000). 5 – *Heliolithus kleinpelli,* Palaeocene: a) ordinary light; b) crossed nicols (x 2000).

2.25 and 2.26 **Pollen and Dinoflagellates**[1] (x 200).

The term "palynology" was created by Hyde and Williams in 1944 for the study of spores and pollen grains. Since then the scope of the term has been widened to cover the study of all fossil micro-organisms contained in sedimentary rocks which can be extracted by palynological techniques (spores, pollen, dinoflagellates, chitinozoans, microforaminiferids, diatoms, etc.).

Continental micro-organisms may be dispersed by winds, water and insects and may fall into lacustrine, lagoonal or marine environments where they may be buried with the indigenous microplankton. Provided that conditions are suitable for their preservation (absence of oxygen or severe diagenesis), they can be used for correlating the sediments formed in a variety of environments.

A. Tertiary Palynology

By Tertiary times, plants were already highly evolved and all the groups extant at the present day were represented. The use of palynology in stratigraphy is largely based on the classical concepts of appearances and extinctions of genera and species and their relative abundance. In addition, it is necessary to recognise the significance of purely geographical factors (linked with the mode of dispersion) and phytogeographical factors (the controls of plant distribution by latitude, altitude, etc.). A full understanding of these factors is essential to ensure the validity of palynological zones. The elimination of local factors makes it possible to recognise the substantial variation in plant populations due to climatic differences, and hence to identify those which have time significance and so can be used for long-distance correlations.

1) Microflora

Early Tertiary (Palaeocene-Eocene) times are characterised by the rapid development of the Angiosperms, which by the late Neogene produced a flora closely resembling that of the present day. In the Paris Basin, it is possible to see a steady change from a tropical or warm-temperate flora to a temperate one which contains elements of a cold-climate flora, the change being most marked in the late Stampian. The microflora of the Palaeocene is very similar to that of the late Cretaceous and is composed mainly of Normapolles, pollen grains of complex structure attributed to plants belonging to the *Amentiferae* or to primitive members of the *Myrtaceae* (Myrtle). The early and middle Eocene yields many species of Normapolles divided between several genera. These are associated with forms indicating a hot and humid climate which are reminiscent of the present-day flora of south east Asia. The warm flora began to disappear in the late Eocene, being replaced by representatives of a cold flora (*Ulmus, Fagus, Alnus*, etc.), a tendency which persisted in the early Stampian. By the late Oligocene the vegetation was dominated by a North European flora (*Picea, Betula, Abies, Alnus, Carpinus, Tsuga*, etc.) but still included a number of warm-climate genera (*Carya, Sequoia, Liquidambar, Engelhardtia*), which had survived successive falls in temperature.

2) Dinoflagellates

The Tertiary epoch is characterised by the appearance or proliferation of a number of genera such as *Wetzeliella, Cordosphaeridium, Lingulodinium, Areospheridium, Cannosphaeropsis, Chiropteridium* and *Hemicystodinium*. In the Paris Basin, several of the major transgressions (Thanetian, Cuisian, Bartonian, Stampian) were associated with an almost complete replacement of the dinoflagellates. Some forms adapted well to abnormal salinities present from time to time and became abundant in these restricted environments. From Miocene times onwards, the dinoflagellates show little evidence of evolution and forms similar to those of the Miocene can be found in present-day seas.

B. The Palynostratigraphy of the Palaeogene of the Paris Basin.
The information given in figures 2.25 and 2.26 must be regarded as provisional in those parts of the succession which have not been studied in detail or where organisms have not been preserved, as in the case of the Montian, the early Thanetian, the middle Cuisian, the early Lutetian, a large part of the Auversian, the Marinesian and the late Stampian. The data for the late Oligocene and the Aquitanian are based on samples collected near Marseilles.

1) Microflora (Fig. 2.25)

The Table shows only the distribution of pollen. The spores of Pteridophytes are omitted since their occurrence is facies-controlled and their evolution is of little significance in the Tertiary.

The Montian is characterised by an abundance of pollen of the group *Stephanoporopollenites*, especially in Belgian deposits. In the Thanetian, certain species of *Complexiopollis, Trudopollis* and *Vacuopollis* are common in the Bracheux sands. The Sparnacian of the Paris Basin is rich in *Plicapollis pseudoexcelsus* and *Triatriopollenites platycarioides*. The microflora of the Cuisian is similar to that of the Sparnacian, though *Anacolosidites* and *Pentapollenites* become more frequent. *Nipa* and *Araucariacites* are abundant at the top of the Cuisian. The Lutetian often contains high proportions of *Pentapollenites* and *Plicapollis* of the group *plicatus*. The flora of the Auversian and the Marinesian is little known, but in the few samples which have yielded pollen, *Corsinipollenites, Ulmus, Liquidambar* and *Pterocarya* have been recorded. Intercalations of marl in the gypsum of the Paris Basin have yielded abundant pollen of Ludian age, including the group *Taxodiaceae-Cupressaceae*, the genus *Carya* and numerous species of *Cornaceae* and *Araliaceae*.

From early Stampian times, the proportion of cold forms increases (particularly *Betula* and *Picea*) and conifers become dominant. The most characteristic species is *Boehlensipollis hohli*, which seems to appear simultaneously in all the North European basins at the end of the early Stampian. A sharp fall in temperature in the late Stampian was responsible for the disappearance of many Eocene species and led to a further increase in cold forms (*Abies, Alnus, Tsuga*, etc.).

---

1. Text and plates by J.-J. Chateauneuf (B.R.G.M., Orleans).

TERTIARY POLLEN AND SPORES

PALAEOCENE		EOCENE						OLIGOCENE			MIOCENE	Figure No.	Taxon
MONTIAN	THANETIAN	SPARNACIAN	CUISIAN	LUTETIAN	BARTONIAN (Auv. / Mar / Lud.)			STAMPIAN (early / late)		late	AQUITANIAN		
								1	Stephanoporopollenites group
								2	Complexiopollis sp.
								3	Trudopollis cf. oculoides
								4	Vacuopollenites sp.
								5	Tricolporopollenites sp.
								6	Nothofagidites sp.
								7	Intratriporopollenites cf. pseudoinstructus
								8	Plicapollis pseudoexcelsus
								9	Triatriopollenites platycariodes
								10	Plicatopollis gr. plicatus
								11	Thomsonipollis magnificus
								12	Caryapollenites circulus
								13	Intratriporopollenites sp.
								14	Anacolosidites pseudoefflatus
								15	Spinizonocolpites sp.
								16	Pentapollenites group
								17	Tricolporopollenites pseudolaesus group
								18	Polyporopollenites eoulmoides
								19	Araucariacites sp.
								20	Inaperturopollenites magnus
								21	Corsinipollenites oculis noctis
								22	Ephedripites group
								23	Ulmus-Zelkova group
								24	Liquidambar
								25	Tricolporopollenites
								26	F. Cornaceae-Araliaceae
								27	Trivestibulopollenites betuloides
								28	Pterocarya
								29	Tricolporopollenites sp.
								30	Stephanocolpites sp.
								31	Sciadopitoid forms
								32	Carya group
								33	Leptocarpus group (in Chanda 1961)
								34	Tricolporopollenites sp.
								35	Tricolporopollenites sp. Gr Cingulum-oriformis (sontag 1966)
								36	Caryapollenites simplex
								37	Fagoid pollen group
								38	Sequoia
								39	Picea
								40	Boehlensipollis hohli
								41	Tsuga
								42	Alnus
								43	Abies
								44	Cedrus
								45	Carpinus

The pollen associations known from late Aquitanian are of Stampian affinities.   In late Aquitanian times, however, a temporary rise in temperature led to a reduction in the number of cold species. During the remainder of the Neogene, a slow fall in temperature caused a progressive modification of the plant cover which led finally to the present-day flora.

2) **Dinoflagellates** (Fig. 2.26)
The lowest horizon of Tertiary age in the Paris Basin which has yielded dinoflagellates is the Thanetian. Its palynflora is strongly differentiated from that of the upper Cretaceous by the arrival of the genus *Wetzeliella* and by an abundance of *Cordosphaeridium* and *Areoligeraceae*. The Sparnacian contains microplankton indicating a restricted environment (*Wetzeliella* cf. *parva, Lingulodinium, Baltisphaeridium funginum*). The Cuisian, which is characterised by the appearance of the short-ranged species, *Lanternosphaeridium bipolare,* contains abundant *Cyclonephelium, Wetzeliella, Hystrichosphaeridium* and *Hystrichokolpoma*.
With the arrival of the Auversian sea there was a slight modification in the dinoflagellate population with the appearance of new species of *Areosphaeridium* and *Wetzeliella* and of the genus, *Surculosphaeridium*. Subsequently there was little change until the early Stampian when at the level of the Sannois limestone new species of *Wetzeliella* appeared. The genus *Chiropteridium* appeared at the base of the Fontainebleau sands, and is characteristric of the Septarian clay in Germany and the Boom Clay in Belgium.

III. – PALAEOGEOGRAPHY

By the end of the Cretaceous the major geographical features of the Earth had begun to resemble those with which we are now familiar. The three great continents of North and South America and Africa (with the Arabian peninsula attached) could have been recognised. North America was still welded to Greenland which in turn was joined to the British Isles. Communication between the Arctic Ocean and the growing Atlantic Ocean only occurred at the end of Eocene times or at the beginning of the Oligocene. The main fracture occurred to the west of Britain, which remained essentially part of Europe. The cooling of the seas of the European area as a result of the influx of Arctic waters, while not provoking a faunal change of the magnitude of that at the end of the Cretaceous, was sufficient to bring about an "evolutionary crisis" in most groups.

In contrast, *the two Americas, temporarily united at the beginning of the Palaeocene, became separated and remained so until the Pliocene*. This resulted in the "most splendid isolation" in all of Phanerozoic time, just as the evolution of the mammals became rapid. It led to the survival in South America of an archaic fauna in which opossums, armadilloes and ant-eaters predominated. But the strait which separated the Americas was situated in the latitude of Venezuela, so that the marine provinces of the Gulf of Mexico and of California were isolated from each other and developed very different faunas and facies.

In Eocene times, Asia had an unfamiliar form, for India, formed by the break-up of Gondwanaland, had not yet linked up with it. However, the Indian Ocean continued to grow and Australia and Antarctica separated. This separation had begun in the Palaeocene and led to a difference between the marine formations of the two areas in this epoch. Asia remained separated from Europe by the Ural Sea and did not become re-united with it until the end of the Oligocene. Nevertheless, Asia was intermittently joined to North America by way of the Bering isthmus.

The southern part of Europe was unrecognisable, the main mass of Europe being separated from Africa by the Tethys Ocean which communicated westwards with the

PALAEOCENE		EOCENE							OLIGOCENE			MIOCENE	Figure No.	
MONTIAN	THANETIAN	SPARNACIAN	CUISIAN	LUTETIAN	Auv.	Mar. BARTONIAN	Lud.	STAMPIAN early	STAMPIAN late	late	AQUITANIAN			

Figure No.	Name
1	Palaeocystodinium sp.
2	Wetzeliella cf. parva
3	Peridinium resistente
4	Diphyes colligerum
5	Baltisphaeridium cf. funginum
6	Wetzeliella articulata
7	Lingulodinium cf. disjunctum
8	Hystrichokolpoma poculum
9	Lanternosphaeridium bipolare
10	Achomosphaera sp.
11	Cyclonephelium sp.
12	Wetzeliella clathrata
13	Areosphaeridium sp.
14	Hystrichosphaera tertiaria
15	Operculodinium centrocarpum
16	Surculosphaeridium sp.
17	Wetzeliella sp.
18	Areosphaeridium diktyoplokus
19	Wetzeliella glabra
20	Hystrichosphaeridium floripes
21	Perisseiasphaeridium sp.
22	Incertae sedis
23	I.s.
24	Cyclonephelium sp.
25	Hystrichosphaeridium paradoxum
26	Geiselodinium sp.
27	Cyclonephelium sp.
28	I.s.
29	Wetzeliella sp.
30	I.s.
31	Wetzeliella gochtii
32	Chiropteridium partispinatum
33	Hemicystodinium sp.

2.26

2.27 The displacement of the evaporite belt during the Cenozoic. In the Eocene, the equator was oblique to its present day position. In Europe and Africa, the belt of evaporites (arid zone) has migrated southwards while in Asia it has moved northwards. (Eocene position shown by solid line; Quaternary position shown by dotted: after Lotze (1964)).

As a consequence of tectonic movements and the evolution of the oceans, epicontinental transgressions spread across the margins of all the continents. Because of the richness of their molluscan faunas and the enormous variety of their benthic foraminiferids (in particular, the nummulites) these neritic deposits of western and southern Europe have, for almost two centuries, been the cradle of historical geology.

This abundant marine life, and correspondingly abundant tropical vegetation on the continents, indicate that the climate was hot and humid. From late Eocene times, however, there were periods of aridity which favoured the deposition of evaporites, such as the gypsum of the Paris Basin, and the rock salt and potash of Alsace, southern Poland and Romania. This "evaporitic belt" had a tendency to move southwards during the Eocene and Oligocene, until in the Miocene it dominated the Mediterranean area (Fig. 2.27).

Although it did not attain the extremes of the Quaternary, the climate became more varied. The nummulites did not extend north of Belgium and Holland and the planktonic foraminiferids of the North Sea have a "cold" character which makes their identification difficult. The vegetation of the northern lands shows that the climate had become temperate. During the Oligocene the *Myricaceae* (bog-myrtle) and the *Betulaceae* (birch), which were becoming more abundant, signalled the cooling of the climate of north western Europe.

As well as the connection by way of the Ural Sea, the Tethys and the northern sea were periodically connected by other routes. In the early and middle Eocene the connection was by way of a forerunner of the English Channel and the developing Atlantic Ocean; in the late Eocene it was by way of the Russian platform, and finally in the Oligocene by this same Russian platform and the Rhine graben. In contrast to the barrier formed by the Ural Sea, the temporary connection between the Rhine graben and the circum-Alpine depressions was no obstacle to the arrival in western Europe, by way of the Asian corridor, of the mammals of Stehlin's "great break" (see p. 76).

At the end of the Oligocene, epicontinental seas disappeared almost completely from the northern areas. In southern England and the Paris Basin, they departed, never to return.

Atlantic by two straits, the northern or Betic and the southern or Rif, between which lay a land mass in the area now occupied by the Straits of Gibraltar and the region to the east. Detrital sediment from this land mass was carried southwards to the Maghreb sea and northwards to the sea of Provence. Further south, the African shield was, in the Palaeocene, traversed by an arm of the sea stretching from Libya and Egypt to the Gulf of Guinea (Fig. 11.13, p. 221).

The precursors of the Alpine chains began to emerge from the Tethys, later to be eroded to form the Flysch.

At the beginning of the Eocene (in the Ilerdian and not in the Lutetian as formerly supposed), the Pyreneo-Provencal chain began to be formed. Over-thrusting began in the Alps and the Dinarides and continued into the Oligocene. In broad terms, *the Eocene was a period of compression followed by a period of relaxation in the Oligocene*. This was a very important period in the western world since it corresponded to an acceleration in the rate of opening of the North Atlantic, the formation of numerous grabens (such as those of the Rhine, Limagne and Bohemia) and the beginning of the great circum-Alpine molasse depression.

The Palaeogene in France | 3

Following the post-Cretaceous regression, the only parts of France still covered by the sea were the Aturian region (Bas-Adour and western Pyrenees) and the internal zones of the Alps (the sub-Briancon, Briancon and probably, the Piedmont zones). A gulf persisted in the western part of the English Channel and another in the North Sea, though there was no communication between them.

The Palaeogene transgressions advanced across France from three directions: the English Channel and North Sea, the Atlantic and the Ligurian Sea. Further inland, numerous lakes and lagoons were formed. During the Oligocene, some of these (Limagne, Alsace) became fault bounded areas of subsidence and were filled with a thick series of detrital sediments.

I. – TRANSGRESSIONS FROM THE ENGLISH CHANNEL AND NORTH SEA: THE PARIS BASIN

The initial advance of the sea (the Dano-Montian advance) was from the direction of the English Channel and not until Thanetian times is there evidence of a link between the North Sea and the Paris Basin. This stage is the first of a series of five cycles of sedimentation (**Thanetian, Ypresian, Lutetian, Bartonian and Stampian**) which followed one another during the Palaeogene (Table VI and Fig. 3.1).

1. The pre-Tertiary surface and the Dano-Montian transgression (early Palaeocene)

The youngest Cretaceous deposits recognised in the Paris Basin belong to the Campanian (*Belemnitella mucronata* chalk). It is possible, however, that thin Maastrichtian sequences are present at depth in synclinal areas, especially as a *remanié* Maastrichtian micro-fauna occurs locally in the clays at the base of the Dano-Montian.

It should not be forgotten that *a large part of the Chalk was removed by erosion* as is shown by the local variations in thickness and by the existence of a fossil topography beneath the marine and continental deposits of the lower Eocene. However, without detailed knowledge of Senonian stratigraphy, it is still difficult to determine the relative importance of Cretaceous tectonism, variations in rate of sedimentation, and post-Cretaceous erosion, in controlling the present thickness of the Chalk.

It was this irregular surface which the Dano-Montian sea invaded from the west. The present distribution of Dano-Montian beds does not indicate a series of separate channels, as might be supposed but rather the gentle flooding of a wide gulf. The extent of this gulf is shown by Fig. 3.2.

It seems probable that there was a direct link between the Paris Basin and the Mons Basin at this time on account of similarities in their foraminiferal and ostracod faunas. The latter was connected to the North Sea, while the Paris Basin opened to the Atlantic by way of the western part of the English Channel, where CURRY (1962) has reported the presence of Danian sediments.

Table VI

Principal Palaeogene formations in the centre of the Paris Basin

Periods	Stages and sub-stages		Principal facies	Lateral facies
OLIGOCENE	Stampian s.l.	Aquitanian	Calcaire de Beauce	Calcaire de l'Orléanais
		Stampian s.s.	Calcaire d'Etampes / Sables et grès de Fontainebleau / Marnes à Huîtres	Meulière de Montmorency
		Sannoisian	Calcaire de Sannois / Caillasses d'Orgemont / Argile verte de Romainville / Glaises à Cyrènes	Calcaire de Brie
EOCENE	Bartonian s.l.	Ludian (= Priabonian)	Marnes blanches de Pantin / Marnes bleues d'Argenteuil / Gypse and associated clays / Marnes à *Pholadomya ludensis*	Limestones of Champigny and Château-Landon
		Marinesian	Sables de Marines / Sables de Cresnes	Calcaire de Saint-Ouen
		Auversian	Sables de Beauchamp / Sables d'Auvers	Calcaire de Nogent-l'Artaud / Argile de Saint-Gobain
	Lutetian	late	Caillasses - Falun de Foulangues with *Discorinopsis kerfornei and Alv. elongata*	
		middle	Calcaire à Milioles / *Nummulites variolarius* / *Orbitolites complanatus*	Calcaires de Provins et de Morancez
		early	Calcaire grossier with *Nummulites laevigatus*	
	Ypresian	Cuisian	Grès de Belleu / Sables de Cuise	Argile de Laon / Sables à Unios et Térédines
		Sparnacian	Argile à lignites du Soissonnais / Argile plastique	Sables de Sinceny / Falun de Pourcy
PALAEOCENE	Thanetian		Sables de Bracheux / Tuffeau de La Fère	Calcaire de Rilly / Conglomérat de Cernay / Travertin de Sézanne
	Dano-Montian		Marnes de Meudon / Calcaire « pisolitique »	
UPPER CRETACEOUS	Campanian		Belemnitella Chalk	

In the Paris Basin, the deposits of the Dano-Montian were mostly biocalcarenites and cream or yellow marls. These marls contain montmorillonite derived from the weathering of the chalk. Lying between the weathered chalk and the Sparnacian clays, they are often not recognised in borehole records, which probably explains the widely-held belief in the limited extent of the Dano-Montian Sea.

Most of the classical localities show limestones of various kinds and include a "pisolite" which is, in fact, a coarse grained, bioclastic calcarenite (Fig. 3.3 and 3.4) formed from fragments of calcareous algae (*Litho-*

thamnion, Lithophyllum), corals, echinoid spines (*Cidaris*) and large molluscs. These include bivalves (*Lima carolina, Crassatella pisolithica, Corbis sublamellosa, Chlamys*), gastropods (*Cerithium, Athleta, Campanile, Turritella*) and cephalopods (*Hercoglossa danica, Nautilus heberti*).

This limestone forms reef-like masses or *bioherms* and is the only formation of this type in the Tertiary beds of the Paris Basin. These reefs were formed in a tropical climate and are banked against steep cliffs of Campanian chalk, as at Vigny. This explains the vertical and overhanging contacts and also the chalk

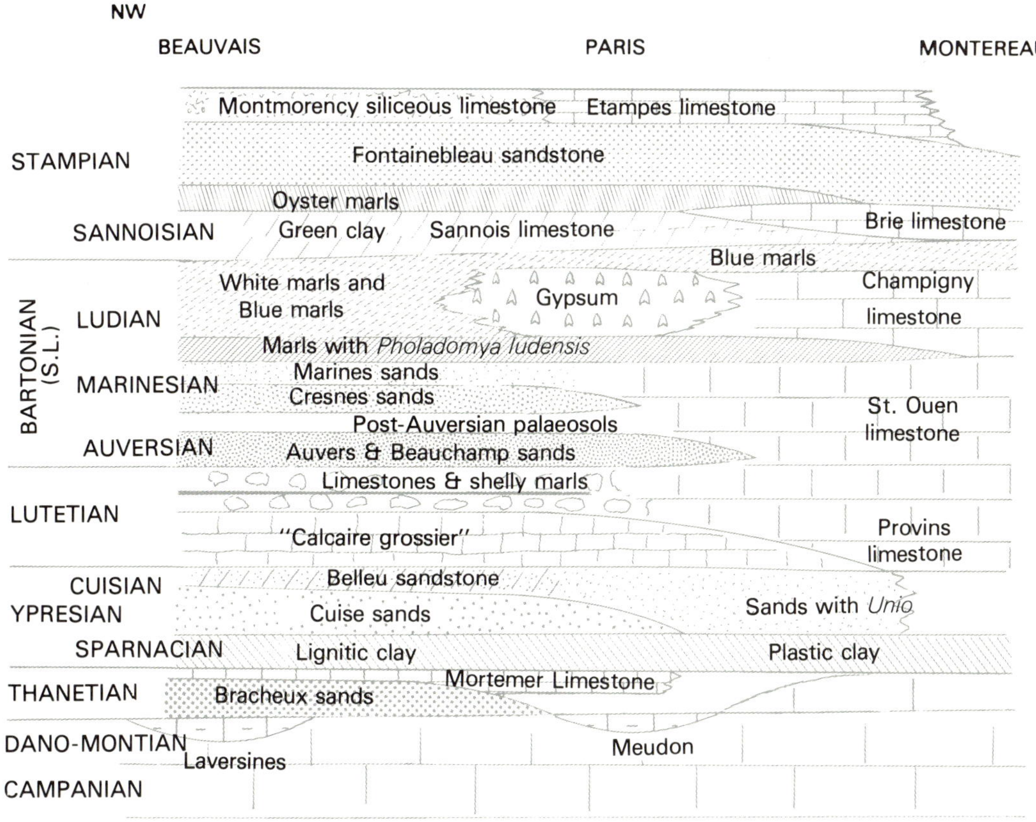

3.1 **The Palaeogene succession in the Paris Basin and its transgressions.** The diagram shows the lateral variation in facies from north-west to south-east. The beds are predominantly marine in the north-west and wholly continental in the extreme south-east. The thicknesses of the continental beds at the right of the diagram are greatly exaggerated in comparison with those of the marine beds. The Thanetian transgression did not reach Paris, though that of the Cuisian did. The other transgressions penetrated beyond (south-east of) Paris with the exception of the Marinesian (Cresnes sands) which stopped close to Paris. The Fontainebleau Sands (Stampian) mark the final and most extensive transgression across the area. The lacustrine Calcaire de Beauce (Aquitanian) which overlies the Calcaire d'Étampes (late Stampian) is not shown. See Table VI (p. 58) for the corresponding formation names in the French language.

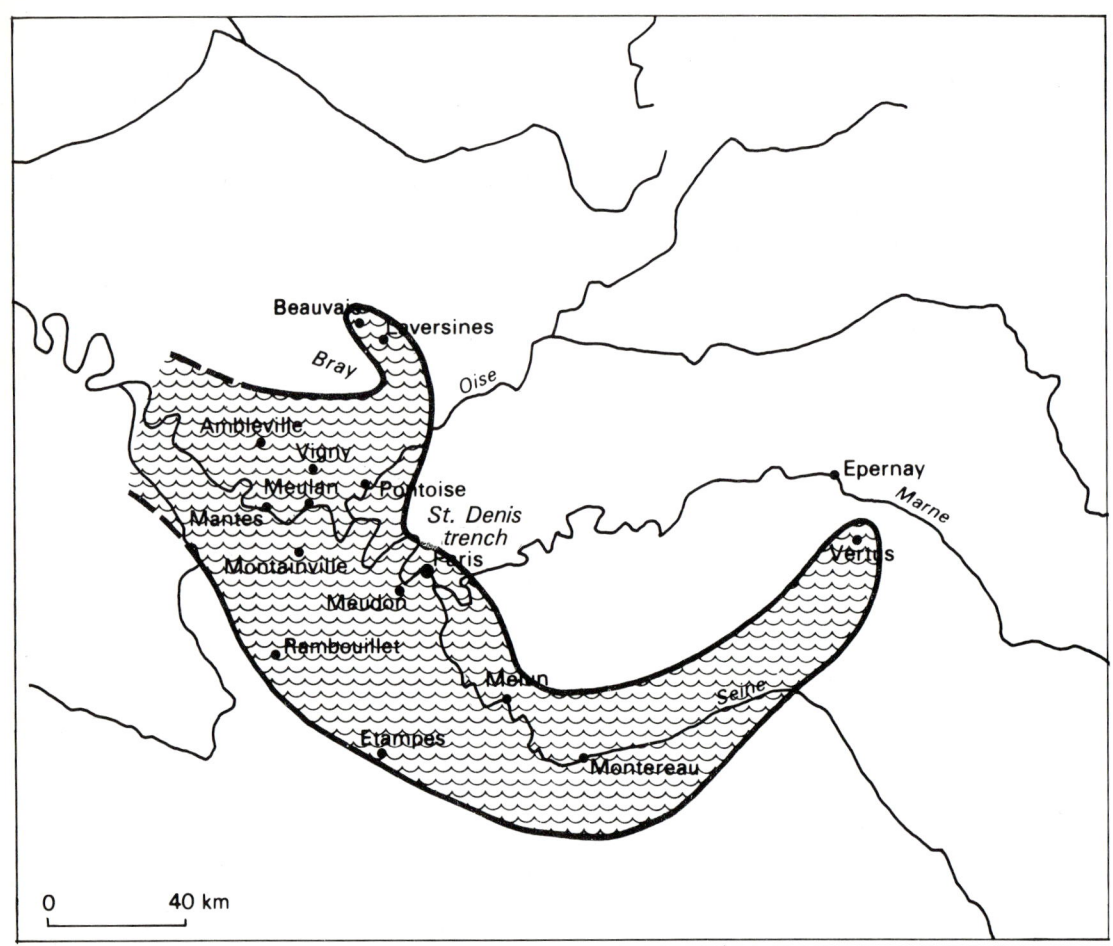

3.2 **The extent of the Dano-Montian in the Paris Basin.** The direction of penetration follows a N.W.-S.E. structural line as far as Melun and then turns towards the north east. There may have been communication with the Mons Basin in Belgium.

boulders incorporated in the limestone. At the contact between the two formations, the chalk is yellowish in colour, hard and bored by marine organisms. When the contact is nearly horizontal, hardened chalk is often overlain by a layer of flint pebbles with a calcareous cement.

The flora and fauna of this limestone have strong Cenozoic affinities but also include Mesozoic elements, especially amongst the corals. There has been discussion of the precise age of these beds for more than fifty years, but the problem was finally solved by DAMOTTE in 1963 who examined the ostracods and by MARIE in 1964 who examined the foraminiferids. The

ostracods from the lowest beds of the Vigny quarry equate these beds with the *Tuffeau de Ciply* (near Mons) of late Danian age (MARLIÈRE's *Cytherelloidea* beds) and with the *Calcaire de Mons* (*Cytherella* beds). Certain genera found at Vigny persist up to the *Calcaire de Mons* s.s. (*Triginglymus* bed). **The "pisolite" is therefore definitely Cenozoic,** as are the other Dano-Montian formations of northern Europe.

At the end of this episode the sea retreated and continental marls with calcareous concretions containing *Physa, Pupa* and *Helix* were deposited locally (Meudon marls).

2. The Thanetian (late Palaeocene): marine invasion from the north.

Whereas the Dano-Montian transgression had come from the west, bypassing Picardy, the Thanetian transgression came from the north east, though it did not reach Paris (Fig. 3.5).

At this time the Paris Basin was a large gulf, open to the north and north-east, in which the Thanetian transgression reached its furthest extent. This transgression is represented there only by the deposits of its terminal phase. The type section of the Thanetian stage, as defined by RENEVIER in 1873, is the outcrop of **Thanet Sands** in the Isle of Thanet on the south side of the Thames estuary (Kent, England). The northern origin of this transgression led to the occurrence of a marine fauna living in cooler waters than those of the Dano-Montian or of the succeeding middle and upper Eocene beds. This does not, however, imply that the climate had undergone any great change, as is witnessed by the tropical flora of the fresh water limestone of Sézanne.

While the lower beds of the Thanetian (or Landenian) Zone I – *Arctica morrisi* zone, were being deposited in Belgium, subaerial erosion still continued to affect the Chalk and, in the centre of the basin, the overlying Dano-Montian deposits. This erosional interval was locally the longest in the Palaeogene and largely explains the absence of Dano-Montian beds over most of the area covered by the Thanetian. The exception is in the extreme north in the region of Beauvais.

The Thanetian sea, as it advanced southwards, received the products of continental erosion and deposited the sandy, glauconitic and phosphatic **Tuffeau de La Fère** and the sandy, glauconitic clay of Vaux-sous-Laon with, at its base, a layer of green pebbles (Zone II – *Pholadomya oblitterata* zone). These pebbles owe their colour to the presence of glauconite in micro-fissures and are accompanied by black pebbles coated with iron and manganese oxides. Such pebbles are found at the base of most of the marine formations of early and middle Eocene age, and are remnants of the erosion of the Chalk in an environment which was sometimes oxidising and sometimes, reducing.

3.3 Subvertical contact between Dano-Montian biogenic limestone of Vigny and hardened and fissured Campanian chalk (Photo: Pomerol).

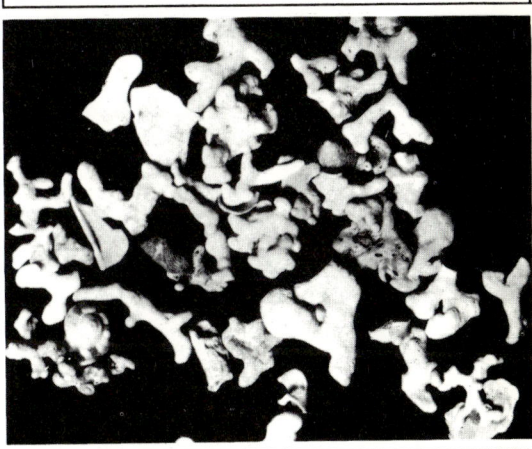

3.4 Calcareous algal debris and small gastropods from recent beach deposits of the **Brittany coast.** Known as "maerl", this debris was formerly used as soil dressing. The Vigny limestone is a cemented deposit of similar character. (Scale: the fragments are about 1cm) (photo: Pomerol).

In the late Thanetian (Zone III – *Arctica scutellaria* zone) the sea spread further south, depositing sands which are glauconitic, especially at the base. These are the **Sables de Bracheux,** which are 30m thick at "Justice Hill" (a mound which formerly carried a gallows!) near the hamlet of Bracheux, 5km east of Beauvais. As well as *Arctica scutellaria,* these sands contain numerous molluscs including *Cucullaea crassatina, Ostrea bellovacina, Cardita pectuncularis, Glycymeris terebratularis* (Fig. 3.6).

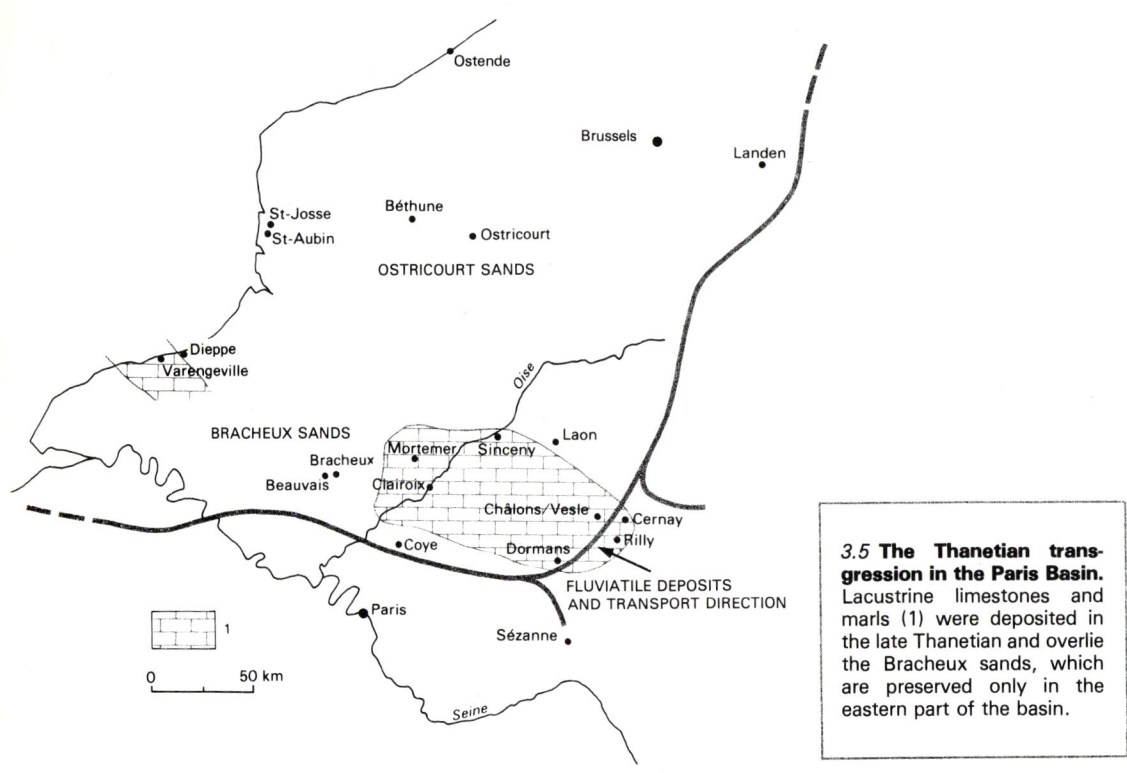

3.5 The Thanetian transgression in the Paris Basin. Lacustrine limestones and marls (1) were deposited in the late Thanetian and overlie the Bracheux sands, which are preserved only in the eastern part of the basin.

The Thanetian sea traversed the Pays de Bray and was terminated a little further south in the Pays de Thelle at the line of the present day escarpment along the northern border of the Vexin (Fig. 3.5). Further east, the coastline was situated 2km north of Luzarches, where there is a conglomerate composed of flint pebbles with a siliceous cement (Poudingue de Coye, Fig. 3.7).

Eastwards, towards Champagne, the marine Bracheux sands give way to the lagoonal **Sables de Châlons-sur-Vesle** and then to the white **Sables de Rilly,** which are thought to be littoral. In the Champagne area the Thanetian sands are estuarine and fluviatile, and especially interesting. The sandy, argillaceous and calcareous **Conglomérat de Cernay** overlying the Rilly sands or even resting directly on the Chalk, outcrops on the side of Mont Berru, east of Rheims. It has yielded a rich vertebrate fauna, studied by RUSSELL (1964). Crocodiles, tortoises and a gigantic bird (Gastornis) are found, with an archaic mammalian fauna (*Arctocyon, Tricuspiodon, Pleuraspido-therium*) which resembles the late Palaeocene faunas of North America. These mark the climax of the evolution of the Mesozoic mammals and can be clearly distinguished from the Sparnacian mammals (*Coryphodon*) which, in the Paris Basin, so far as the mammals are concerned, indicate the true beginning of the Tertiary era. *This is a good reason for placing the Sparnacian in the Ypresian cycle and for putting the Palaeocene-Eocene boundary between the Thanetian and the Sparnacian.*

South of Épernay, the **Travertin de Sézanne** (formed in Thanetian times by the emergence of a spring from the Chalk) contains beautiful imprints of vegetation, cases of caddis fly larvae, crustaceans including crayfish, *Helix, Physa* and *Viviparus.* The flora, of a hot and humid climate, includes alders, laurels, willows, walnuts, magnolias, vines, and a profusion of ferns, including tree-ferns.

The Thanetian cycle ended with a continental episode in which localised cementation of the higher sands occurred, as for example near Dieppe and in the region of Laon. At the same time lakes were formed in which fresh-water limestones and marls were deposited. These include the Calcaire de Rilly, the Dormans marl, the Sinceny marl to the west of Laon and the Varengeville limestone near Dieppe. This regression is seen in the north of France (Ostricourt sands, near Béthune) and Belgium (Landen sandstone with plant remains, the type locality of the Landenian). The limestones of Clairoix and Mortemer, near Compiègne, traditionally ascribed to the late Thanetian, are now thought to be possibly Sparnacian on the basis of their charophytes.

3. The Ypresian cycle (early Eocene): Paris Basin, and region flanking the Massif Central.

At the beginning of the Ypresian (DUMONT, 1849), the Paris Basin was invaded by a sea which was situated to the north and northwest (and not to the north and northeast as in the Thanetian). It is the last cycle of the Tertiary era in which the Paris Basin was fully open to the North Sea. It began with a lagoonal episode (early Ypresian or Sparnacian), which was followed by a marine one in which the Cuise sands (late Ypresian or Cuisian) were deposited.

a) **Sparnacian:** While the Ypresian was already fully marine in Belgium (**Ypres clay**) and in the north of France (**Orchies clay**), *a series of lagoons developed in the Paris Basin, south of the Artois axis, which reappeared for the first time since the Thanetian transgression.* These lagoons extended a little way south of Paris (Fig. 3.8) overlying the Dano-Montian or the Chalk, and encircling the "axes" of Beynes and Meudon. On the edge of these uplifted areas was deposited the **Conglomérat de Meudon,** formed from fragments of chalk and of pisolithic limestone, which has yielded a large mammal fauna including *Coryphodon* and *Hyracotherium* (see Fig. 5.9, p. 146).

It was probably at this time that the **Argile Plastique** was formed. This brightly coloured often variegated clay is rich in kaolinite and is extensively exploited between Mantes and

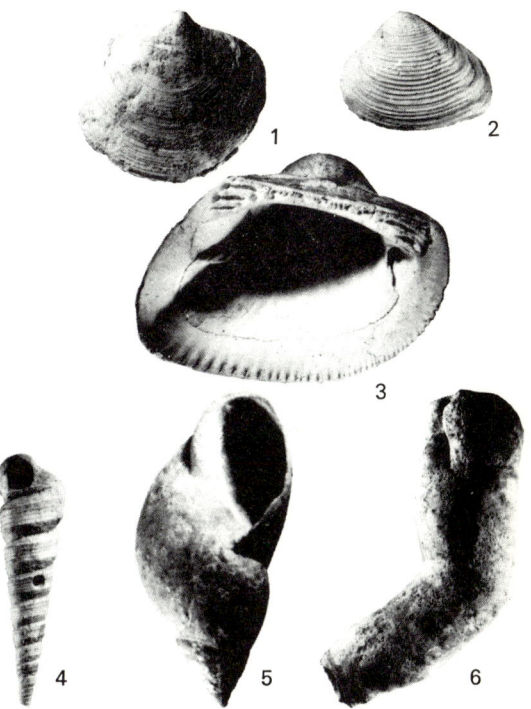

3.6 **Thanetian fossils** (x ½): 1 – *Phacoides contortus*; 2 – *Crassatella bellovacensis*; 3 – *Cucullaea crassatina*; 4 – *Turritella compta*. (x ⅔): 5 – *Physa gigantea* (sinistral (left-handed) spiral, compare with *Turritella*); 6 – *Teredina personata* (Continental Sparnacian-Cuisian) (photo: Fay).

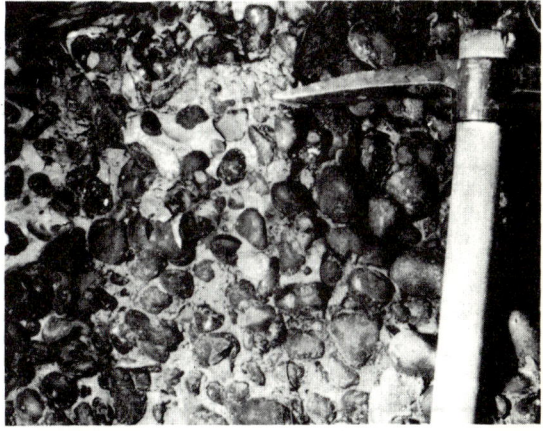

3.7 **Coye Conglomerate.** Erosion of the Chalk of the Pays de Bray by the Thanetian sea led to the formation of a beach composed of flint pebbles. Cementation of the pebbles within sandstone (pebbles in relief, right side of the photograph) or quartzite (harder than the flint, left side of photograph) has resulted in the formation of this conglomerate (photo: Pomerol).

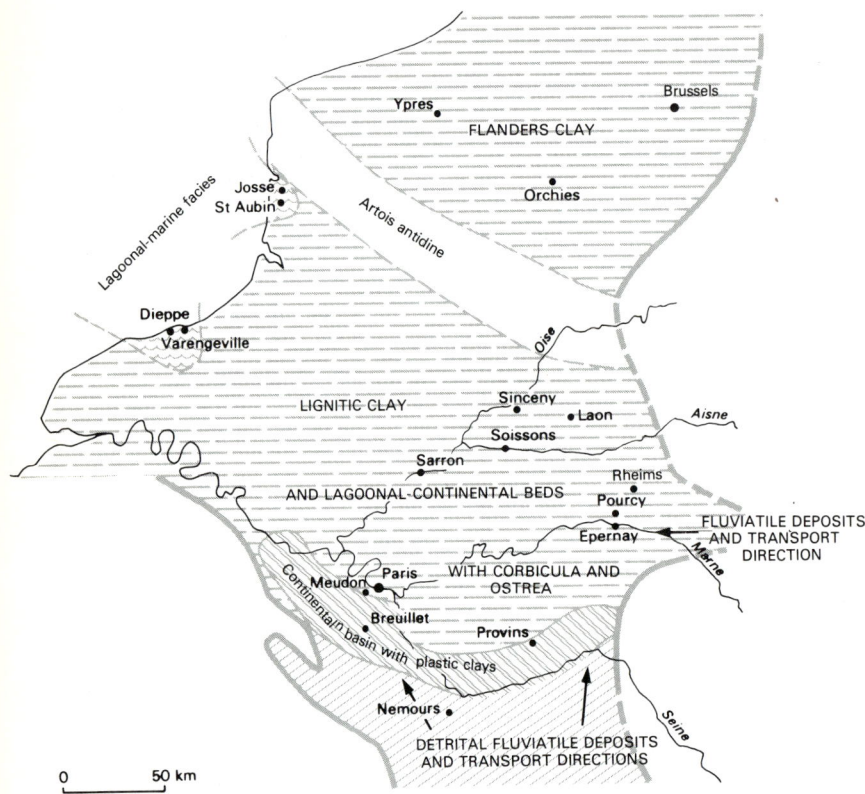

3.8 **The extent and principal facies of the Sparnacian in the Paris Basin.** The beds become more continental from north to south. For the first time in the Palaeogene, the Artois anticline acts as a sill (or partial barrier) between the marine formations in northernmost France and Belgium and the lagoonal and continental beds of the Paris Basin (after Feugueur, 1963).

3.9 **Sparnacian lignitic clays which, mixed with chalk, are used for the manufacture of cement.** Guitrancourt (Val d'Oise). The clays are about 10m thick and are overlain by a thin (2-3m) bed of Cuise sand. The uppermost bed seen in the photograph is a bioclastic limestone of middle and late Lutetian age about 8-10m thick (photo: Pomerol).

Provins. It does not extend northwards much beyond the southern flank of the Bray axis and its eastward extension. This continental clay probably originated from the Massif Central. It is thought to have been brought by streams which, at about the same time or a little later, had deposited sands and ferruginous clays further south (the most northerly evidence of the *"siderolitic" facies*). These sands are succeeded by felspathic sands, locally consolidated into sandstone and arkose (Breuillet arkose).

It must not be assumed, however, that all the continental sands resting on the Chalk and underlying well-dated marine formations, whether Lutetian, Bartonian or Stampian, are necessarily Sparnacian. Thus the greater part of the Nemours conglomerate, formerly thought to be Sparnacian, was probably deposited at different times in the early and middle Eocene.

To the north of the plastic clays, the pyritic **Argiles et Lignites du Soissonnais** were deposited (Fig. 3.9). These were formerly

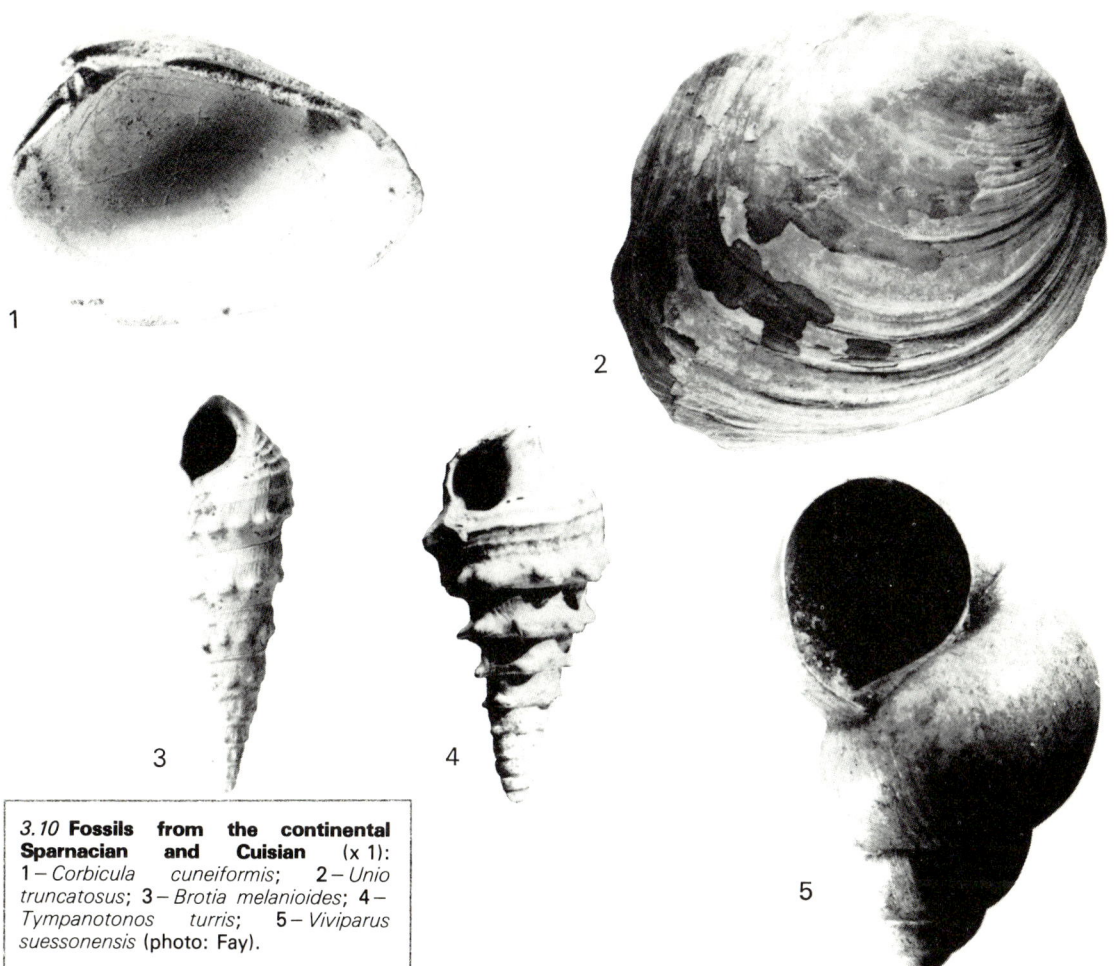

3.10 Fossils from the continental Sparnacian and Cuisian (x 1): 1 – *Corbicula cuneiformis*; 2 – *Unio truncatosus*; 3 – *Brotia melanioides*; 4 – *Tympanotonos turris*; 5 – *Viviparus suessonensis* (photo: Fay).

exploited in the valleys of the Oise, Aisne and Marne for the production of alum and ferrous sulphate and for dressing the land. This last usage continues in the champagne vineyards, particularly at Épernay, the type-locality of the Sparnacian stage (DOLLFUS, 1880).

A more marine episode (the Sinceny and Auteuil sands and Sarron clay), correlated with the estuarine facies of Pourcy (*Falun de Pourcy*) to the east, was followed by a lagoonal phase which extends southeastwards towards Provins and Épernay. This phase is represented by clays with *Corbicula cuneiformis* (Fig. 3.10), *Brotia melanioides*, *Tympanotonos funatus* and shell beds with *Ostrea bellovacina* var. *sparnacensis*.

In southern Europe, where marine sedimentation was continuous from the Thanetian to the Cuisian, the marine stage corresponding approximately to the Sparnacian is the *Ilerdian* (see p. 128).

b) **Cuisian:** At the beginning of the late Ypresian (=Cuisian, DOLLFUS, 1880), the Artois anticline disappeared (Fig. 3.11) beneath the transgressive sea in which the **Sables de Cuise** were laid down. These sands contain *Nummulites planulatus*, *Turritella solanderi* and *Velates schmiedeli* Fig. 3.12).

In their lower part nummulites are absent, however (Sables de Cuise inférieurs, equivalent to the Formation de Varengeville, near Dieppe (Fig. 3.8) and to the London Clay).

The Cuise sands correspond to the *Mons-en-Pévèle sands* and the *Roubaix clay* (which overlies the Orchies clay) of northern France and to the Paniselian sands of Belgium (FEUGUEUR, 1958).

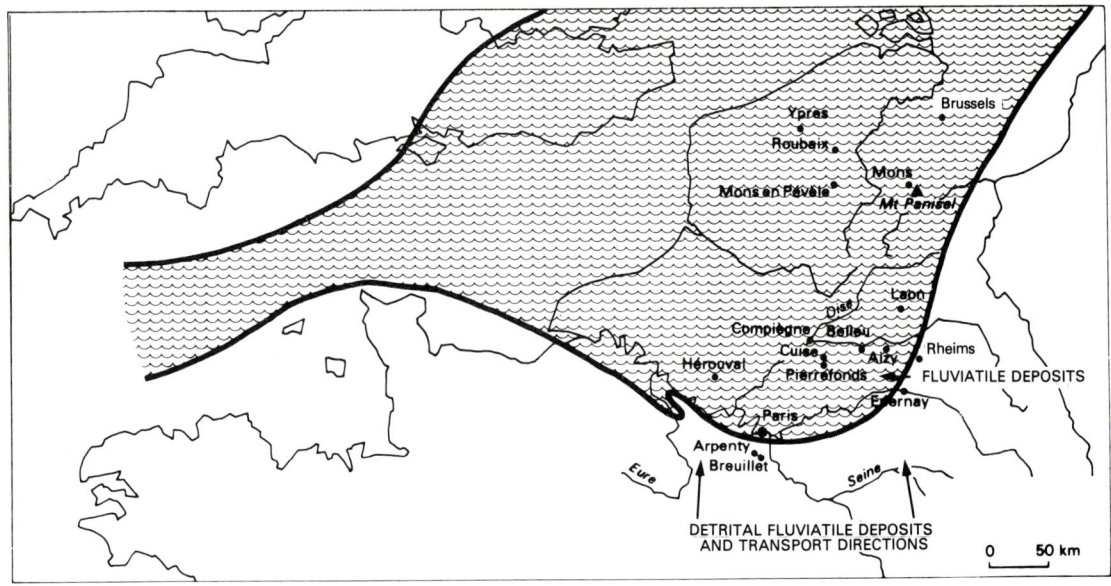

3.11 The Cuisian sea in the Anglo-Paris Basin, showing connections with the North Sea and the Atlantic (which provided a migration route for *Nummulites planulatus*).

The Sables de Cuise (= Aizy sands overlain by Pierrefonds sands) were laid down in a sea which hardly reached south of the Seine valley, being halted by the Meudon anticline (Fig. 3.11). During its retreat the Cuisian sea deposited the **Argile de Laon** in the region of Soissons and Laon. In the continental episode which followed, the upper part of the Cuise sands, particularly in the region of Soissons, became indurated (Belleu sandstone). To the east, in the region of Épernay and Rheims, beyond the limits of the transgression, the Sparnacian was overlain by the Cuisian fluviatile "sands with *Unio* and *Teredina*" (Fig. 3.10), in which the first *Lophiodon* appears.

To the west, in the Vexin, the final deposits of the Cuisian sea were the Hérouval sands. The retreat of the sea seems to have occurred only shortly before the onset of the Lutetian transgression.

4. The Lutetian cycle (middle Eocene): the isolation of the Paris Basin.

The Lutetian cycle (stage defined by A. De Lapparent, 1883, from Lutèce = Paris) *corresponds nearly to the classical concept of a sedimentary cycle.* It begins with a bed of small flint pebbles, rich in glauconite (*Glauconie Grossière*), overlain by a less sandy calcareous facies (**Calcaire Grossier**). Towards the end of the cycle, the deposits become lagoonal and evaporites replace the organo-detrital sediments. The total thickness of the Lutetian is about 40m.

After the partial retreat of the Cuisian sea and the development of a continental facies in the eastern part of the Paris Basin, the Lutetian sea transgressed from the north and west, *crossing the Artois dome for the last time* and extending beyond the limits of the Cuisian sea. In the middle Lutetian, the present course of the Seine was crossed and marine deposits reached the Rémarde anticline (Fig. 3.13). Lakes were formed in the area south of the Lutetian sea in which calcareous sediments were deposited. The lime in these rocks came from the Chalk, the source being indicated by the presence of titaniferous corundum, a heavy mineral characteristic of the Senonian. These Lutetian limestones include the Morancez limestone near Chartres, the Darvault limestone near Nemours and those of Provins and the border of the Champagne (Fig. 3.14). Since the discovery of *Discorinopsis kerfornei* in the Creil area by Cavelier and Le Calvez in 1965 it has become apparent that *the Lutetian of the Paris Basin is equivalent to the whole of the Middle Eocene. D. kerfornei* is a foraminiferid occurring in the youngest Lutetian of Aquitaine (the Biarritzian "stage" of Hottinger and Schaub, 1960).

At the Colloquium on the Eocene in Paris, 1968, Pomerol proposed that the Lutetian should be divided into three parts:

a) **Lower Lutetian** with *Nummulites laevigatus* (see Fig. 3.15, p. 69) and various echinoids, especially *Maretia*. This horizon is a sandy limestone, locally indurated, which is well exposed in the Soissonnais and the Laonnois. The basal beds are generally, in a transgressive facies, the **Glauconie Grossière.** This is a poorly-sorted, coarse, quartz sand with glauconite and flints, containing many shark's teeth, bivalves (*Ostrea, Chama, Crassatella, Cardita = Venericardia planicosta*) and an abundant coral, *Eupsammia trochiformis* (Fig. 3.16). Also present are forms reworked from the Cuisian, including *Nummulites planulatus*. Radiometric dating of the glauconite has given an age of 47-49Ma. It should be noted that the *glauconie grossière* is a facies linked to the southward progress of the Lutetian transgression. As a result it is early Lutetian near Soissons, but middle Lutetian around Mantes, that is, it is a *diachronous deposit*.

b) **Middle Lutetian** with *Nummulites variolarius, Orbitolites complanatus* and abundant miliolids (Fig. 3.18). This facies, known locally as the *banc royal*, is well-developed in the Soissonnais, Valois and Vexin areas and extends thence to Paris (Fig. 3.17). This limestone has been quarried and mined for a very long time as a fine building stone (Notre Dame and the Sorbonne in Paris). The Paris catacombs and the mushroom caves of the Oise Valley are former limestone mines. The limestone was laid down in a calm, tropical sea where the waters were sufficiently shallow to

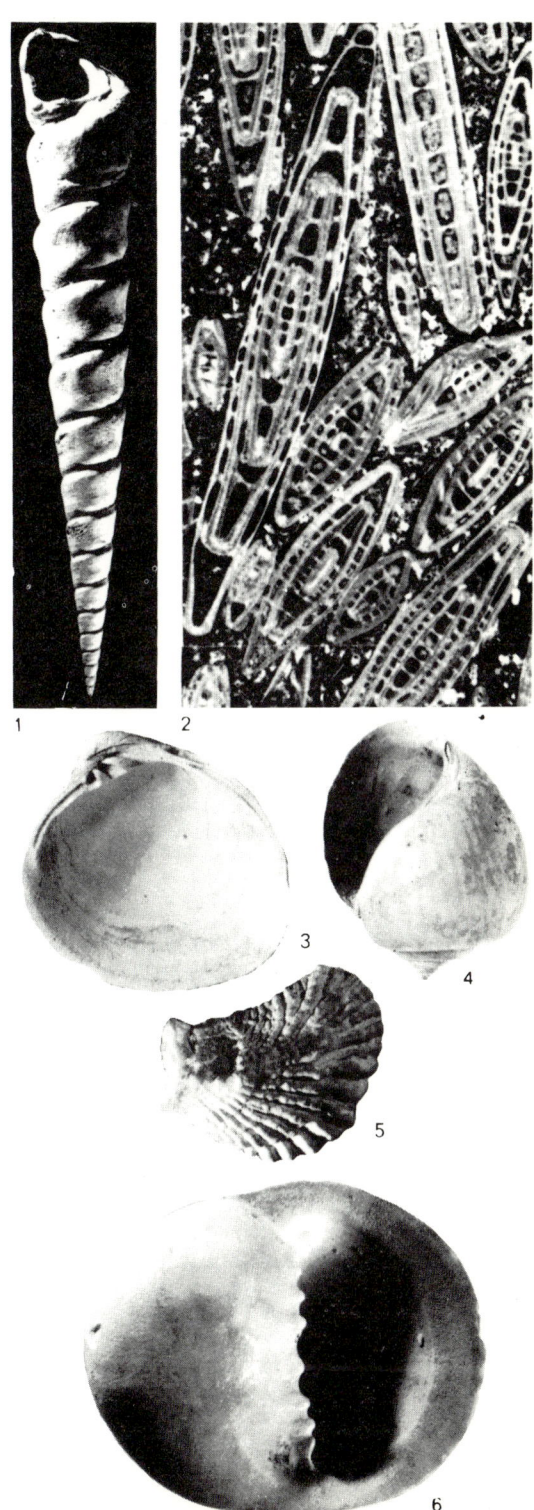

3.12 **Cuisian fossils** (x 1): 1 – *Turritella solanderi*; 2 – Sections of *Nummulites planulatus* (x 20); 3 – *Cyrena gravesi*; 4 – *Ampullina intermedia*; 5 – *Ostrea multicostata*; 6 – *Velates schmiedeli* (photo: Fay).

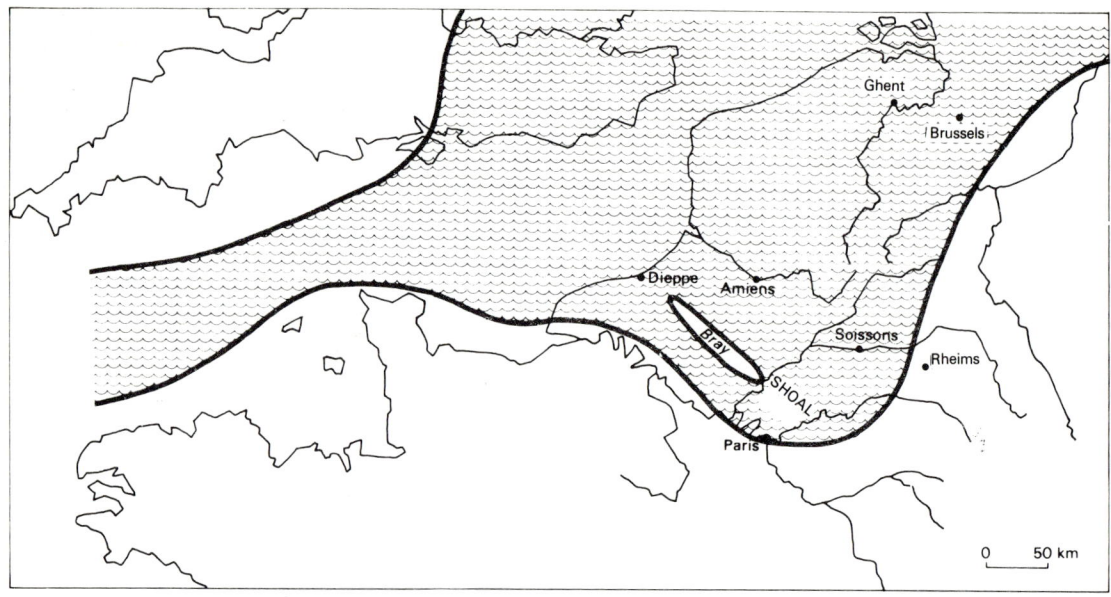

3.13 **The early Lutetian sea** (zone of *Nummulites laevigatus*) in the Anglo-Franco-Belgian basin; calcareous facies in the Paris Basin; sandy and sandy limestone facies (Bruxellian) in Belgium.

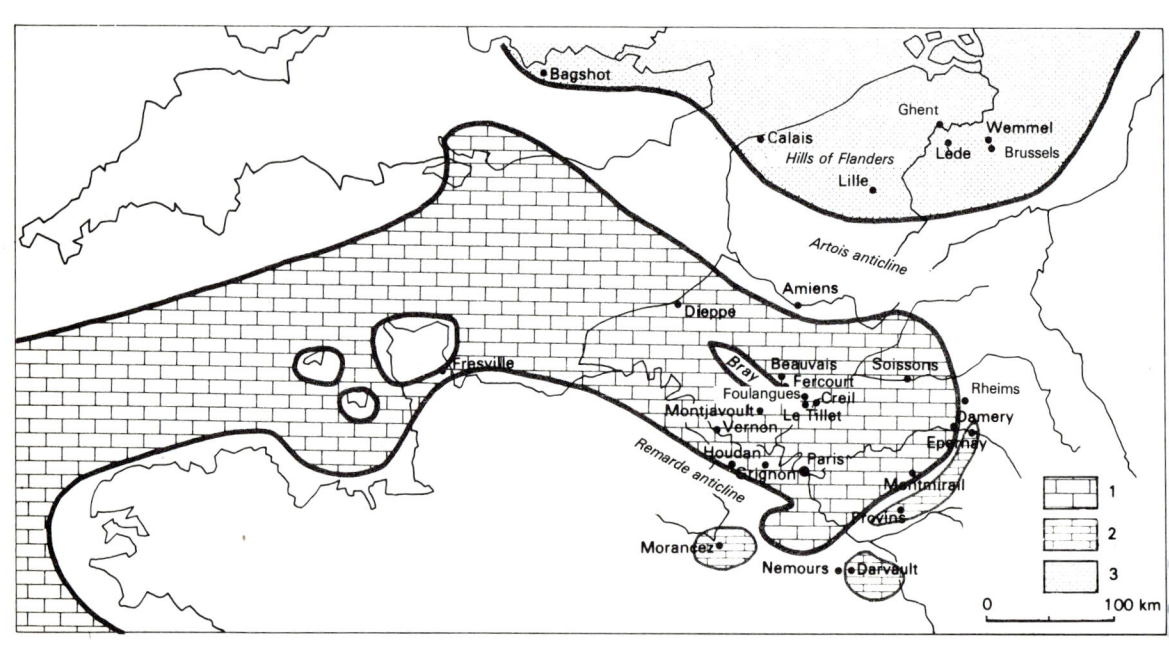

3.14 **The middle and late Lutetian seas in the basins of Paris and of Belgium, separated by the Artois and Wealden anticlines.** 1 – Marine limestones and laguno-marine limestones with marls; 2 – lacustrine limestones; 3 – Ledian and Wemmelian sands and sandy limestone, etc.

allow the growth of sea grasses. A similar environment can be seen in the Bay of Florida at the present time (*Thalassia* fields, Fig. 11.7, p. 216). In the Soissonnais, this limestone commonly contains many tubes of the worm, *Ditrupa strangulata*, and is locally dolomitised. It then becomes friable and forms a sand of minute rhombohedra with calci-dolomitic concretions, known as "cat's heads".

The Middle Lutetian limestone remained unconsolidated in the areas where anticlines were active at the time of deposition, as, for instance, around the Bray and Beynes anticlines, and near the eastern shore of the Lutetian sea near Épernay. In these areas it yields a magnificent fauna of well-preserved molluscs which are to be found in collections the world over. Genera include *Natica, Cerithium* (*C. giganteum* is more than 1m in length), *Turritella, Voluta, Murex, Rimella, Sycum, Arca, Cardita, Corbis* and *Meretrix* (Fig. 3.16).

This horizon sometimes also contains an abundance of *Nummulites variolarius*, a species now known to appear in the Lower Lutetian, but whose presence had escaped attention until recent years. This nummulite had previously been considered to be characteristic of the Bartonian and its presence in the Ledian of Belgium had led to the supposition that the Bartonian Stage was to be regarded as synonymous with the Ledian. Nowadays, however, it is considered that *the Ledian is equivalent to the Middle Lutetian of the Paris Basin* (POMEROL, 1965). By that time, the Paris and Belgian basins had become separated by the Artois anticline, which terminates to the east against the Ardennes-Rhenish massif (Figs. 3.13 and 3.14).

This separation had a profound influence on the palaeogeography of the Paris Basin, which had previously had a direct link with the North Sea from Thanetian times (60Ma), and had been linked at least temporarily with the Atlantic Ocean, by way of the English Channel, from Cuisian times (Fig. 3.11). By this latter route *Nummulites planulatus* had migrated from Aquitaine to the Paris Basin in the Cuisian, to be followed by *N. laevigatus*, and probably *N. variolarius*, in the early Lutetian. In the middle Lutetian (45Ma), the sea in which the miliolid limestone was deposited was

3.15 **Fragment of nummulitic limestone (pierre à liards)** showing equatorial sections of *Nummulites laevigatus* (diameter 1cm) (photo: Pomerol).

restricted to the gulf south of the Artois anticline and for the first time in the Eocene a limestone was formed which resulted directly from the erosion of the Chalk.

c) The **upper Lutetian** comprises both a formation (the *caillasses* or rubbly limestones) and a biozone (Biarritzian, see below). The caillasses commence with beds containing *Cerithium*, a lagoon dweller (abundant individuals representing few species, Fig. 3.20). They are overlain by marly or dolomitic beds containing magnesian clays (sepiolite, attapulgite), authigenic quartz grains, and gypsum in lenses, or in beds, several metres thick in places. This evaporitic and chemical sedimentation marks the *increasing isolation of the Parisian gulf* which had begun in the middle Lutetian. This phase was interrupted locally by a brief transgression which flooded across the centre of the lagoons of the caillasses. It deposited a shelly, calcareous sand (*Falun de Foulangues*) containing the foraminiferids *Discorinopsis kerfornei, Linderina brugesi* and *Alveolina elongata*, at several places in the western part of the basin (Biarritzian biozone, see p. 128). But this time the nummulite, *N. brongniarti*, known from Aquitaine and found in Lutetian beds on the Breton coast, does not seem to have penetrated further east.

5. The Bartonian cycle (Upper Eocene): the further isolation of the Paris Basin.

As explained earlier (p. 26) the term Bartonian is here used to include also beds of Priabonian (= Ludian) age.

3.16 Lutetian fossils (x 1, except where indicated): 1 – *Nummulites gizehensis* (Egypt, Senegal) (x ⅔); 2 – *Alveolina elliptica* (x 12); 3 to 5 corals: 3 – *Acropora* (*Madrepora*) *ornata*; 4 – *Eupsammia trochiformis*; 5 – *Stylocaenia* (*Astrea*) *emarciata*; 6 to 9 Echinoderms; 6 – *Echinolampas calvimontana*; 7 – *Maretia grignonensis*; 8a and 8b – *Scutellina lenticularis*; 9 – *Echinanthus issyavensis*; 10 – *Phacoides concentricus*; 11 – *Corbula gallica*; 12 – *Meretrix laevigata*; 13 – *Anomia tenuistriata*; 14 – *Cardita planicosta* (x ½); 15 – *Chama calcarata*; 16 – *Corbis lamellosa*; 17 – *Terebratula bisinuata*; 18 – *Lithocardium aviculare*; 19 – *Volutilithes muricinus*; 20 – *Athleta spinosus*; 21 – *Rimella fissurella*; 22 – *Natica cepacea*; 23 – *Cassidaria nodosa*; 24 – *Potamides lapidum*; 25 – *Turritella carinifera*; 26 – *Mesalia fasciata*; 27 – *Cerithium lamellosum*; 28 – *Bayania lactea*; 29 – *Rhinoclavis striatus* (photo: Fay).

3.17 **The Saint-Vaast-les-Mello quarry (Oise).** This quarry is worked for building stone and shows (in the foreground) the Saint-Leu limestone (8m) overlain by the ''banc royal'', a miliolid limestone (11m) in which swallow holes of karstic origin occur. These cavities are filled with Auversian sands. The overburden, seen in the back wall of the quarry, is marl and marly limestone of late Lutetian age overlain by a thin layer of Auversian sand, with loess at the top of the section (photo: Pomerol).

After the Lutetian lagoons had dried up, a new sandy transgression invaded the Paris Basin at the beginning of the Bartonian cycle (Auversian) which began about 43Ma. Two later regional oscillations of this major transgression are recognised and named the Marinesian and the Ludian. The type locality of the Bartonian is Barton cliff in Hampshire, England (MAYER-EYMAR, 1857).

a) **Auversian episode** (type locality, Auvers-sur-Oise, Val d'Oise, France, DOLLFUS, 1907).

Slightly smaller than the Lutetian sea, the Auversian sea was also more isolated. It was separated from the Belgian basin by the Artois dome and surrounded the emergent area of the Pays de Bray. It crossed the Meudon anticline, but did not pass beyond that of the Rémarde (Fig. 3.21).

Lateral variations in facies are more numerous than in any other stage of the Eocene. Only two examples will be given. The first is the **Sables d'Auvers,** a beach deposit in a gulf with strong tides. It contained pebbles and rolled fossils. The second example, the **Sables de Beauchamp,** occurs laterally to, or above, the Auvers sands. The Beauchamp sands were formed in more tranquil conditions and have a better-preserved fauna with bivalves still in the position of growth (Fig. 3.22 and 3.23). Molluscs and foraminiferids were a little less abundant than in the Lutetian. In this semi-isolated gulf, *Nummulites variolarius* proliferated (Fig. 3.28), but, as previously stated, this form cannot be considered as a good stratigraphical indicator. Elsewhere, forms derived from this nummulite occur at the same stratigraphical level and include *N. prestwichianus* in the south of England and *N. orbignyi* in Belgium.

At the end of the Auversian, and somewhat earlier in the east of the basin, a temporary emergence led to the formation of **soils of**

3.18 **The microfacies of the "banc royal" (middle Lutetian).** The rock contains many miliolids and the flattened *Orbitolites complanatus*. The oval form of *Fabularia discolithes* can be seen in the top right hand corner (x 3) (photo: Blondeau).

tropical podzol type, locally indurated, well seen (Figs. 3.24 and 3.25) at a number of localities (POMEROL, 1964). Concurrently, aeolian reworking affected a variable thickness of the Beauchamp sands. This is the Fleurines facies of the Valois and the Tardenois, which is characterised by fine and very pure white sands which have the same industrial uses as the sands of Fontainebleau (foundry sands, glass making and the manufacture of silicones).

b) **Marinesian episode** (Marines, Oise valley, DOLLFUS, 1907 and POMEROL, 1965).

An inundation, rather than a transgression, flooded the Auversian fossil soils without destroying them, and deposited the laguno-marine **Formation d'Ézanville.** This was followed in the centre of the basin by the laguno-lacustine limestone and marls of Ducy, the laguno-marine formation of Mortefontaine (limestones, marls and especially sands) and the laguno-lacustrine beds of St. Ouen (limestones and marls with *Lymnaea longiscata*). Chemically deposited sediments, and notably the magnesian clays attapulgite and sepiolite, occur throughout these sequences.

At the same time, in the Vexin area to the west, the open sea was depositing the **Sables de Cresnes** in a facies similar to that of the Auvers sands. The Cresnes sands are overlain by more lagoonal, though still glauconitic sands, the **Sables de Marines.** Further south in the immediate neighbourhood of Paris the Monceau sands overlie the St. Ouen limestone and are probably coeval with the Marines sands. This is a good example of the lateral variation of facies: marine in the west, continental in the east (Fig. 3.26). The palaeogeography of Marinesian times differs from that of the Auversian by the lesser extent of the sea, which did not cross the Bray anticline to the north, and by its more strongly continental aspect, particularly to the east of the Oise, where the St. Ouen limestone is well-developed and has yielded the remains of *Palaeotherium magnum* (cf. Fig. 2.8).

This lagoonal character became still more pronounced in the succeeding Ludian episode, so that, in the Paris Basin, the boundary between the Eocene and the Oligocene is difficult to define.

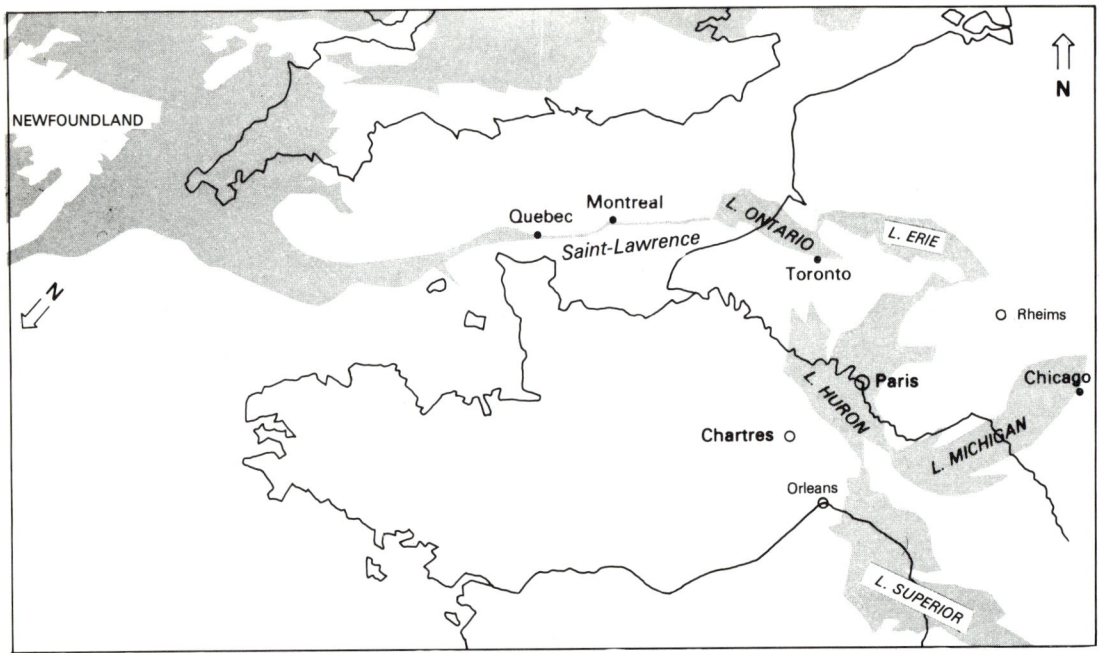

3.19 An approximation to the palaeogeography of the Paris Basin in the late Eocene. Note the rotation and change of scale of the map of North America. In this comparison, the St. Lawrence river corresponds approximately to the River Seine and the western part of the English Channel, while the Great Lakes represent the lagoons of the Paris Basin. The comparison would be more realistic if the present day map of North America were replaced by the palaeogeographic map of the Würm (= Wisconsin) phase at the time when the Champlain sea penetrated as far as Montreal (see Fig 12.6, p. 228).

c) The Ludian episode (Ludes, near Rheims, MUNIER-CHALMAS and DE LAPPARENT, 1893).

This episode began with a marine transgres-

3.20 Cerithium limestone at the base of the calcareous marls of the upper Lutetian, showing internal and external moulds (photo: Pomerol).

sion which extended eastwards (Fig. 3.27) as far as Rheims, in which the **Marnes à Pholadomya ludensis** were deposited. In fact, the change in sea level needed to produce this transgression was small since the surface which it traversed was flat and low-lying. The salinity of the sea was much reduced in the centre and the east of the basin as is shown by the paucity of marine microfauna together with *an abundance of charophytes* unequalled in any other Tertiary horizon in the Paris Basin (see fig. 3.33, p. 84).

This marine episode was brief, and in the still-subsiding centre of the Paris Basin was replaced by a laguno-lacustrine regime. Alternating beds of gypsum (**Gypse**) and attapulgite clay, often dolomitic, pass laterally into the lacustrine limestones of Champigny in the east and southeast, and into the Château-Landon limestone in the Gâtinais. After long controversy (see Fig. 3.29) recent geochemical studies based on the determination of the isotope ratio O^{16}/O^{18} and on the boron content (FONTES, 1968) have shown that the gypsum was

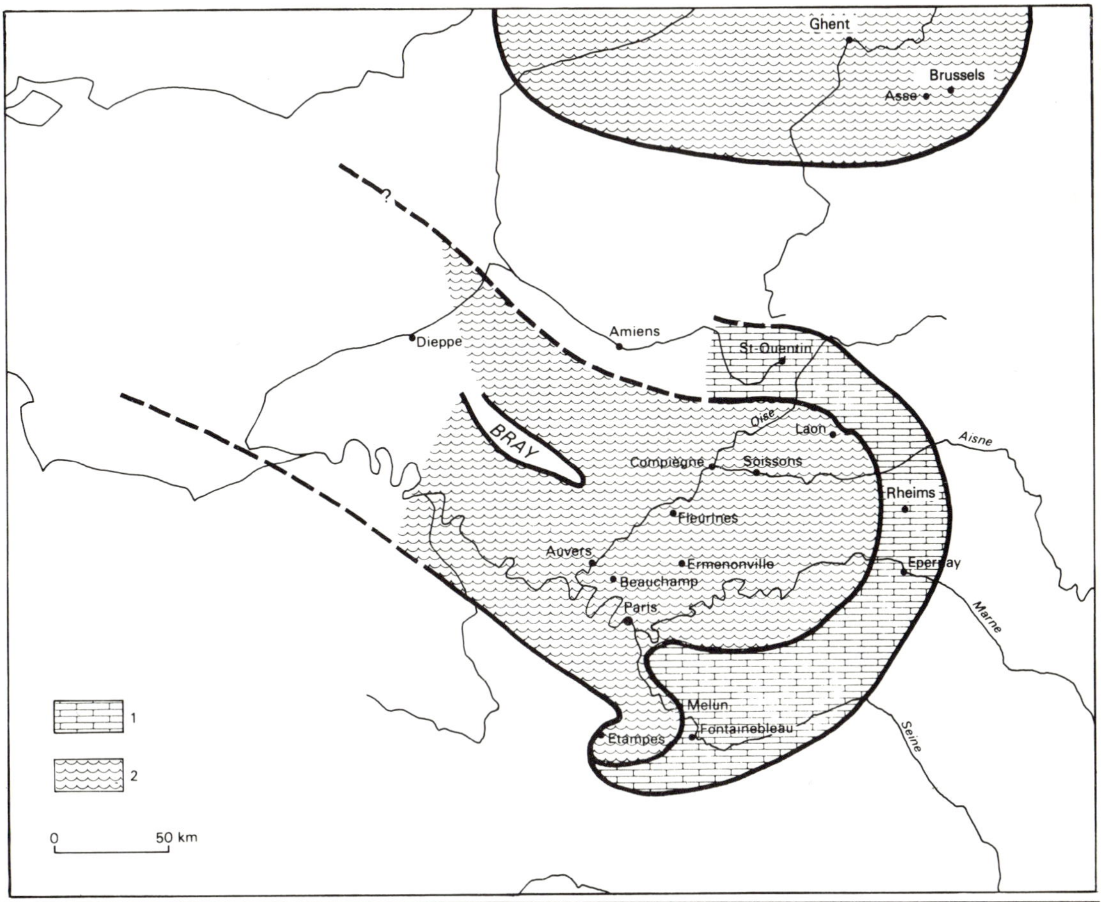

Ghent

Brussels
Asse

?

Dieppe
Amiens
St-Quentin

BRAY
Oise
Laon
Compiègne
Soissons
Rheims
Aisne

Fleurines

Auvers
Ermenonville
Epernay
Beauchamp
Paris
Marne

Melun
Fontainebleau
Etampes
Seine

1

2

0   50 km

3.21 **The Auversian sea and its lagoonal border:** 1 – lagoonal formations; 2 – marine formations.

3.22 **Aerial view of quarry at Champlâtreux near Luzarches (Val d'Oise) showing the Auvers sands overlain by the Beauchamp sands and the Saint-Ouen limestone** (photo: Bouhot).

3.23 **The "sand sea" at Ermenonville:** a barren area in the Beauchamp sands where the mobility of the sand has prevented afforestation (photo: Bouhot).

3.24 Fossil podsol at the top of the Beauchamp sands near Ermenonville (photo: Pomerol). The Marinesian marls (M) were deposited in a lagoon formed by gentle flooding of a land area without disturbing the humus layers of the soil (A1). Below this is the leached horizon (A2) which rests on the horizon of accumulated humus (B1) and the ferruginous layer (B2). The soils then grades down through a transitional level (B3) to the parent rock, the Beauchamp sands (C). The letters refer to the conventional classification of soil horizons.

3.25 A cemented fossil soil forming an overhang at the top of the Beauchamp sands at Attainville (Val d'Oise) (photo: Pomerol).

brought in by fresh water which had leached it from the saline hinterland of Lorraine.

The marls overlying the Gypse, that is, the blue marls of Argenteuil and the white marls of Pantin (Fig. 3.30), have an impoverished fresh water fauna consisting mainly of ostracods (Fig. 3.31 and 3.32). They also contain the same mammalian fauna as the Gypse of Montmartre from which CUVIER described *Palaeotherium magnum*, *Anoplotherium commune* and *Xiphodon gracile*. These species are considered to be Eocene, an important argument in favour of the view that the Oligocene in the Paris Basin begins with the green clays of Romainville which rest on the white marls of Pantin. This junction is STEHLIN's ''great break'' in the mammalian faunas (see p. 38).

6. The Stampian cycle (Oligocene): the last marine incursion into the centre of the Paris Basin.

The Stampian (named after the town of Étampes, by D'ORBIGNY, 1852) is equated with the whole of the Oligocene period in the Paris Basin. It begins with a slow but progressive change from the laguno-lacustrine environment of the preceding cycle (white marls of Pantin) to a marine lagoonal environment in which the green clays of Romainville were formed (Fig. 3.30 and 3.34). At the centre of the Paris Basin, this horizon begins with the **Glaises à Cyrènes,** in which there is a marked change of fauna heralding the Stampian cycle. At the same time, attapulgite disappears and kaolinite replaces it.

These earliest beds of the Stampian comprise the Sannoisian stage (from Sannois, Val d'Oise, MUNIER-CHALMAS and DE LAPPARENT, 1893), the palaeogeography of which has been studied by D'ALBISSIN (1956). The lagoons with *Cyrena convexa* were limited to the north by the Bray anticline and extended southwards, across the Meudon anticline, towards the Beynes and Roumois domes. The green clay itself extends appreciably further to the north and east. It is overlain by a rubbly limestone (**Caillasses d'Orgemont**), almost devoid of fossils, and by the **Calcaire de Sannois,** which contains a characteristic Stampian fauna.

Eastwards the Orgement and Sannois lime-stones are replaced by a lacustrine deposit, the

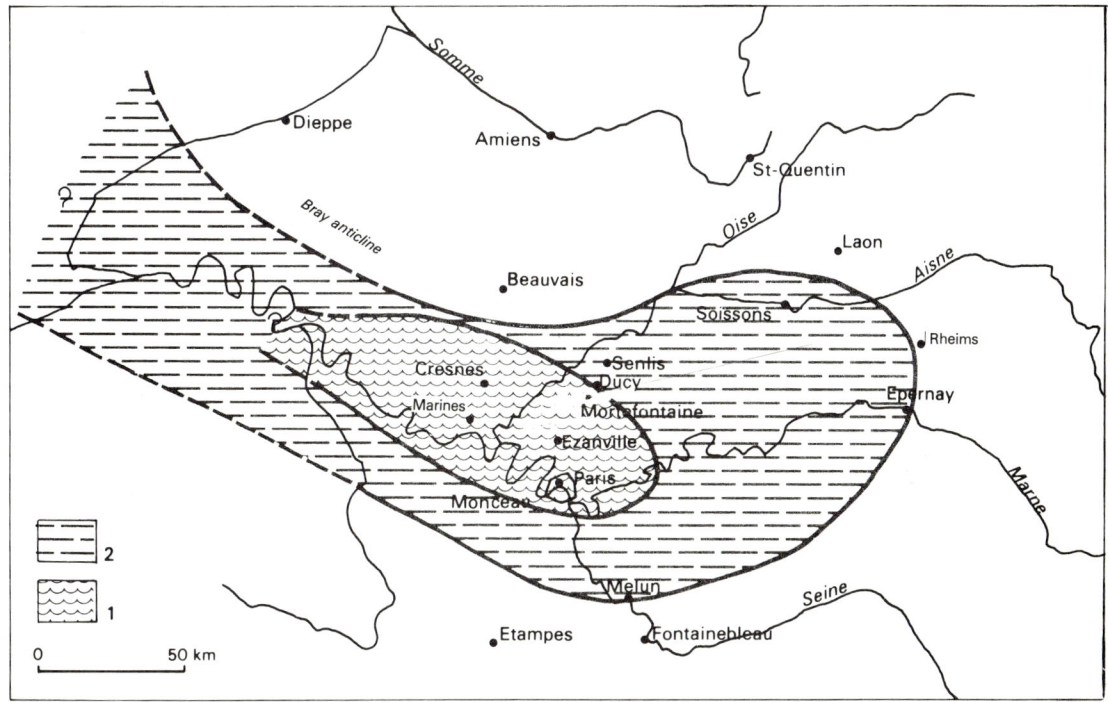

3.26 **The Marinesian palaeogeography of the Paris Basin.** 1 – Marine Cresnes sands and lagoonal sands of Marines; 2 – Lagoonal-lacustrine limestones of St. Ouen. The Ézanville, Ducy and Mortefontaine formations of the Marinesian are not shown. The hypothesis of a closed sea (with a laguno-lacustrine margin), at least until the late Marinesian, is supported by the recent discovery of the St. Ouen limestone in the English Channel.

Calcaire de Brie, where mammals of the Ronzon fauna (cf. p. 38) and in particular *Entelodon magnum,* appear. This species is one of the chief indicators of the arrival of Oligocene forms (STEHLIN, 1909). The most abundant species of charophyte, *Gyrogona medicaginula* (Fig. 3.33, p. 84), persisted throughout the Stampian (GRAMBAST, 1962).

In the centre of the Paris Basin, the **Marnes à Huîtres** overlie the Sannois limestone. This is a shell bed in which clay was probably trapped due to filtration by oysters and deposited as faecal pellets. Two species are present, *Ostrea cyathula* and *Crassostrea longirostris.* The latter is of some stratigraphical importance, as it has a short range in time both in the Paris Basin and Aquitaine, England, Belgium and Germany.

The laguno-marine episode in which the oyster marls were formed, was followed by a transgression which extended further to the east and to the south in the Paris Basin than any

previous one (Fig. 3.34). It deposited the **Sables de Fontainebleau,** which have a thickness of 60m. Outcrop and borehole evidence shows that the transgression probably extended eastwards beyond Rheims and Épernay, and reached Orléans in the south. To the north it apparently did not cross the Bray anticline or the valley of the Aisne (the most northerly Stampian outlier is at Villers-Cotterêts).

It seems probable that, in the west, *there was now communication with the Loire gulf in the region of Orléans,* where a Stampian fauna with *Archiacina armorica* in known from boreholes at Chaingy and Darvoy (Fig. 3.34).

This link with the Loire Gulf was *a precursor of the Miocene transgressions. However, positive epeirogenic movements (swells), which had begun in the Artois in the middle Lutetian and slowly spread southwards,* prevented the Miocene sea from reaching the centre of the Paris Basin.

Around the type locality of Étampes the

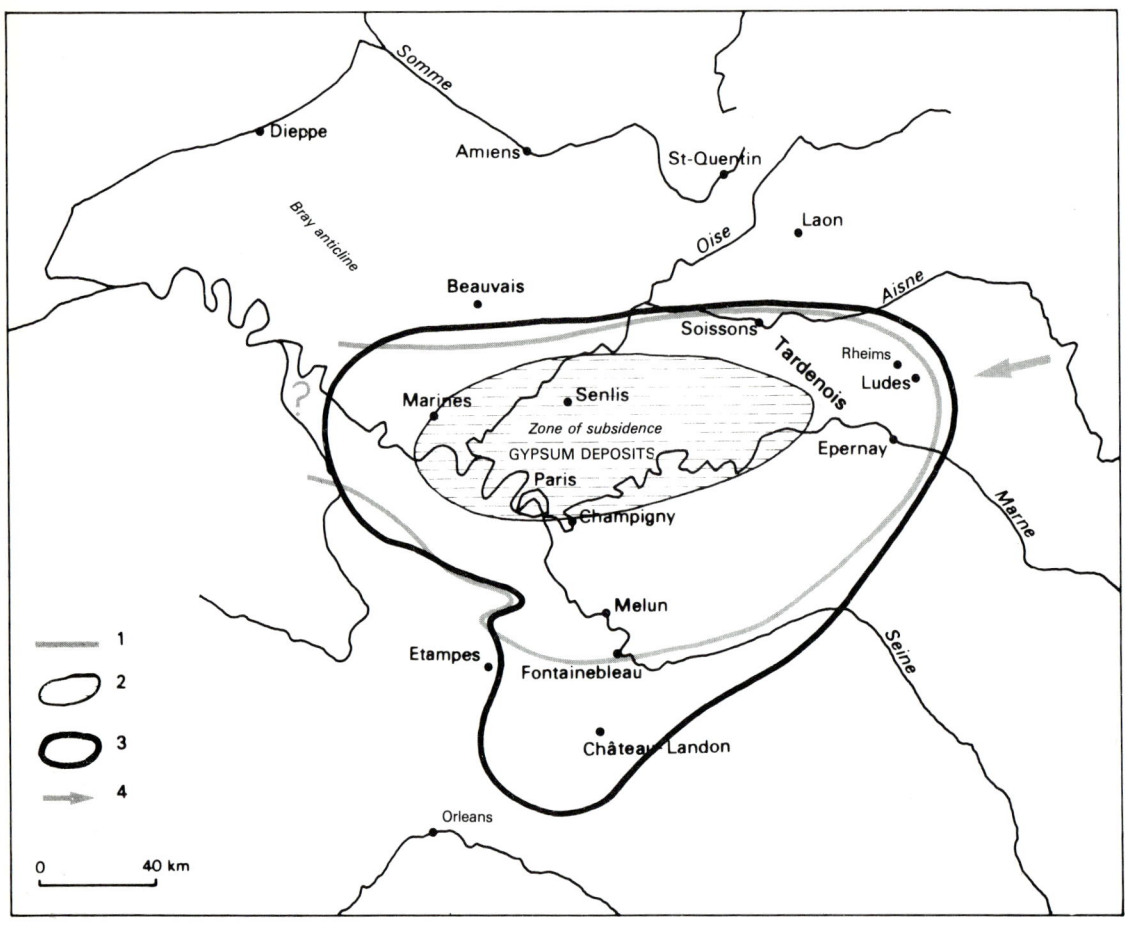

3.27 The Ludian in the Paris Basin. 1 – limit of the *Pholadomya ludensis* marls; 2 – limit of the gypsum mass; 3 – lacustrine limestone and marls: Champigny limestone, Château-Landon limestone, Tardenois marls; 4 – direction of sediment transport into the Ludian lake.

Fontainebleau sands are highly fossiliferous, and have yielded *Ampullina crassatina, Glycymeris angusticostata, Potamides plicatus, P. lamarcki, Cytherea* and *Cardium* (Fig. 3.35 and 3.36).

The Stampian sea retreated from the north of the Paris Basin before it left the southern part, where the successive stages of its retreat were marked (according to ALIMEN, 1936) by a series of strongly indurated, parallel dunes with a W.N.W.-E.S.E. orientation (Fig. 3.37 and 3.38). In this region, the fresh water **Calcaire d'Étampes** was deposited in late Stampian times. This is the lowest subdivision of the Calcaire de Beauce, *sensu lato*. To the north, between Hurepoix and Valois, clays were

deposited which have been partly silicified to form a hard, tough stone suitable for millstones (the Meulière de Montmorency).

At the end of the Stampian (or Oligocene) cycle the sea abandoned the Paris Basin leaving behind, between Orléans and Normandy, a swampy region which persisted into the Aquitanian. The **Calcaire de Beauce** *sensu stricto* (Pithiviers limestone, Orléans limestone), was deposited in these swamps and yields a mammalian fauna like that of St. Gérand (see Table V), of Aquitanian aspect but with some Oligocene characteristics. Thus, in the Aquitanian, there was a contrast between the Mediterranean regions (a transgressive phase with a Miocene marine fauna) and the

3.28 Bartonian fossils. 1 – *Nummulites variolarius* (Form B, microspheric) (x 30); 2 – *N. variolarius* (Form A, megaspheric) (x 30); 3 to 5 Algae (x 10): 3 – *Acicularia pavantina* (x 15); 4 and 5 – *Dactylopora cylindracea* (x 6); 6 and 7 Corals (x 1): 6 – *Trochoseris* (*Anthophyllum*) *distorta*; 7 – *Lobopsammia* (*Dendrophyllia*) *cariosa*; 8 to 19 Molluscs (x 1): 8 – *Belosepia blainvillei*; 9 – *Avicula defrancei* (Mortefontaine horizon, Marinesian); 10 – *Arca biangula*; 11 – *Pholadomya ludensis* (base of Ludian); 12 – *Perna lamarcki*; 13 – *Ostrea cubitus*; 14 – *Ampullina parisiensis*; 15 – *Cerithium tricarinatum*; 16 – *C. tuberculosum*; 17 – *Clavilithes longaevus*; 18 – *Melongena minax*; 19 – *Delphinula lima* (photo: Fay).

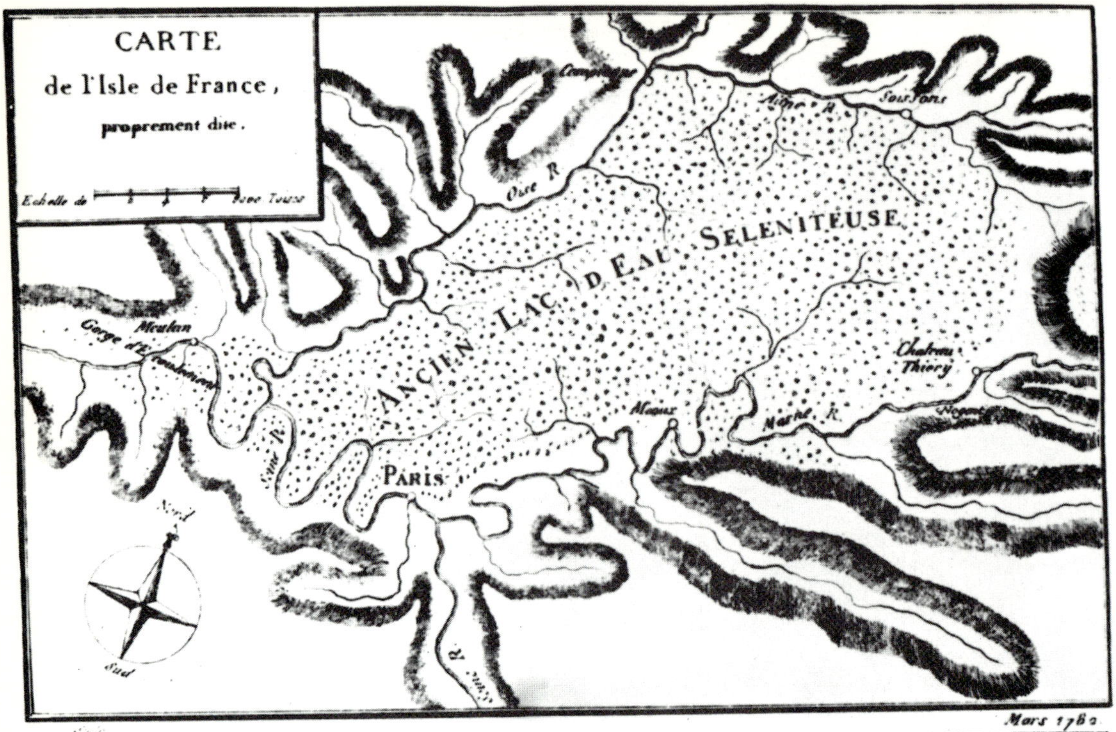

3.29 Sketch map illustrating one of the earliest hypotheses for the formation of gypsum (Lamanon, 1782 *in* Fontes, 1968). The gypsum was supposed to have been formed by the conversion of calcium carbonate to calcium sulphate by the action of rivers containing sulphuric acid which flowed over the chalk and limestone to the east of the Paris Basin.

3.30 **Lambert Quarry at Cormeilles-en-Parisis** (photo: Bouhot).
Ludian: 1 – Marl; 2 – Upper gypsum (17m); 3 – Argenteuil blue marl; 4 – Pantin white marls, with a thin gypsum horizon (5) at the top, marking the final evaporitic phase of the Eocene. Lower Stampian = Sannoisian: 6 – Plastic clay with *Cyrena* forming the base of the Oligocene; 7 – Green clay of Romainville with, at the top, the "white band", a lacustrine limestone, 50cm thick but extending for 150km; 8 – Orgemont calcareous marls and Sannois limestone. Middle Stampian: 9 – Oyster marls; 10 – Fontainebleau sands, white, locally cemented; 11 – Fontainebleau sands, stained yellow by infiltration during the Quaternary. Upper Stampian: 12 – Thin veneer of clay derived from the Montmorency sandstone, associated with loess (Quaternary).

PALAEOGENE OSTRACODS OF THE PARIS BASIN

Scale |———— 1 mm ————|
except where indicated

Stratigraphic stages (left to right): MONTIAN, THANETIAN, SPARNACIAN, CUISIAN, LUTETIAN (early, middle, late), Auversian, BARTONIAN (Marinesian, Ludian), SANNOISIAN, STAMPIAN

1	Limburgina bilamellosa bilamellosa
2	Bairdia montensis
3	Cytheretta multicostata
4	Krithe rutoti
5	Schizocythere tessellata
6	Vetustocytheridea lignitarum lignitarum
7	Trachyleberidea prestwichiana
8	Pokornyella ? ventricosa
9	Pterygocythereis cornuta
10	Schuleridea perforata
11	Occultocythereis mutabilis mutabilis
12	Cytheretta grignonensis
13	Cytheretta costellata
14	Idiocythere lutetiana
15	Cypridina homoedwardsiana
16	Uroleberis parnensis
17	Xestoleberis subglobosa
18	Paijenborchella eocaenica eocaenica
19	Leguminocythereis dumonti
20	Krithe bartonensis
21	Neocyprideis apostolescui
22	Pokornyella limbata
23	Cyamocytheridea punctatella
24	Eocytheropteron grekoffi
25	Cytheridea gypsi
26	Hemicyprideis montosa
27	Cypridopsis soyeri
28	Cytheridea pernota
29	Hemicyprideis helvetica
30	Loxoconcha nystiana
31	Cytherella jonesiana
32	Cytheromorpha zinndorfi
33	Pterygocythereis ceratoptera
34	Hornibrookella macropora
35	Cytheretta tenuistriata
36	Pterygocythereis retinodosa

for legend see p. 82.

3.31 Stratigraphy and Palaeoecology of Tertiary Ostracods[1].

The great diversity of ostracod faunas in the Palaeogene of the Paris Basin (more than 150 species) reflects the depositional environment: generally shallow water, sometimes even emergent. Ostracods occur in a variety of aqueous environments (lacustrine, hypo- or hyper-saline, shallow or deep marine) but the greatest variety of species is found in the shallow, marine environment (0-50m). Deeper shelf deposits and those of the continental slope show a rapid decrease in both numbers of species and individuals. The hypo-saline lagoonal environment (which may become hyper-saline in warm periods) is also poor in species, though the numbers of individuals may be very great. Lagoons and estuaries are environments which are relatively hostile to many organisms, including predators, but certain species of ostracods (*Neocyprideis, Vetustocytheridea, Hemicyprideis* and *Cytheromorpha*, for example) thrive there. At the present time, as well as in the past, considerable concentrations may occur, often consisting of a single species.

Even without specialist knowledge of the morphology of ostracods, it is possible to distinguish a number of environments on the basis of general characters. For example, the lagoonal environment is characterised by few species but many individuals; shallow marine environments have many species (often highly ornamented), while deep water marine environments have few species and few individuals which may be smooth or ornamented. (Ostracods of this last category are not known in the Palaeogene of the Paris Basin, though they are present in the upper Cretaceous). The fully lacustrine environment (such as occurs in the Sannoisian) generally contains few species and very few individuals and these have thin, and predominantly smooth, valves.

The evolution of ostracod species (as is well shown by their changes in morphology) was rapid and they are therefore of considerable stratigraphic value. This is particularly true of the ornamented forms, in which evolution has left clearer traces than on the smooth forms. Species which survive for more than one stage are rare and even in these cases detailed study has sometimes shown that what was considered to be a single species, can, in fact, be divided into two or more species. Some species are restricted to a small part of a stage. The figure opposite shows only a few of the more important species.

The genus *Cytheretta* has shown spectacular development from the Palaeogene to the present day and at least 20 species can be recognised. Equally varied are the genera *Cytherella* and *Cytherelloidea* (about 12 species), the group of similar genera *Vetustocytheridea, Hemicyprideis, Neocyprideis* and "*Clithrocytheridea*" (about 25 species) and the genus "*Leguminocythereis*" (at present being revised) (about 20 species).

1.   Text and figures by H.-J. Oertli (S.N.P.A., Pau)

3.32 **Palaeogene ostracods of the Paris Basin** (photo: Oertli).
1 – *Pokornyella limbata* (x 100) – Stampian – internal view of left valve.
2 – *Pokornyella limbata* (x 500) – Stampian – detail of hinge (anterior end, left valve).
3 – *Schizocythere tessellata* (x 100) – Lutetian – exterior, right valve.
4 – *Schizocythere tessellata* (x 1000) – Lutetian – detail of surface, alveolus.
5 – *Leguminocythereis ? striatopunctata* (x 50) – Lutetian – right valve, exterior.
6 – *Cytheridea pernota* (x 100) – Stampian – right valve, exterior.
7 – *Pterygocythereis cornuta* (x 100) – Lutetian and Bartonian – left valve, exterior.
8 – *Paijenborchella eocaenica inornata* (x 200) – Sannoisian – carapace viewed from right.

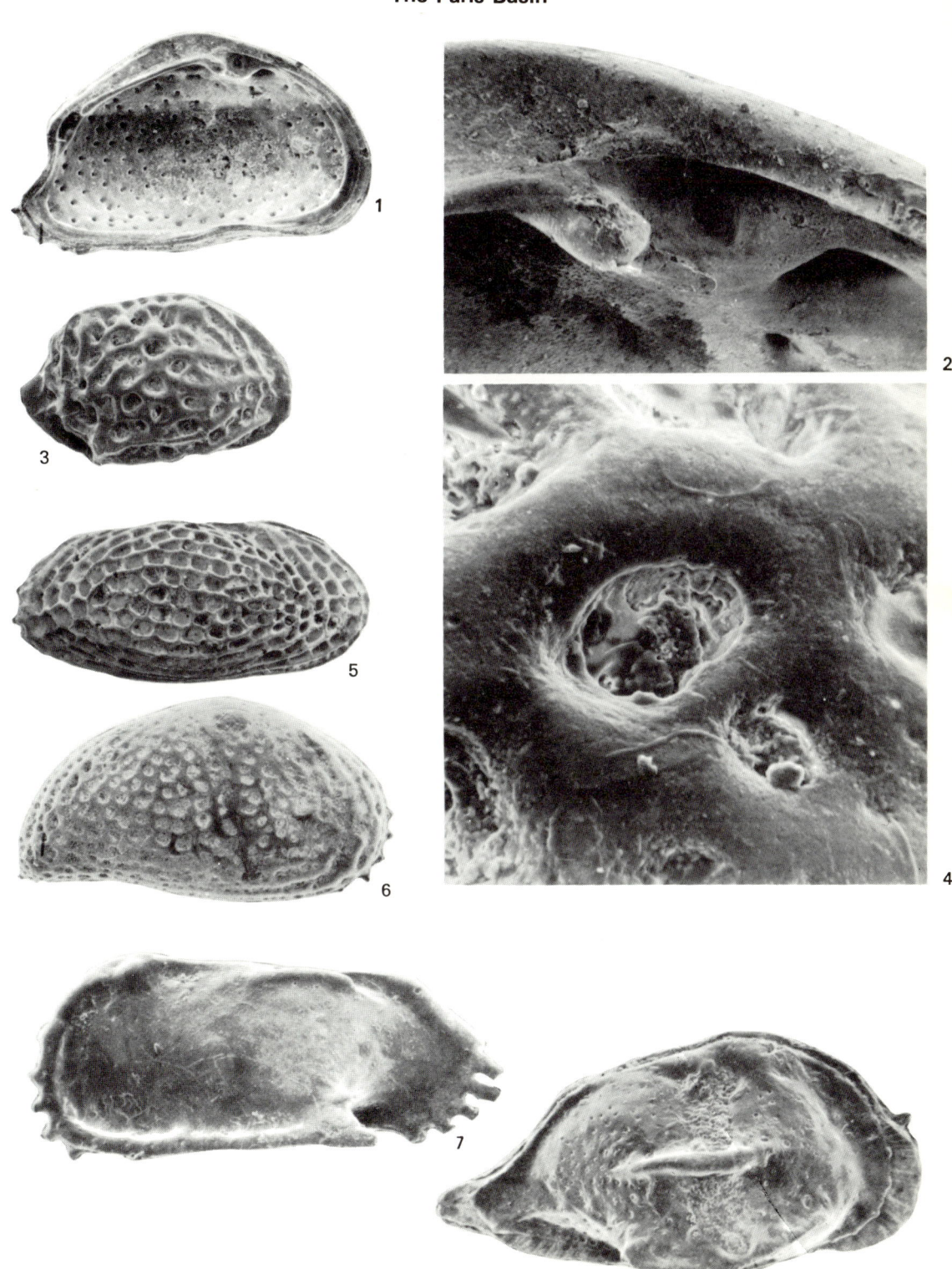

PALAEOGENE CHAROPHYTES OF THE PARIS BASIN

THANETIAN	SPARNACIAN	LUTETIAN	Auversian	Bartonian s.s.	EOCENE-OLIGOCENE boundary	STAMPIAN	Figure No.		
■							1	Tectochara dutemplei	
■	*Disappearance of Peckichara*						2	Peckichara disermas	
	■						3	Peckichara varians	
	■						4	Microchara hystrix	
		■					5	Psilochara undulata	
		■ *Appearance of Gyrogona and Psilochara*	■				6	Maedleriella mangenoti	
			■				7	Gyrogona lamarcki	
			■				8	Gyrogona morelleti	
			■				9a	Raskyella pecki (side view)	
			■				9b	Raskyella pecki (profile)	
				■			10	Chara friteli	
				■	■		11	Gyrogona wrighti	
				■			12	Maedleriella amblyodon	
				■			13	Psilochara repanda	
				■			14	Harrisichara vasiformis	
				■			15	Grovesichara distorta	
					■		16	Harrisichara tuberculata	
					■		17	Rhabdochara stockmansi	
						■	18	Gyrogona medicagulina	
						■	19	Psilochara acuta	

3.33 **Charophytes**[1] (x 20).

The Charophytes appear to have been neglected until about 1959 because their stratigraphic usefulness seemed to be very limited. These are plants which developed in brackish, marginal littoral or lacustrine environments. Their fossil remains consist of calcified fructifications ranging in size from about 200 microns to about 1700 microns together with

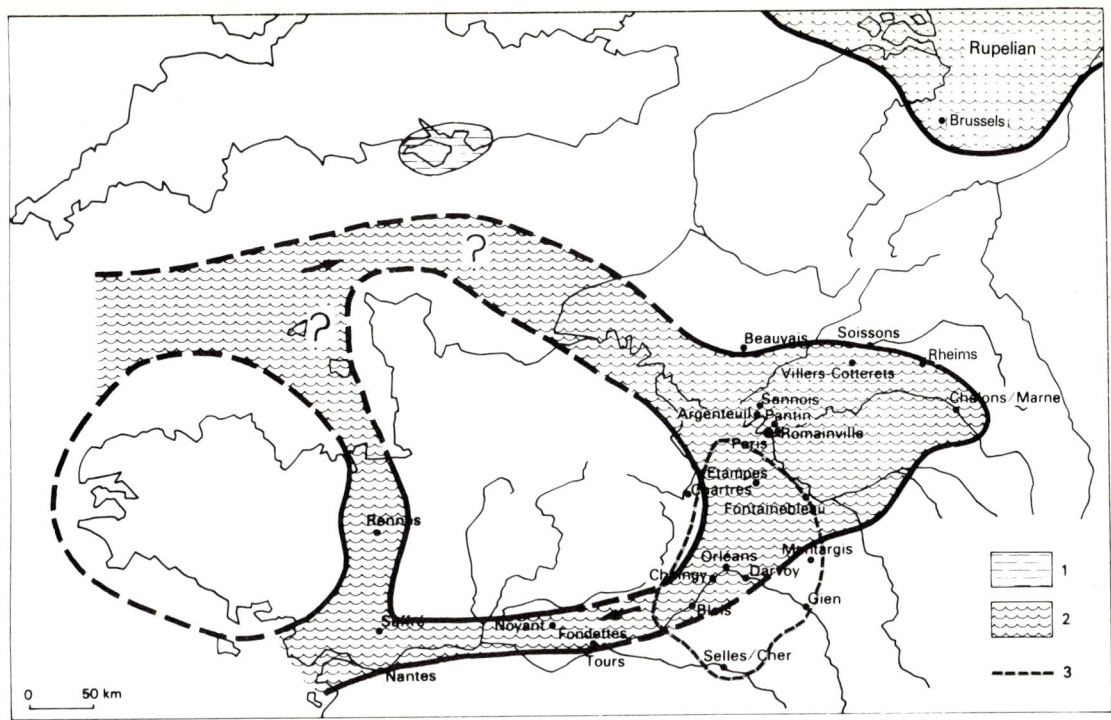

3.34 **The Stampian Sea** (2) **and the boundary of the lake of Beauce** (3). It appears that the Stampian sea retreated initially from the northern part of the Paris Basin towards the south and finally left the area through the Loire channel. The Rennes strait divided the Armorican massif into two islands. In the Isle of Wight (see Fig. 4.12, p. 122) the deposition of Sannoisian and lower Stampian lagoonal beds (1) was followed by emergence of the area.

fragments of stems. The fruiting bodies ("gyrogonites" or "oogonia") are usually well-preserved and detailed study has allowed Grambast to establish a classification based not only on the ornamentation (which is relatively unimportant in classification) but also on several other characters, such as the general form and dimensions, the nature of the apex and the base, the nature and form of the basal plate and the number, dimensions, appearance and ornamentation of the spirals. The Charophytes, known from the Silurian to the present day, are in some cases the only group by which a formation can be dated.

The Thanetian and Sparnacian floras are similar. In fact, most of the Thanetian species (*Peckichara disermas, Tectochara dutemplei* . . .) are present also in the Sparnacian. However, some species, such as *Peckichara varians* and *Microchara hystrix* seem to be restricted to the Sparnacian.

The genus *Peckichara* was extinct by Lutetian times while the genera *Gyrogona* and *Psilochara* appear for the first time. The easily recognised *Psilochara undulata* is restricted to the Lutetian. It is often associated with *Maedleriella mangenoti* and *Gyrogona lamarcki* which, however, persist into the Auversian where *Raskyella pecki* and *Gyrogona morelleti* make their appearance.

The flora of the St. Ouen limestone (Marinesian = middle Bartonian) is distinct from that of the underlying beds. It includes *Chara friteli, Maedleriella amblyodon, Gyrogona wrighti* and *Grovesichara distorta*. The last two species extend into the overlying beds.

The Ludian marls above the *Pholadomya ludensis* bed are characterised by the persistence of some elements of the St. Ouen limestone flora and by the appearance of new forms including *Harrisichara vasiformis* and *Psilochara repanda*. *Harrisichara tuberculata* and *Rhabdochara stockmansi* occur in the Bembridge Beds of the Isle of Wight, which are equivalent to the gypsum horizons which have yielded the Montmartre mammal fauna. These forms are, however, found also at higher levels and occur, for example, in the "white band" at the top of the green marls of Cormeilles-en-Parisis (cf. Fig. 3.30).

The Charophytes are of little palaeogeographic importance, because they are not endemic to particular areas. However they are valuable for establishing inter-continental correlations because of the very widespread distribution of individual genera throughout the history of the group.

1. Text and figure by J. Riveline-Bauer (Paris University). The drawings are taken from the publications of L. Grambast.

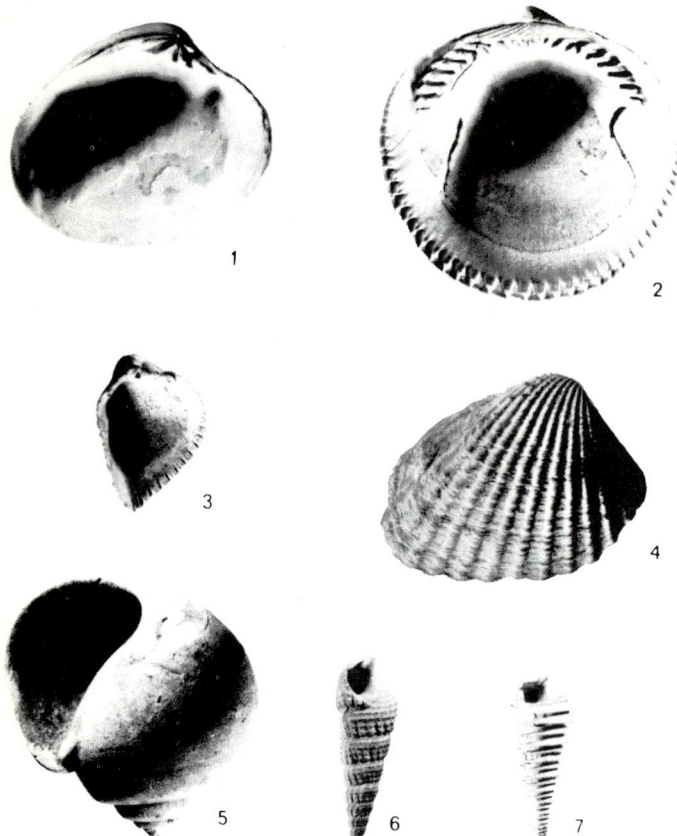

3.35 Stampian fossils (x 1). **1** – *Meretrix splendida*; **2** – *Glycymeris obovata*; **3** – *Cardium stampense*; **4** – *Cardita bazini*; **5** – *Ampullina crassatina*; **6** – *Potamides plicatus*; **7** – *Potamides trochlearis* (photo: Fay).

3.36 Benthic Foraminiferids[1] (x 20).

The benthic foraminiferids are both good stratigraphic fossils and precise indicators of facies. It is rare to find marine formations which do not contain these tiny fossils and even small amounts of sediment may contain large numbers of them. For these reasons, they are of great importance in the study of borehole cores and in oilfield research. However, not all foraminiferids are of equal stratigraphic value. The Table opposite shows that while some characterise a particular zone and allow it to be dated with precision, others range through a long period of time. Thus, in the Paris Basin, *Protelphidium rolshauseni* is confined to the Thanetian, *Spiroloculina morelleti* to the upper Cuisian, *Rotalia trochidiformis* to the middle Lutetian, *Halkyardia minima* to the Marinesian and *Ammonia propingua* to the lower Oligocene. On the other hand, *Angulogerina muralis* is present from the Sparnacian to the Marinesian and *Elphidium hiltermanni* from the base of the lower Cuisian to the top of the lower Stampian.

Geological formations therefore contain a succession of genera and species of foraminiferids which, in many cases, enable correlation to be made between formations of different basins and even of different continents. As well as their stratigraphic significance, foraminiferids are indicators of facies. For example, *Rosalina bractifera* and *Ammonia propingua* indicate a salinity which is different from normal and, in particular, the presence of brackish water. In contrast, *Miliola saxorum*, *Quinqueloculina crassa* and *Spiroloculina morelleti* indicate a fully marine environment. As a group, the foraminiferids illustrated are typical of a littoral facies of a warm sea.

1. Text and figures by Y. Le Calvez (Institute of Palaeontology, Museum of Natural History, Paris). Stereoscan photographs by the Geological Laboratory of the Museum.

The Paris Basin

PALAEOGENE BENTHONIC FORAMINIFERIDS OF THE PARIS BASIN

THANETIAN	YPRESIAN – SPARNACIAN	CUISIAN early	CUISIAN late	LUTETIAN early	LUTETIAN middle	LUTETIAN late	BARTONIAN Auversian	BARTONIAN Marinesian	BARTONIAN Ludian	STAMPIAN early	STAMPIAN late	Figure No.	Species
■												1	Protelphidium rolshauseni
■	■											2	Cibicidina cunobelini
■	■	■	■	■								3	Cibicidoides proprius
	■	■	■	■	■	■	■					4	Angulogerina muralis
	■	■	■	■	■	■	■	■	■			5	Elphidium hiltermanni
		■	■	■	■							6	Spiroloculina morelleti
			■	■	■	■	■	■	■			7	Miliola saxorum
			■	■	■	■	■	■	■			8	Epistomaria rimosa
		■	■	■	■	■	■	■	■			9	Rosalina bractifera
		■	■	■	■	■	■	■	■			10	Pararotalia armata
			■	■	■							11	Pararotalia curryi
				■	■	■	■	■	■			12	Rotalia trochidiformis
				■	■	■	■	■	■			13	Neocribrella globigerinoides
				■	■	■						14	Boldia lobata
				■	■	■	■	■	■			15	Linderina brugesi
					■	■	■	■				16	Quinqueloculina crassa
							■	■	■			17	Halkyardia minima
										■		18	Ammonia propingua

1. Protelphidium rolshauseni
2. Cibicidina cunobelini
3. Cibicidoides proprius
4. Angulogerina muralis
5. Elphidium hiltermanni
6. Spiroloculina morelleti
7. Miliola saxorum
8. Epistomaria rimosa
9. Rosalina bractifera
10. Pararotalia armata
11. Pararotalia curryi
12. Rotalia trochidiformis
13. Neocribrella globigerinoides
14. Boldia lobata
15. Linderina brugesi
16. Quinqueloculina crassa
17. Halkyardia minima
18. Ammonia propingua

3.37 Fontainebleau Sands overlain by a thin sandstone layer forming an overhang. The tree-covered slope above the quarry is formed by the Montmorency siliceous limestone (millstone) containing pockets of Lozère sands (grassy talus slopes). Near Villejust (Essonńe) (photo. Bouhot).

3.38 Ridge of tumbled sandstone blocks (aligned W.N.W.-E.S.E.) near Fontainebleau (photo: Bouhot).

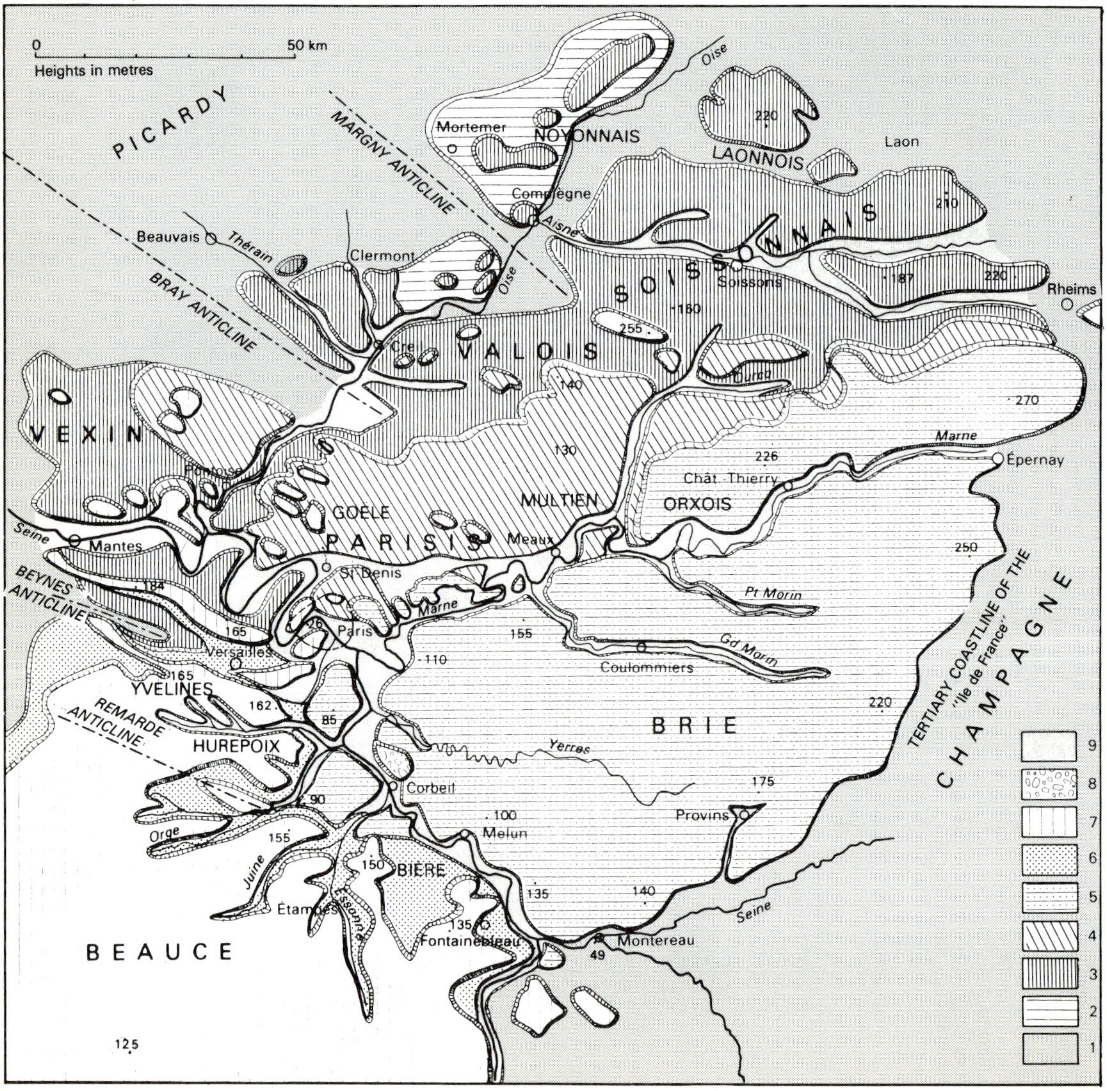

0   50 km

Heights in metres

PICARDY

MARGNY ANTICLINE

BRAY ANTICLINE

Mortemer

NOYONNAIS

Oise

220

Laon

LAONNOIS

210

Compiègne

Aisne

Beauvais

Thérain

Clermont

Oise

S O I S S O N N A I S

Soissons

187

220

Rheims

255

180

V A L O I S

140

270

VEXIN

Cre

130

Marne

Épernay

GOELE

MULTIEN

Chât Thierry

226

ORXOIS

250

Seine

Mantes

P A R I S I S

Meaux

184

St Denis

Pt Morin

Gd Morin

BEYNES
ANTICLINE

165

Marne

Paris

155

Coulommiers

220

Versailles

110

165

162

YVELINES

REMARDE
ANTICLINE

85

HUREPOIX

Yerres

B R I E

90

Corbeil

175

Provins

Orge

155

100

Melun

TERTIARY COASTLINE OF THE
"Ile de France"

C H A M P A G N E

Juine

150

BIÈRE

135

140

Étampes

Essonne

135

Fontainebleau

Seine

B E A U C E

49

Montereau

125

9
8
7
6
5
4
3
2
1

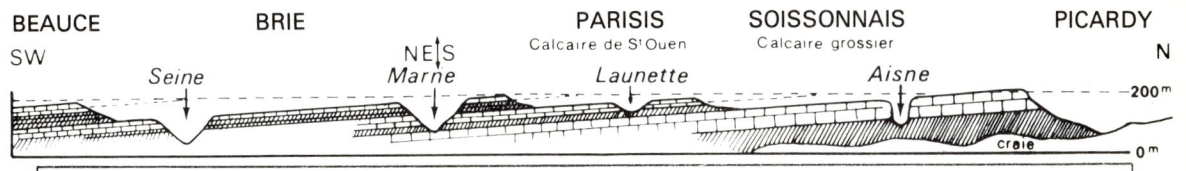

BEAUCE   BRIE   PARISIS   SOISSONNAIS   PICARDY

SW   Calcaire de St Ouen   Calcaire grossier   N

NE|S

Seine   Marne   Launette   Aisne

craie

3.40 Diagrammatic section across the structural platforms of the Tertiary rocks of the Paris Basin. The slope of the platforms and the escarpments is greatly exaggerated. The regional dip of the beds is much less than 1°. The valleys cutting through the *Calcaire grossier* (some 30 to 40m thick) are steeper-sided than those in the other platforms. Oblique shading; unconsolidated horizons; small rectangles; limestones.

northern regions (a regressive phase with a relict Oligocene terrestrial fauna).

After the Lake of Beauce had dried up in the Aquitanian, the central part of the Paris Basin remained above sea level. The Neogene transgressions were, therefore, unable to penetrate any further than the English Channel (p. 209) and the synclinorium of the Loire (p. 153). The palaeogeographic evolution of the Paris "cuvette" was thenceforth essentially continental and is marked by four successive stages, as follows[1]:

1. In the **Burdigalian,** the deposition of the sands of the Orléans region, Sologne and Lozère.

2. In the late **Miocene,** the initiation of the present Paris Basin and the formation of the highest erosion surface.

---

1. See Pomerol and Feugueur, Guide géologique du Bassin de Paris. Paris, Masson.

3. In the **Pliocene,** the erosion of structural platforms and the isolation of outliers (Figs. 3.39 and 3.40).

4. In the **Quaternary,** formation of the existing valleys; periglacial phenomena, such as the deposition of loess.

The Quaternary period will be discussed later (p. 223) and our purpose in referring here to the above sequence of events is to draw attention to its effect in eroding valleys and structural platforms bordered by escarpments where the Palaeogene formations outcrop. The dip of the Palaeogene beds is so low, that it is necessary to travel considerable distances to cover the whole succession. By contrast, in Alum Bay and Whitecliff Bay in the Isle of Wight, England, where intense folding has turned the beds to a vertical position, it is possible to see the whole of the Palaeogene succession in a distance of only a few hundred metres (see p. 121).

II. – THE ATLANTIC COAST: THE AQUITAINE BASIN AND THE ARMORICAN BASINS

The evolution of the Aquitaine basin in the Palaeogene is not an exact parallel of that of the Paris basin. Although the transgressions in the Gironde area in the north show a similarity with the marine incursions of the Paris Basin,

in the Adour region to the south deposition was continuous from the Cretaceous into the Tertiary. This latter area was part of the Pyrenean gulf which covered much of northern Spain (see Fig. 4.20) in which the beginning of the Tertiary (Danian) is marked by the

3.39 Map of the Paris region showing the succession of structural platforms.
1 – Mesozoic formations; 2 – Structural surface of the Mortemer limestone (Thanetian); 3 – Structural surface of the *Calcaire grossier* (Lutetian); 4 – Structural surface of the St. Ouen limestone (Marinesian); 5 – Structural surface of the Brie limestone (Sannoisian); 6 – Erosion zone in the Fontainebleau Sands (Stampian); 7 – Structural surface of the Beauce limestone (late Oligocene); 8 – Pliocene gravel of Senart; 9 – Alluvium.
Note the dip of each of the platforms towards the south and west. The platform of Vexin is more an erosion platform than a structural one. In the Hurepoix region, the Fontainebleau Sands have been stripped off by erosion beneath their clay cover, which is less resistant than the Étampes and Beauce limestones to the south.
A section across the area is shown in Fig. 3.40.

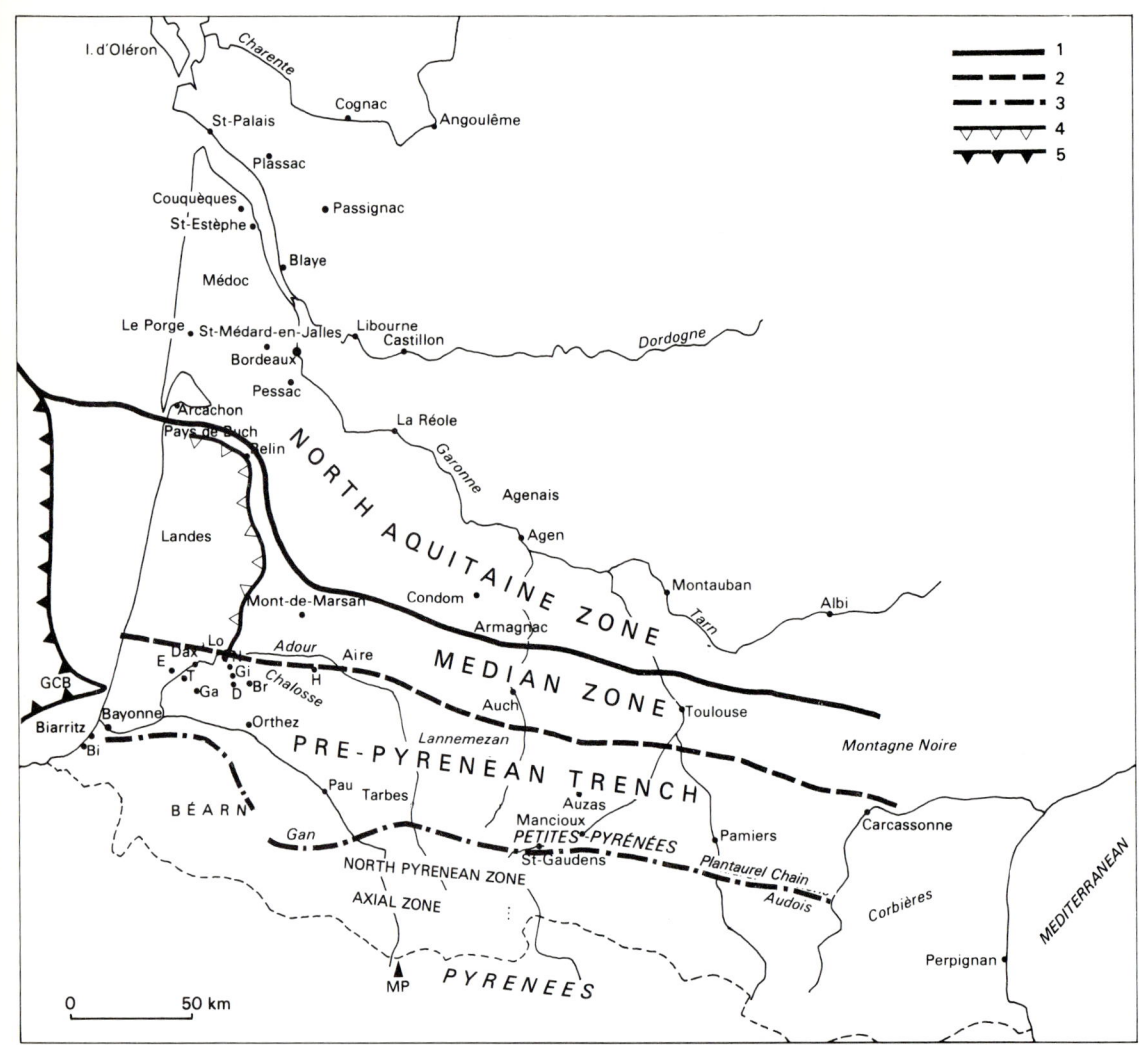

3.41 The structure of the Aquitaine basin and the locations of Palaeogene type-localities. For mammaliferous deposits see Fig. 2.6, p.35. Bi – Bidart; Br – Brassempouy; D – Donzacq; E – Escornebéou; Ga – Gaas; GCB – Gulf of Cap Breton; Gi – Gibret; H – Horbaziou; Lo – Louer; M – Mancioux; MP – Mont-Perdu; N – Nousse. 1 – Celtaquitaine line; 2 – northern limit of pre-Pyrenean trench; 3 – present day North Pyrenean front; 4 and 5 – the Landes slope (continental slope) in the Palaeocene (4) and at present (5) (after Kieken (1973)).

appearance of globorotalids in flysch-like sediments.

In the Palaeogene this gulf was *dominantly marine in the west and continental to the east*. It was bounded to the N.E. by the Massif Central which remained emergent, undergoing denudation, until the Oligocene. The southern boundary was formed by the land mass of Asturias and Ebro, and the funnel-shaped gulf was closed to the east by a land embracing

Provence, Languedoc, the Gulf of Lions and Catalonia. In France, the head of the gulf (the region of the Petites Pyrénées and Corbières) was reached by the sea on two occasions, the first was in the late Palaeocene (Thanetian) and the second in the earliest Eocene (Ilerdian). Uplift of the eastern and central parts of the Pyrenees chain in later Ilerdian times led to the formation of a great thickness of molasse and conglomerates (the *Poudingue de Palassou*) in the region of the Petites-Pyrénées, south of Toulouse, which separated the Basque and Aragon seas and isolated the Corbières region.

The Aquitaine basin can be divided, structurally, into a *northern zone separated by* *a median zone from the pre-Pyrenean trench* (an extension of the Aturian gulf). The **northern zone** is bounded on the north by the Poitou sill, and on the south by a sinuous line of folds extending from Arcachon to Toulouse. This line extends into the Atlantic where it forms the southern edge of the Breton platform (see Fig. 1.4) from which it takes its name – the Celtaquitaine line of WINNOCK (1971).

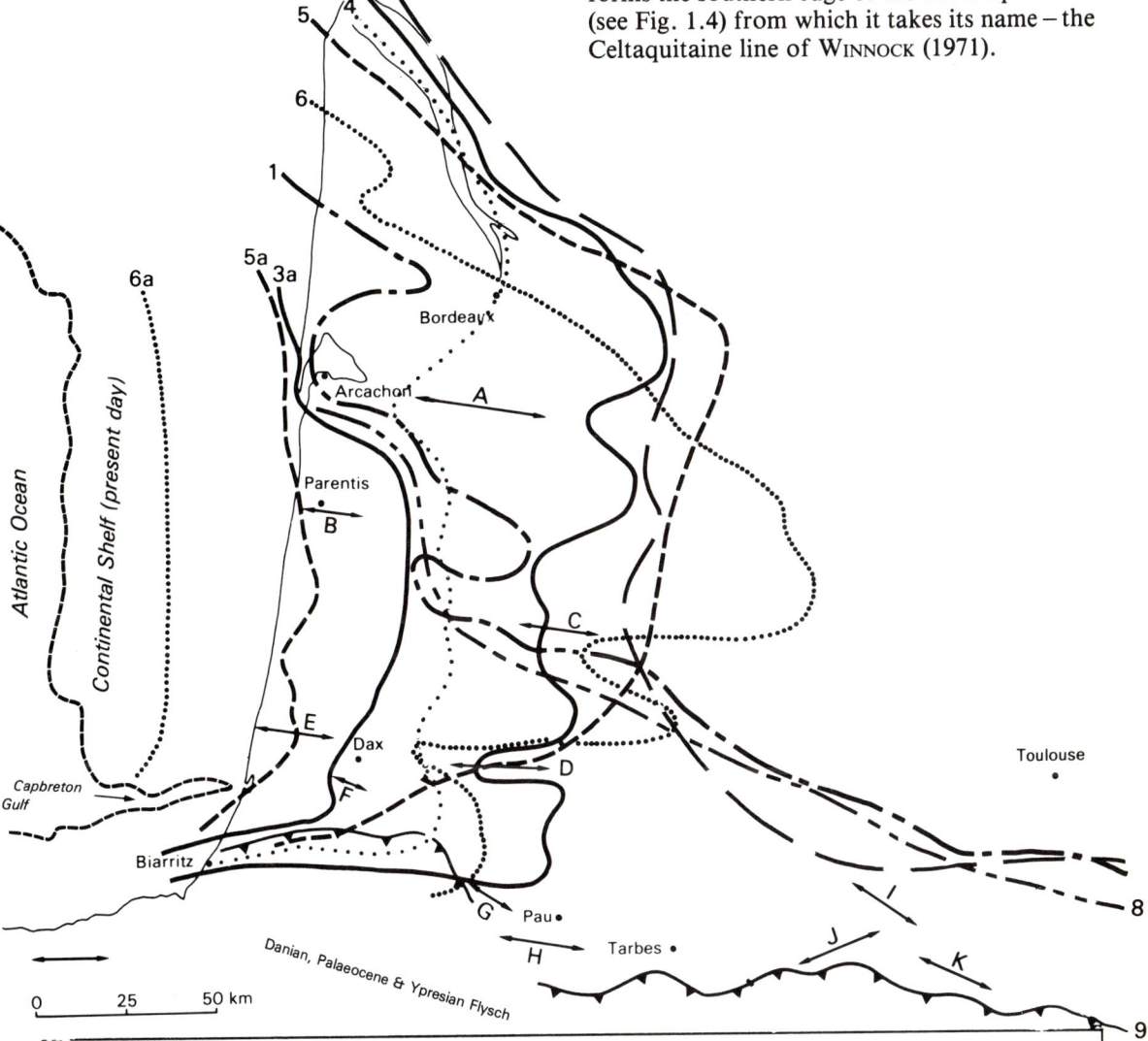

3.42 **The position of shorelines and of the edge of the continental shelf in Aquitaine during the Tertiary** (after Kieken (1973), modified). 1 – Palaeocene; 2 – early Eocene (Ilerdian); 3 – middle Eocene; 3a – edge of the continental shelf in the middle Eocene; 4 – late Eocene; 5 – Oligocene; 5a – edge of the continental shelf in the Oligocene; 6 – Miocene; 6a – edge of continental shelf in the Miocene. Note the westward accretion of the continental shelf from the Eocene to the present day (7). See Fig 3.41 for position of type-localities.
The position of the southern limit of the north Aquitaine plateau (8) and that of the north Pyrenean front (9) are shown.
Anticlines: A – Villagrains-Landiras; B – Parentis; C – Roquefort-Créon; D – Audignon; E – Magescq; F – Tercis; G – Lacq; H – Pau; I – Puymaurin; J – Gensac; K – Saint-Marcet-Plagne.

Neritic beds were deposited on the western edge of this platform (in the Médoc), while in the east, continental deposits derived from the Massif Central were formed. These latter included iron ores of early Eocene age and the Périgord sands of Lutetian to Miocene age. The total thickness of the Tertiary here is less than 600 metres.

To the south, the **pre-Pyrenean trench** extended as far as the Corbières from late Palaeocene to middle Ilerdian times, but by the late Cuisian it extended no further east than Pau. This trench always subsided more in the west than in the east and the submarine gulf of Cap Breton (20km N of Biarritz) is a present day relict. *North to south barriers*, more or less linked to *diapiric movements*, divided the trench into a series of steps, exemplified by those of Pau, St. Marcet, Orthez and Dax. These barriers hindered the transport of detritus towards the Atlantic and caused it to be deposited in a string of basins where the thickness of the Tertiary may reach 3000m.

The **median zone** lay between the north Aquitaine zone and the subsiding trench, which today has been overriden by the north Pyrenean front (Fig. 3.41). The western part of this zone was subsiding rapidly, while the eastern part was gently uplifted between Mont-de-Marsan and Toulouse. These two areas are separated by the *Landais talus*, a north-south fold extending from Belin to Louer. The axis of this fold moved progressively towards the Landes (Figs 3.41 and 3.42) in step with the westward progression of Palaeogene sedimentation.

The microfauna of the Aquitaine Palaeogene is richer than that of the Paris Basin in large foraminiferids (numerous species of *Nummulites, Assilina, Alveolina* and *Discocyclina*), planktonic foraminiferids and nannoplankton and this facilitates correlation within the Mediterranean region.

Recent studies have substantiated classical ideas concerning the separation of the Gironde and Aturian Basins by east-west folds, such as those of Villagrains-Landiras and Roquefort-Creon between the Garonne and the Adour. This intermediate region is mantled by the Quaternary sands of the Landes, 2 to 20m thick, and the underlying rocks can only be studied in borings or in deep river valleys.

The succession in the Gironde area consists mainly of marine sediments but they contain a rich sequence of mammal faunas known also from the continental molasse of the Agenais. These permit excellent correlation between the marine and continental beds across the area. The continental beds were derived from the Massif Central until middle Lutetian times after which their source was the Pyrenees, a situation which continued until the end of the Oligocene.

The stratigraphy of the Aquitaine basin therefore has a two-dimensional pattern starting from the neritic region of the Gironde; southwards towards a pelagic domain and eastwards towards the land.

1. A model neritic basin: the Gironde area[1].

The pattern of well-marked belts displayed by the Jurassic and Cretaceous rocks of the Charente and Périgord areas is not repeated in the Tertiary, in which there is a Palaeogene plateau east of the Garonne-Gironde river (continental Eocene and marine middle Oligocene) and a Neogene plateau to the west of the river. The marine Palaeogene reappears in the Blaye and Médoc districts as a result of gentle anticlines with an Hercynian trend.

After the post-Maastrichtian transgression and during the Palaeocene, unfossiliferous red clays and quartz sands were deposited, up to 12m in thickness. These formations are continental and pass laterally into lacustrine limestones with charophytes (Pessac). The Passignac plant horizon, SSE of Jonzac, which can be compared with that of Sézanne in the Paris Basin, is of Thanetian, or possibly, Sparnacian, age. However the Gironde area experienced a *brief marine episode* during the Palaeocene (KLINGEBIEL and PUECHMAILLE, 1966) and traces of this can be seen at Porge and St.-Médard-en-Jalles. The position of these marine beds at the foot of the Villagrains-Landiras anticline suggests the presence of a rise separating a southern zone of marine sedimentation from a northern littoral zone.

Thus it will be seen that during a period in which the Paris Basin experienced two transgressions (in the Dano-Montian and in the Thanetian), the Gironde remained almost free

---

1. This section and the later section on the Neogene of Aquitaine have been reviewed by MM. Vigneaux and Alvinerie and their colleagues of the "Institut de géologie du bassin d'Aquitaine" at Bordeaux.

of them. The first major marine penetration did not occur until the beginning of the early Eocene, when the Paris Basin was, by contrast, a continental area with (Sparnacian) lagoons, separated from the Belgian Ypresian sea by the Artois swell.

Marine deposits of this age from Southern Europe which correspond more or less to the Sparnacian have been placed in a new stage, the Ilerdian, by the Swiss geologists, HOTTINGER and SCHAUB (1960). The stratotype for this stage was defined in the north of Spain near the town of Lerida (= Ilerda, in Latin) (see p. 128).

In general terms, *the Tertiary marine transgressions entered the hinterland of Aquitaine from areas around Arcachon and Bayonne, regions which remained permanently marine* (Fig. 3.42).

a) **Early Eocene: Ilerdian and Cuisian** (Fig. 3.43).

The Ilerdian, which marks the first important

Palaeogene transgression locally, has been recognised in borings in the Médoc. At its base in the Pays de Buch (near Arcachon) is a limestone with *Alveolina cucumiformis* which is overlain by a marly limestone with *Nummulites exilis* and *N. globulus*. These two species are found also in sandy limestone in the Médoc and mark the wide spread of the sea across the region of the Gironde. The transgression reached the line of the right bank of the Gironde estuary in Cuisian times and cut an irregular surface in the Maastrichtian beds, on which pockets of glauconitic sand containing *Nummulites planulatus* and *Alveolina oblonga* were deposited. These are overlain in the Médoc by sands containing *Nummulites burdigalensis* and *N. aquitanicus*, of late Cuisian age.

b) **Middle Eocene: Lutetian.**

Following a brief regression, the detrital sands known as the **"Sables inférieurs"** (Lower

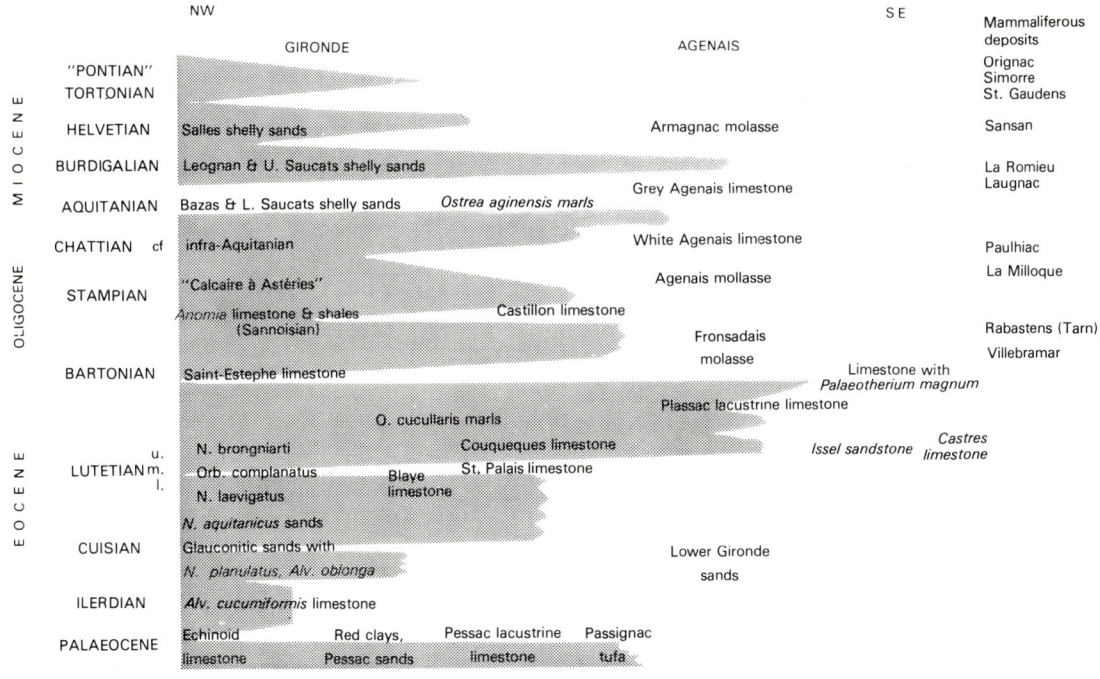

3.43 **Schematic outline of the stratigraphy of the principal Tertiary formations of the Aquitaine Basin from the Gironde to the Agenais.**
The term "Blaye limestone" has been the source of some confusion among geologists. It has been variously applied either to wholly Lutetian beds, or to Lutetian plus some late Eocene beds, or just to a part of the upper Eocene. It is more correct to avoid the use of the term and to refer simply to "Lutetian limestone", etc. as appropriate.
To the west the formations are marine (shaded); to the east they are continental.

Sands), with *Nummulites laevigatus* and *N. aquitanicus*, mark the base of the middle Eocene. This basal formation reflects a *reaction to the Pyrenean movements* which provoked uplift and renewed erosion in the Massif Central, to provide a supply of detritus to the deltas of the Bordeaux region. North of the estuary this unit is overlain by thick Lutetian limestones (the **Calcaire de Blaye** *s.s.*) (Fig. 3.43). These massive, sandy limestones, with intercalations of sand, yield a rich echinoid fauna (*Echinolampas, Lenita*).

As in the Paris Basin, the Lutetian of Aquitaine can be divided into three sub-stages identified by their microfaunas:
- **lower Lutetian:** sandy limestone with *N. laevigatus*.
- **middle Lutetian:** sandy limestones and marls with *Orbitolites complanatus*, *N. millecaput* and *Assilina*.
- **upper Lutetian:** ("Biarritzian") with *N. brongniarti*, *N. variolarius*, *Alveolina elongata* and *Linderina brugesi*.

These three episodes, well marked in borings and recognisable within the Blaye limestone are separated by slight regressions. The late Lutetian transgression is the most widespread of the Atlantic coastal area, penetrating deep into Brittany, the English Channel and the Cotentin peninsula. It also briefly entered the Paris Basin to deposit the Falun de Foulangues (see p. 69).

The **Calcaire de Saint-Palais** on the right bank of the Gironde begins with sands yielding *N. planulatus* and *Alveolina oblonga* derived from the Cuisian and is of early to middle Lutetian age. According to FABRE (1939), the **Calcaire de Couquèques,** south of the estuary, the Marnes à *Ostrea cucullaris* (which overlie the Calcaire de Blaye) and the **Calcaires lacustres de Plassac** belong to the late Lutetian. In the east, these units pass laterally into a wholly continental sequence: the Issel sandstone with *Lophiodon isselensis* and the Castres limestones with *Planorbis pseudoammonius*.

The Jurassic and Cretaceous rocks bordering the Massif Central were covered at the beginning of the Palaeogene by a mantle of kaolinite and ferruginous clays (known as the "*sidérolithique*", formerly exploited as iron ore). These deposits were derived, in part, from the altered rocks of the Massif Central.

c) Late Eocene = Bartonian + Priabonian.

The late Eocene is represented in the Médoc by the **Calcaire de Saint-Estèphe.** This rests discordantly on the Couquèques limestones, which had been gently folded during the final phase of the Lutetian regression (pre-Pyrenean or Illyrian phase). As in the Paris Basin, the late Eocene of the Gironde area can be subdivided into three episodes in ascending order:
- the Saint-Estèphe clays with *Ostrea bersonensis* overlain by the Saint-Estèphe limestone with miliolids.
- the Saint-Estèphe limestone *sensu stricto* with *Echinolampas ovalis* and *Sismondia occitana*.
- the Saint-Estèphe clays with *Anomia* overlain by the Blaignan white marls with *Sismondia occitana*.

Nummulites are rare, but occur in borings as follows (from bottom to top):
- *Nummulites praefabianii* and *N. variolarius;*
- *N. fabianii* and *N. incrassatus;*
- *N. chavannesi* and *Operculina*.

This succession is comparable with that of the Priabonian in northern Italy.

Eastwards between Bordeaux and the borders of the Agenais, the marine formations of the upper Eocene are replaced by continental deposits, the **Molasse inférieure du Fronsadais.** In the Agenais itself, this unit passes into the Ondes limestone containing *Palaeotherium magnum* (Ludian).

d) Oligocene: Stampian.

The Stampian of the Gironde commences with a gentle transgression (Sannoisian episode) which deposited lagoonal limestones with *Anomia oligocenica*, which are more argillaceous and more marine in character in the synclinal troughs (Bordeaux, Pauillac). Their microfauna however has late Eocene affinities and so the unit may alternatively be regarded as marking the final regressive phase of the Eocene. (ANDREIEFF and MARIONNAUD, 1972). Around Fronsac and Libourne the deposits are continental sands and micaceous clayey sandstones, the **Molasse supérieure du Fronsadais,** derived from the Massif Central. At about this time (38 to 26Ma) the karst topography of the Causses du Quercy (region around Cahors and Montauban) provided a trap for the rich mammal fauna associated with the phosphate deposits of the **Phosphorites du Quercy** (see Fig. 2.12, p. 39).

A brief regression, represented to the east of Libourne by the Calcaire lacustre de Castillon (equivalent to the Calcaire de Brie of the Paris Basin) was followed by the **major Stampian transgression which gave rise to the Calcaire à Astéries** ("starfish limestone"), which contains the remains of several species of echinoderms. This was the most important transgression in the Gironde during the Cenozoic and also the most obvious because the Calcaire à Astéries is a prominent morphological and pedological feature and has influenced the economy of the area east of Bordeaux.

The transgression occured in two phases (Fig. 3.42). Marls with *Ostrea longirostris* and the foraminiferid, *Archiacina armorica* (as in the Paris Basin), are succeeded by coral and algal reef limestones which locally contain characteristic Oligocene nummulites (*N. intermedius* and *N. incrassatus*).

The Calcaire à Astéries passes eastwards into the **Molasse de l'Agenais.** This continental formation is rich in mammals including *Anthracotherium* and *Entelodon* and, at Milloque, 25km north east of Agen, the carnivores, *Hyaenodon* and *Amphicyon* (Fig. 2.12).

At the end of Stampian times, the sea withdrew completely from the Gironde area as a result of the Pyreneo-Provencal orogenic phase and, in the Bordeaux region, marls with *Unio*, attributed to the Chattian, were deposited. This unit passes eastwards into the **Calcaire blanc de l'Agenais,** in which the Paulhiac mammal bed occurs.

From the Eocene up to the present day there has been a westward shift in the position of the continental margin in step with the progressive filling-up of the shelf sea to its east (Fig. 3.41 and 3.42). The width of the zone of progradation is around 200km and it seems that the coastline of the Landes has had its present shape since Eocene times. During the Eocene, detritus from the rising Pyrenees was carried

into a prolongation of the Aturian gulf (the subsiding pre-Pyrenean trench), where it was trapped. However, in Oligocene times and subsequently, this debris was carried into the Landes area where it mingled with sediment coming from the north east (Fig. 3.41).

2. A temporary connection between the Atlantic and the Mediterranean: the Aturian gulf (Fig. 3.44)

The Aturian gulf spread over the southern part of Aquitaine and the north of Spain (Aragon) and along its axis the Palaeogene formations are generally thicker than those in the Gironde. Sometimes they become flysch-like in the median trench of the gulf in the region now occupied by the Pyrenean chain. Changing depths in the trench during the Cenozoic explain the lateral variations of sedimentary facies in this arm of the sea. During the periods of maximum trangression (in the Thanetian and the Ilerdian) the sea extended as far as the Corbières, At the present time the last remnant of the Aturian gulf survives as the Cap Breton marine canyon, which in Pleistocene times was the continuation of the River Adour before its diversion to the south.

The flysch-like marly limestone exposed in Bidart cliff, south of Biarritz, and in the Tercis anticline, near Dax was deposited in the Aturian gulf in the Dano-Montian and is continuous with that of the upper Cretaceous. The limestone contains *Ananchytes* and *Nautilus danicus.* It also contains the first Palaeogene globigerinids (*Globigerina daubjergensis*) and globorotalids (*Globorotalia eugubina*). North of Pau, there are algal limestones, which are sometimes reefal and dolomitised. Southwards, towards the Pyrenean trough, the beds become more marly, while to the east in the Plantaurel and the Petites Pyrénées, south of Toulouse, the regression which began in the late Maastrichtian is continued into the Dano-Montian. The passage from the Cretaceous into the Tertiary there is continental and is represented by multicoloured lacustrine limestones and red clays, the equivalent of the Rognacian of Provence and the middle

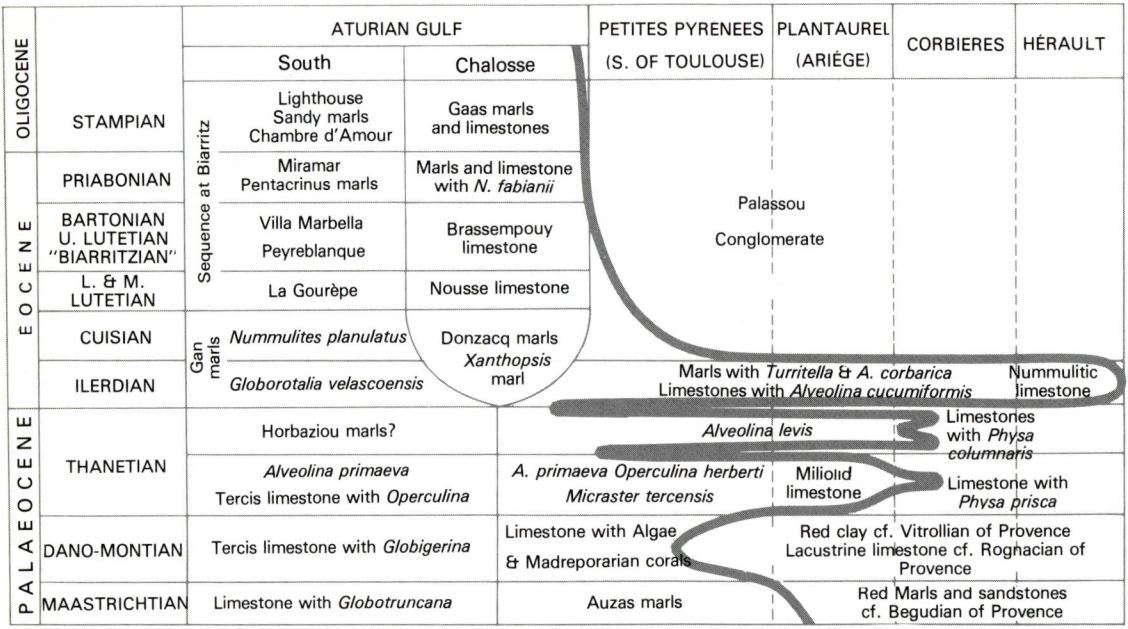

		ATURIAN GULF		PETITES PYRENEES (S. OF TOULOUSE)	PLANTAUREL (ARIÉGE)	CORBIERES	HÉRAULT
		South	Chalosse				
OLIGOCENE	STAMPIAN	Lighthouse Sandy marls Chambre d'Amour	Gaas marls and limestones				
EOCENE	PRIABONIAN	Miramar Pentacrinus marls	Marls and limestone with *N. fabianii*	Palassou Conglomerate			
	BARTONIAN U. LUTETIAN "BIARRITZIAN"	Villa Marbella Peyreblanque	Brassempouy limestone				
	L. & M. LUTETIAN	La Gourèpe	Nousse limestone				
	CUISIAN	*Nummulites planulatus*	Donzacq marls *Xanthopsis* marl	Marls with *Turritella* & *A. corbarica* Limestones with *Alveolina cucumiformis*			Nummulitic limestone
	ILERDIAN	*Globorotalia velascoensis*					
PALAEOCENE	THANETIAN	Horbaziou marls?		*Alveolina levis*		Limestones with *Physa columnaris*	
		Alveolina primaeva Tercis limestone with *Operculina*	*A. primaeva Operculina herberti* *Micraster tercensis*	Miliolid limestone	Limestone with *Physa prisca*		
	DANO-MONTIAN	Tercis limestone with *Globigerina*	Limestone with Algae & Madreporarian corals	Red clay cf. Vitrollian of Provence Lacustrine limestone cf. Roghacian of Provence			
	MAASTRICHTIAN	Limestone with *Globotruncana*	Auzas marls	Red Marls and sandstones cf. Begudian of Provence			

(Sequence at Biarritz; Gan marls)

3.44 The Palaeogene stratigraphy of the southern part of the Aquitaine Basin showing the maximum extent of transgressions towards the east (in part after J. Villatte). Marine formations in the west; continental in the east.

Garumnian[1] of Languedoc. It is only in the extreme western part of the Petites Pyrénées that the Dano-Montian is marine, being represented by an algal (*Lithothamnion*) limestone. During the whole of the Palaeocene, detritus was derived from the Massif Central.

In **Thanetian** times, the sea continued to occupy the western part of the Aturian gulf, from which it advanced eastwards on two occasions (Fig. 3.44). The first (early Thanetian) was accompanied by the deposition of limestones with *Micraster tercensis, Operculina heberti, Discocyclina seunesi* and precursors of the nummulites. These are overlain by *Lithothamnion* limestones, on which rest miliolid limestones with *Alveolina primaeva* (Montian or upper Garumnian of Leymerie). A lagoonal-lacustrine regression followed and is marked by the deposition of marls with oysters and charophytes which have yielded molluscs known from the Thanetian Calcaire de Rilly, near Rheims. This regression was succeeded by a second marine advance, that of the limestones with *Alveolina levis* (Fig. 3.45 and 3.46). The deposits of both the regression and the second transgression were formerly believed to be Sparnacian.

Throughout the Thanetian, the area of the Corbières (to the east) was one of continental deposition where red clays with intercalations of gastropod (*Physa*) limestones (analogous to those of the Vitrollian of Basse-Provence) were formed (Fig. 3.47). In the Chalosse, the marls of Horbaziou (a farm about 10km west of Aire-sur-l'Adour) are possibly the equivalent of the *Alveolina levis* limestones of the late Thanetian (Fig. 3.46).

The stratigraphy of these sequences has long been a source of controversy but has been unravelled by HOTTINGER & SCHAUB (1960) of Basle, VILLATTE (1962) and TAMBAREAU (1972) of Toulouse, and PLAZIAT (1966) of Paris in recent years. In particular, it has been shown that formations previously attributed to the Ypresian and Lutetian are, in reality, of **Ilerdian** age. The early Ilerdian transgression in which the limestones with *Alveolina cucumiformis* were formed was probably the most extensive in the Pyrenees in Tertiary times. According to KROMM (1968) it crossed the Corbières and reached the area of Hérault (west of Montpellier) in the middle Ilerdian.

1. From Garonne: a regional stage created by Leymerie (1862) which included all the horizons from the Maastrichtian to the Thanetian; that is, it straddles the Cretaceous-Tertiary boundary.

3.45 Panoramic view of the southern flank of the Plagne anticline (= the northern limb of the Cassagne-Fabas syncline), Petites-Pyrénées, taken from Pédegas-d'en-Haut (photo: Pomerol). *Maastrichtian:* M – Auzas marls. *Dano-Montian:* D₁ – sublithographic limestone; D₂ – red marls. *Thanetian:* T₁ – Miliolid limestone; T₂ – marly limestone with *Natica*; T₃ – Biones limestone with reef-building corals; T₄ – Sands and sandstone with *Ostrea bellovacina*; T₅ – *Micraster tercensis* beds; T₆ – *Alveolina primaeva* limestones; T₇ – *Alveolina levis* marls. Lower Ilerdian: I₁ – *Alveolina cucumiformis* marls; I₂ – Pink *Nummulites globulus* limestone, similar to Mancioux marble (after J. Villatte and Y. Tambareau, original).

The lower Ilerdian (ex-Ypresian) beds with *A. cucumiformis* are initially marly, but then become calcareous and finally glauconitic. The upper beds include a limestone with *Melobesia* which is used as an ornamental stone (Mancioux marble). The middle Ilerdian (ex-Lutetian) is marly in its lower part, but becomes sandy in its upper layers and contains *Turritella trempina, Nummulites globulus, N. atacicus* and *Alveolina corbarica* (Fig. 3.44).

The sea, which had spread far to the east during the early Ilerdian, rapidly retreated westwards at the beginning of late Ilerdian times and did not advance eastwards again during the remainder of the Eocene. This regression was caused by the initiation of the uplift of the Pyrenean mountain chain, an event formerly dated as Lutetian. A great expanse of boulders, pebbles, sands and red clays began to accumulate in the central part of the southern side of the gulf of Aquitaine and separated the Ilerdian sea of Western Aquitaine from that of the Languedoc region, which was in a less neritic facies, and which may have had a link with the Mediterreanean. This continental deposit, called the **Poudingue de Palassou,** is named after an 18th century Pyrenean geologist. Of middle Ilerdian to Bartonian age, these conglomerates pass laterally and vertically into the Molasse of Aquitaine, which was largely derived from the Massif Central. Thus from late Ilerdian times the eastern part of Aquitaine remained emergent. The southern part of this area was folded in the late Eocene and Oligocene.

During this time the western part of the area (Aturian gulf) remained marine. Flysch was deposited along the axis of the gulf from Maastrichtian to Cuisian times. In the Gan syncline, south of Pau, the Thanetian is overlain by marls (the Gan marls) of Ilerdian and Cuisian age. The latter contain *Nummulites planulatus, Alveolina oblonga* and *Globorotalia subbotinae*, which permit precise correlation with the Paris Basin. Further north, in the Chalosse, sedimentation was more calcareous. Here were deposited the *Marnes à Xanthopsis* (a crab) with *Globorotalia subbotinae*), and the Donzacq marls (with *G. aragonensis* and *Discoaster lodoensis*, both units attributed to the Cuisian. These are overlain by the limestones of Nousse, Gibret and Donzacq (Fig. 3.41) with *Assilina exponens, Nummulites atacicus, N. millecaput,* and *N. laevigatus*, though this last is rare and untypical (BOULANGER, 1968). Towards the end of the Lutetian, a brief transgression occurred in which the Brassempouy limestone (containing *Nummulites brongniarti, Alveolina elongata, Orbitolites complanatus, Discorinopsis kerfornei* and *Fabiana cassis*) was deposited. It is followed by the Priabonian (marly

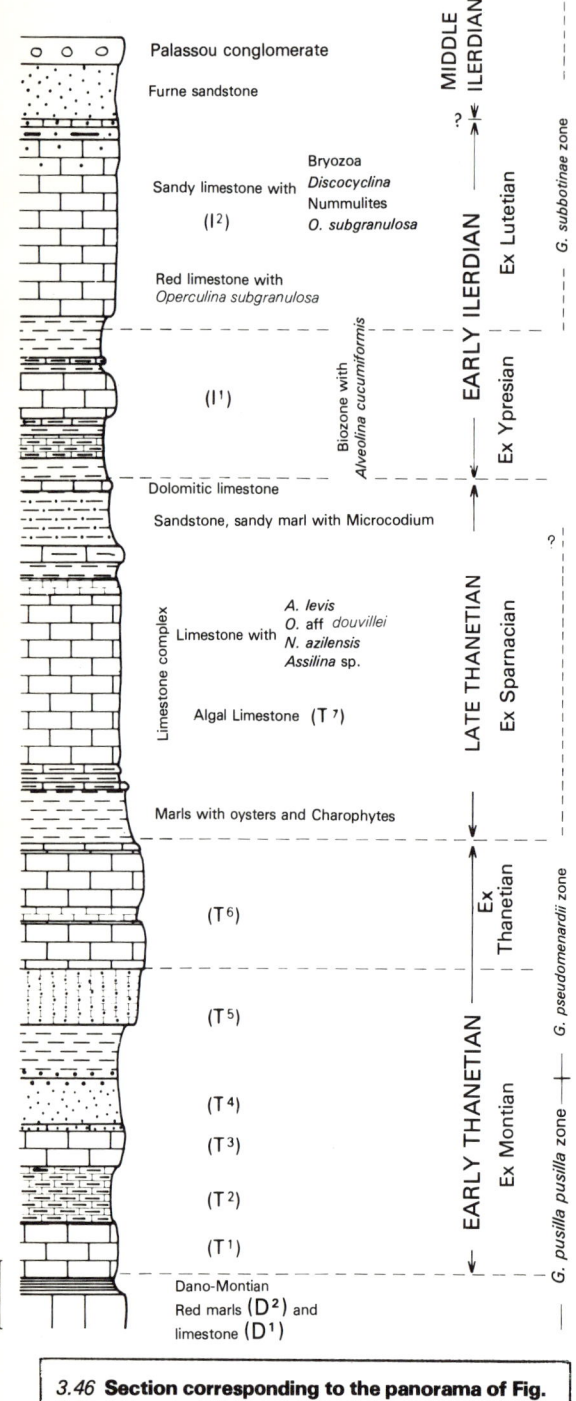

3.46 **Section corresponding to the panorama of Fig. 3.45** (after J. Villatte and Y. Tambareau, original). A – *Alveolina*; G – *Globorotalia*; N – *Nummulites*; O – *Operculina*.

limestone with *N. fabianii*), and the Stampian (sandy marls of Gaas with molluscs comparable to those of the Paris Basin).

The middle and upper Eocene beds are particularly well seen in the cliff section of the Côte-des-Basques at Biarritz, where over 1500m of beds are exposed (Fig. 3.48). An anticline towards the south western end of the section is pierced near its axis by a diapir (the Peyreblanque disturbance, exposed on the beach) which contains rocks of Cretaceous, Liassic and even Triassic age. Similar diapirs can be traced from west to east along the length of the Aturian gulf (Dax, Bastennes, Maubourget). Mid-Eocene movement of this diapir modified the local topography and caused lateral variations of facies associated with changes in seawater depth. This and later movements have led to the disruption of the middle and upper Lutetian beds (the earliest Tertiary beds exposed in this part of the section) and has made it difficult to unravel the details of their stratigraphy.

The **middle Lutetian** is exposed in the Peyre-qué-Bève rocks south west of Peyreblanque and in the Gourèpe rock, a little to the north east. The former is a limestone containing *Nummulites millecaput* while the latter is a marly limestone rich in echinoids and discocyclinids.

Late **Lutetian** marls form the cliff at Handia and contain crustaceans, *Nummulites striatus* and *N. aturicus*. Inland these marls are seen in a ravine at Ilbarritz. The Peyreblanque rocks (Fig. 3.49) are a more neritic lateral facies of the upper Lutetian, and yield, in addition to the nummulites noted above, *N. brongniarti*, *N. millecaput* and *Alveolina elongata*. HOTTINGER and SCHAUB (1960) have designated this unit as the stratotype of the Biarritzian stage. It is, however, particularly badly chosen because the rocks are isolated and displaced and without visible contacts with the nearby beds.

The Biarritzian stage corresponds to a transgression of short duration which, because of its brevity, is of considerable chronostratigraphic interest and, because of its extension into the northern basins, affords a useful horizon for correlation. It occurs in the following localities (Fig. 3.50):

Chalosse: Brassempouy beds
Gironde: upper Blaye limestone and
    Couquèques limestone

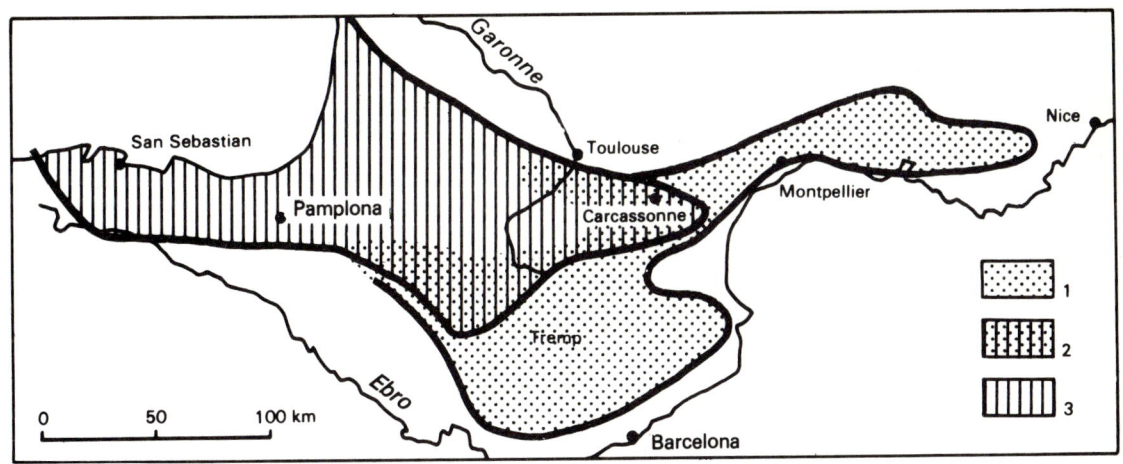

3.47 Thanetian palaeogeography of part of the Pyrenees and Provence (after Plaziat (1973)).
1 – area of continental sedimentation; 2 – zone of fluctuating shoreline; 3 – area of permanent marine sedimentation (see also Fig. 3.43).

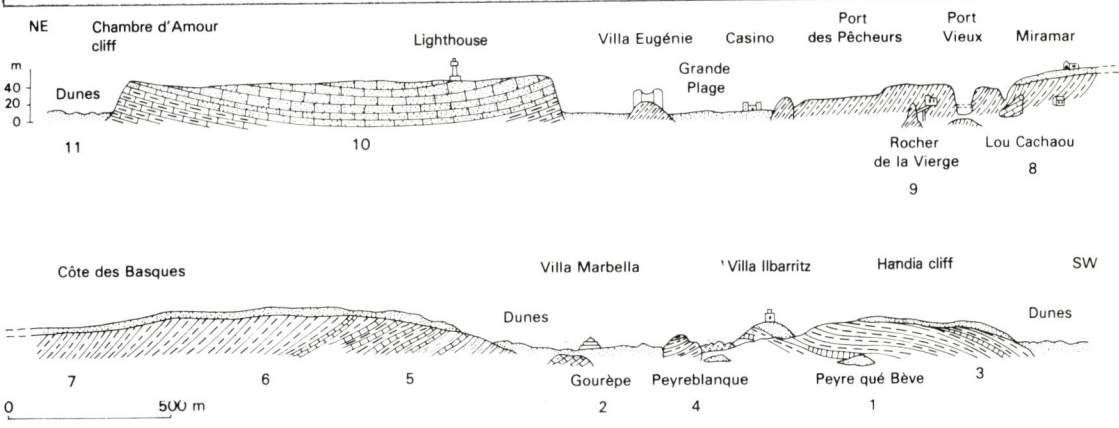

3.48 The Palaeogene section at Biarritz (after Alvinerie, Mayeux and Rechiniac (1964)).
Middle Lutetian: 1 – Peyre qué Bève limestones; 2 – Gourèpe marly limestone.
Upper Lutetian: 3 – Handia marls; 4 – Peyreblanque limestone (see Fig. 3.49); 5 – Villa Marbella marls.
Bartonian: 6 – *Pentacrinus* marls; 7 – Côte des Basques marls; 8 – clays and sandy limestones below the Miramar viewpoint and of the Lou Cachaou rock (see Fig. 3.51).
Oligocene: 9 – marls and sandstones of the Rocher de la Vierge, the port area and the Villa Eugénie; 10 – marls and flyschoid limestones of the Chambre d'Amour.
Quaternary: 11 – Landes sands.

The lower Loire: Bois-Gouët
Off the south coast of Brittany
The sea bed of the western Channel
Hampshire: Upper Bracklesham beds of the Isle of Wight
Cotentin peninsula: Fresville limestones
Paris Basin: Falun de Foulangues, found near Vernon and in borings at Tillet and Montjavoult (Fig. 3.14)

This transgression extends far across Europe. In particular, communication was established between Hungary (Transdanubian Mountains, north west of Budapest) and the central Carpathians (Slovakia). The deposits of this transgression are characterised by neritic limestones rich in *Nummulites brongniarti* and other middle and late Lutetian species (*N. perforatus* and *N. millecaput*.

This is an additional reason for regarding the Biarritzian merely as an episode in the late Lutetian.

The Handia marls (of the Biarritz section) are succeeded by the marls of the Villa Marbella which contain abundant discocyclinids and *Nummulites aturicus, N. striatus* and *N.*

3.49 **The Peyreblanque rocks at Biarritz** (upper Lutetian) (photo: Pomerol).

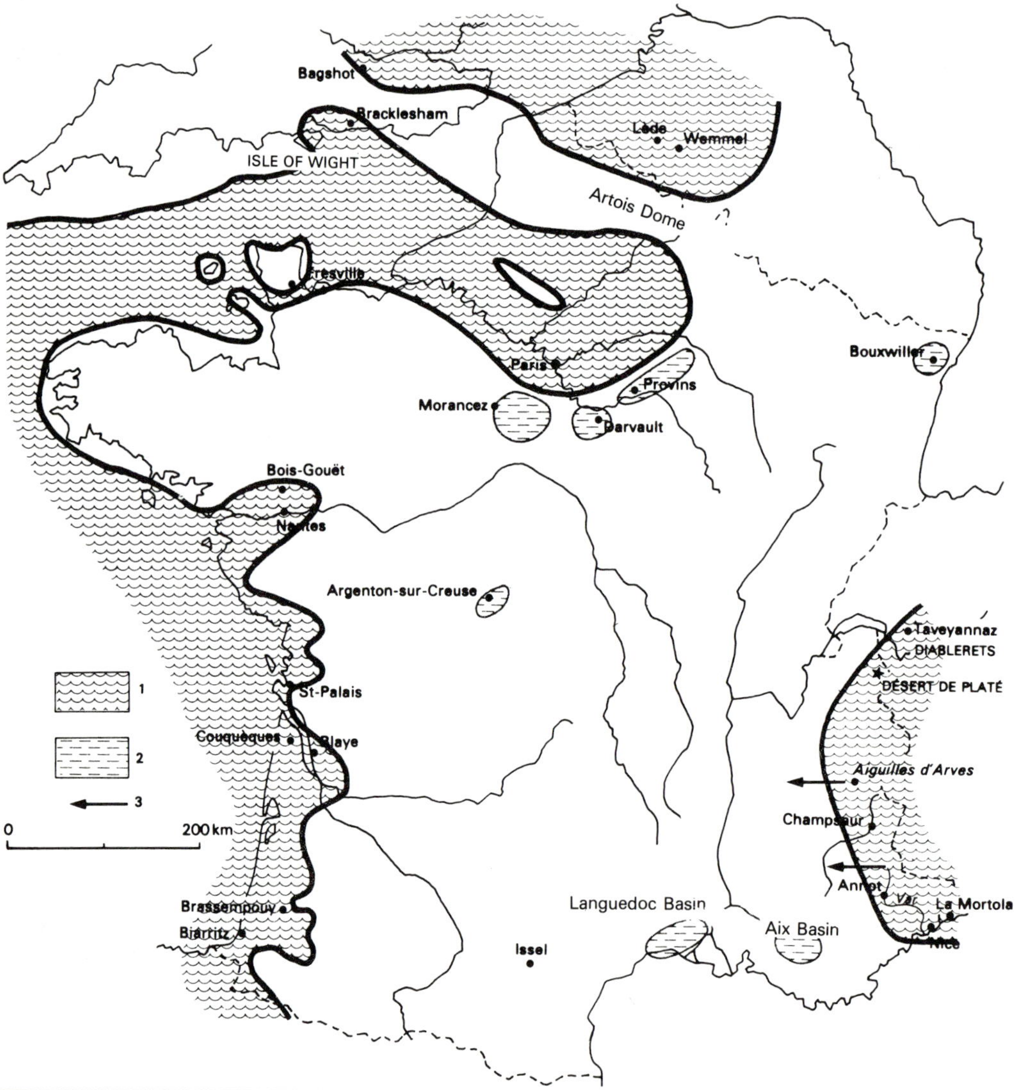

3.50 **France in the middle and late Lutetian.** 1 – marine formations; 2 – lacustrine formations; 3 – direction of the Palaeogene transgression in the Alps.

variolarius. Boussac (1911) referred this to the Auversian stage, but it seems that this unit, together with the grey marls at the top of the cliff at Handia, belongs to the Bartonian (s.s.) because, amongst the planktonic foraminifera, *Truncorotaloides rohri* is present, whereas *Globigerapsis semiinvoluta* is not (Table III, p. 24).

Priabonian: Above the beds of the Villa Marbella is the Priabonian, which begins with *Pentacrinus* marls rich in echinoids and now includes the planktonic foraminifer, *Globi-gerapsis semiinvoluta*. It continues with the marls and marly limestones with *Nummulites fabianii* of the Côte des Basques, then with blue-grey clays and calcareous sandy marls (Miramar and Lou Cachaou) (Fig. 3.51). These beds pass conformably upwards into the Oligocene. Some 30km east of Biarritz, a

3.51 **The Lou Cachaou rock at Biarritz:** Upper Eocene (Priabonian). The Eocene-Oligocene boundary is to the right of the rock (photo: Pomerol).

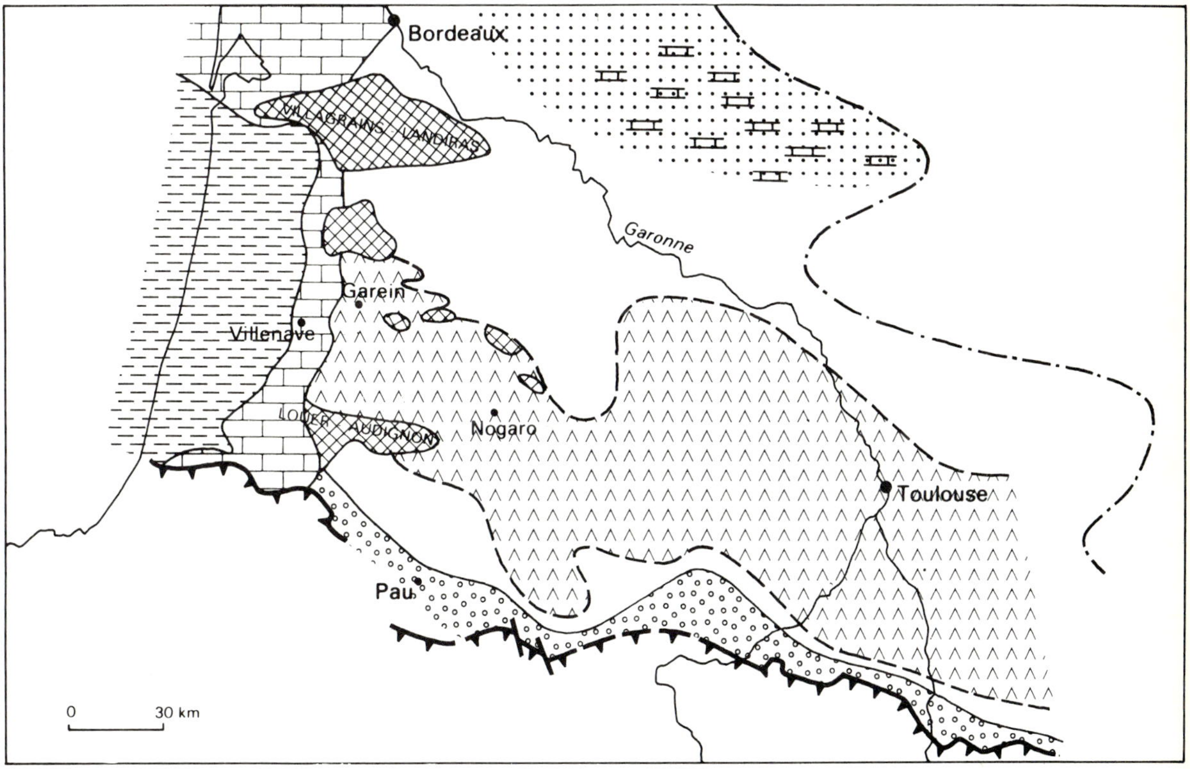

3.52 **Palaeogeography of the Aquitaine Basin at the end of the Eocene** (after Bea and Kieken (1971)) 1, 2, 3; *Lagoons:* 1 – fluviatile sands and lacustrine limestones; 2 – gypsum; 3 – conglomerates; 4 – *coastal bars:* carbonates; 5 – *open sea:* marls and clays; 6 – *Late Eocene beds subsequently removed*; 7 – limit of evaporites; 8 – limit of the Bartonian-Priabonian; 9 – North Pyrenean thrust.

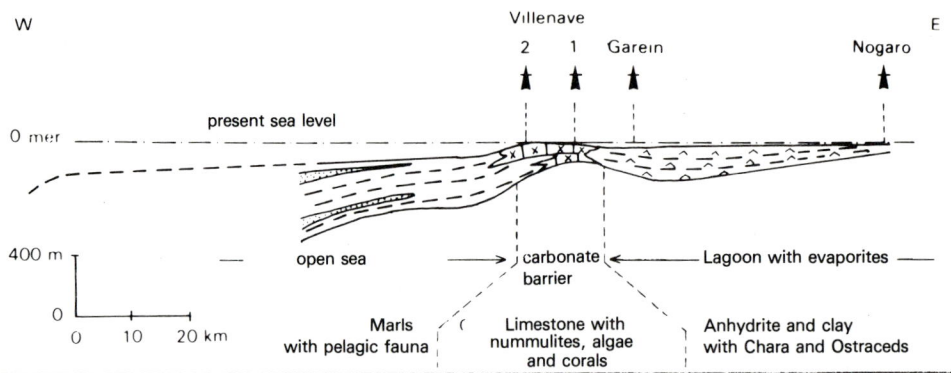

3.53 **Interpretative section of Fig. 3.52:** sedimentary styles in Aquitaine during the late Eocene (after Bea and Kieken (1971)).

carbonate barrier, discovered in boreholes, separated the open sea from a vast lagoon in which gypsum was deposited (Figs. 3.52 and 3.53).

Oligocene: The lowest beds of the Oligocene are marls and sands with *Nummulites intermedius, Chlamys biarritzensis*, and *Clypeaster biarritzensis* (below the Villa Eugénie, Fig. 3.48), and a thin pebble bed at their base (Port des Pêcheurs, Rocher de la Vierge). The succession continues with the sandy flyschoid marls of the Chambre d'Amour (Fig. 3.55) and the sandy and gritty marls of the Phare (Lighthouse) beds, which are covered by the Quaternary sand of the Landes.

The facies of the cliffs of Biarritz (with the exception of the diapiric rocks of Peyreblanque) indicate deposition in deeper water than the beds of the Gironde and Chalosse. They correspond to the axial part of the Aturian gulf and pass inland into more neritic facies. South of Dax, at Gaas, in the Chalosse, there are Stampian beds very rich in molluscs, whilst the shelly marls of Escornebéou are now attributed to the late Oligocene since they contain the last nummulites (*N. bouillei*) and the first *Miogypsinoides* and *Globigerinoides*. The latter is classically considered to appear at the base of the Miocene but it seems that, in fact, it appeared somewhat earlier. Further east, the still rising Pyrenees were continuing to act as the source of the Poudingue de Palassou, which accumulated on their piedmont slopes (Figs. 3.54, 3.55 and 3.56).

3. The link between the Paris Basin and the Aquitaine Basin – the Armorican gulfs.

Connection between the Paris Basin, which was linked to the North Sea, and the Aquitaine Basin (opening to the Atlantic) was established at various times during the Palaeogene. This connection was sometimes by way of the English Channel, sometimes by the valley of the Loire.

The first link between the Paris Basin and the Atlantic via the English Channel was established in the early Senonian (Coniacian-Santonian). This link remained open during most of the Campanian but closed in the Maastrichtian, a stage recognized locally only in the Cotentin and the Western Approaches, whilst the Paris Basin became emergent.

It is possible that the link was re-established in the Dano-Montian, but evidence is lacking in the area between the Western Approaches and the Paris region. There was no communication in Thanetian times, but *a new connection occurred in the late Ypresian* (Cuisian) which, for the first time allowed the passage of Aquitanian foraminiferids (such as *Nummulites planulatus* and *Cuvillierina eocenica* (see p. 207), into the Paris Basin. However the sea had reached the Vendée coast in the Bay of Bourgneuf (30km south west of Nantes) by late Sparnacian times.

The opening to the Atlantic continued through the early Lutetian (zone of *Nummulites laevigatus*). From the middle

Lutetian the uplift of the Artois anticline disrupted the link between the North Sea and the Paris Basin and the only link with the open sea was by way of the English Channel. At the end of the middle Eocene (late Lutetian) there was a *brief transgressive episode* (*the Biarritzian*) which has been referred to earlier. Its effect was to isolate the northern part of the Cotentin (Fig. 3.50) and to deposit marly limestone containing *Discorinopsis kerfornei, Alveolina elongata* and *Linderina brugesi* in the lagoons of the Paris region. However, *Nummulites brongniarti* did not reach the Paris Basin, but only travelled as far as the south east coast of Brittany, from which the sea penetrated small embayments in the shoreline and the region of the Lower Loire near Angers and deposited the shelly marls of Bois-Gouët. At the same period the sea occupied the gulf of Challans-Commequiers, the depression of Machecoul (where it deposited sands and limestones) and the depression of Grand-Lieu. It invaded the Arthon basis along rias, rising to about 35m above present sea-level (TERS, 1961).

In the Bartonian, *Nummulites fabianii* is known only in the Western Channel and off the coast of Brittany. The Breton Lutetian basins were drained, the central Channel emerged above sea level and the Paris Basin became a closed sea in which *Nummulites variolarius* (which had arrived in the early Lutetian) became trapped. In Belgium and in Hampshire the descendants of *N. variolarius* proliferated. These include *N. prestwichianus, N. rectus* and *N. orbignyi* (ex *N. wemmelensis*).

A new opening occurred in the Stampian, probably through the English Channel, which allowed the sea to return to the centre of the Paris Basin, bringing with it *Ostrea longirostris, Archiacina armorica* and a molluscan fauna similar to that found at Gaas. The sea retreated by another route, abandoning the north of the basin before it left the south. Traces can be recognised from Orléans as far as the Loire estuary in limestones and sandy limestones at Noyant-sous-le-Lude, in borings at Chaingy and Darvoy (containing *Archiacina armorica*) and in the Fondettes limestones (with *Potamides lamarcki*) west of Tours (see Fig. 3.34, p. 85). This suggests that a seaway had been established between Beauce and the Loire

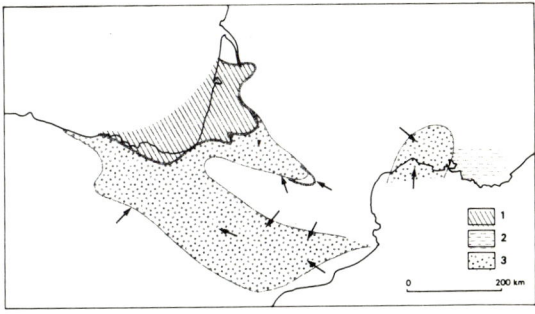

3.54 The Pyrenean region in the early Oligocene (after Mattauer and Henry (1971)). 1 – marine formations; 2 – continental clayey limestones and evaporites; 3 – continental detrital formations.

channel. At the same time there was a gulf running from the lower Loire (**Saffré Basin**) to the outskirts of Rennes where it deposited a coarse limestone with *Archiacina armorica*. However, the Vendée coast (south of the Loire Estuary) was not reached by the Oligocene sea and its deposits.

During the Palaeogene, the Atlantic coastline and that of the western part of the English Channel were near their present day positions. The low-lying continental area was subjected to repeated transgressions and regressions resulting either from changes of sea level or slight epeirogenic movements, associated with the Pyrenean and Alpine orogenic phases. It only needed a small rise in sea level (or a small downward movement of the continent) for the sea to sweep some hundreds of kilometres across the land. Advances of the sea were preceded by the development of coastal lagoons which, when each transgression had passed its maximum, persisted in the hinterland.

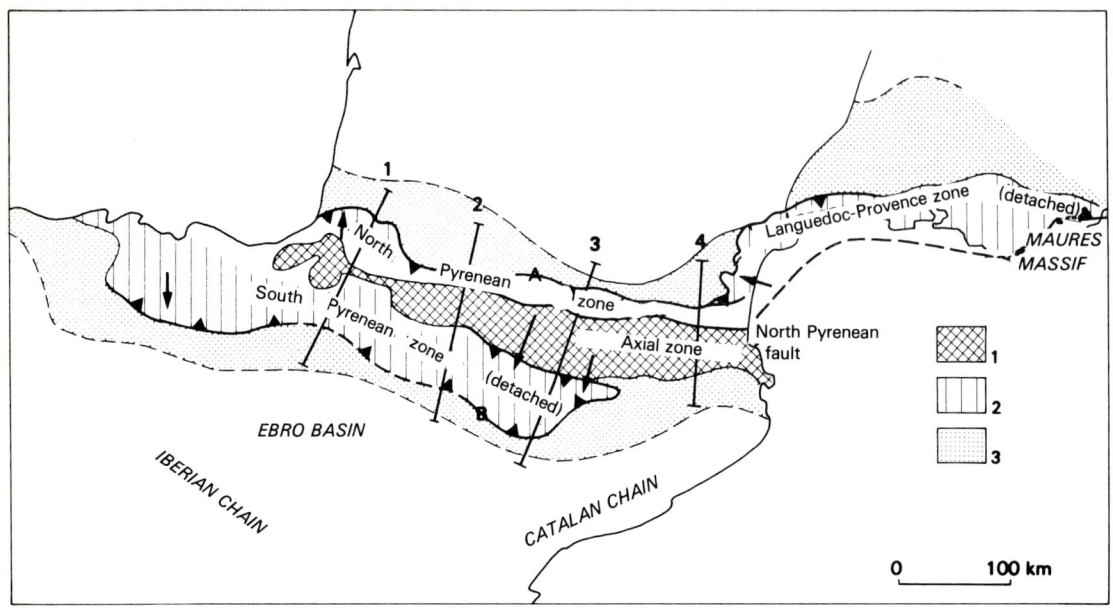

3.55 **The major structural subdivisions of the Pyrenean chain** (after Mattauer and Henry (1971)). 1 – axial zone; 2 – zones of Mesozoic-Eocene nappes; 3 – folded foreland. The lengths of the arrows give a rough indication of the displacement of the nappes with respect to their roots: A – north Pyrenean thrust; B – south Pyrenean thrust. The sections on lines numbered 1, 2, 3, 4 are shown in Fig. 3.58.

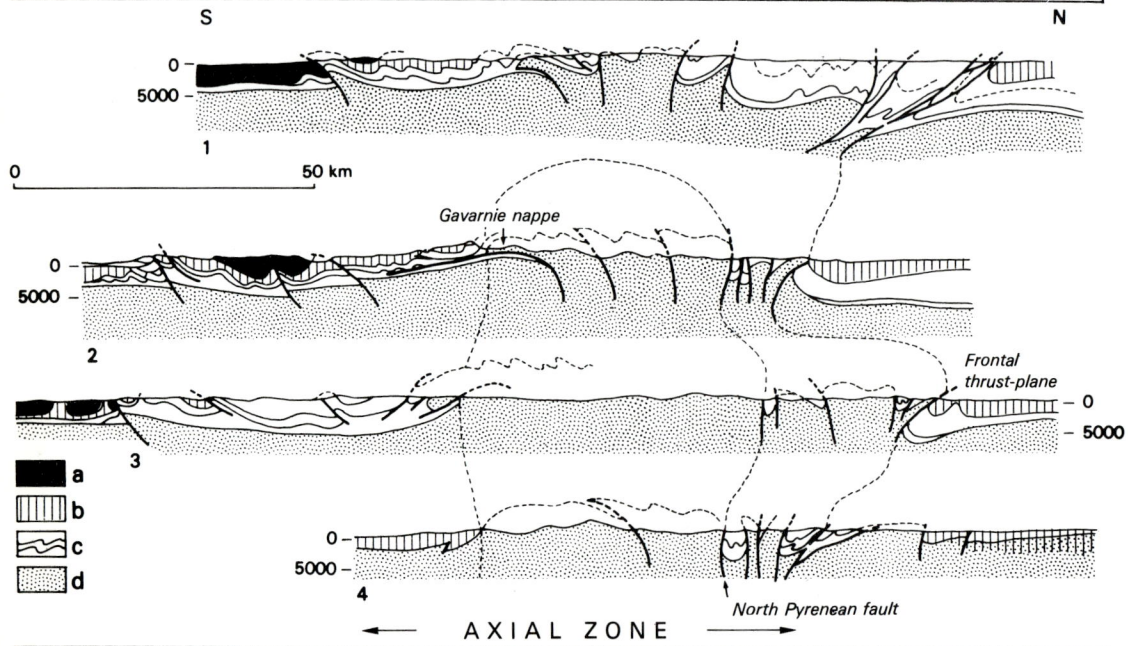

3.56 **Structural sections across the Pyrenean chain** (for positions, see Fig. 3.55) (after Mattauer and Seguret (1971)). a – Oligocene; b – Eocene; c – Mesozoic; d – Hercynian basement.

III. – THE ALPINE AREA

During the Palaeogene, there was an east to west transgression across the area of the western and maritime Alps before the main uplift of the chain at the beginning of the Oligocene[1].

In the **Palaeocene,** the sea occupied only the Sub-brianconnais and Brianconnais zones, where flaggy limestones continued to be deposited from the late Cretaceous until the early Eocene. It is probable that at this stage, the Piedmont zone had already emerged.

In the **middle Lutetian,** sedimentation in the Brianconnais zone changed from limestone to shales and sandstones with rare large nummulites. These beds have been called the "black flysch" by Alpine geologists, although this description is hardly appropriate. But at the same time the transgression continued westwards across the area of the Dauphiné and the Maritime Alps (see Fig. 3.57). In the Savoy area, sandstones and limestones with large nummulites (*N. aturicus, N. millecaput*) and discocyclinids occur at Chatelard, Entrevernes-en-Bauges, Roc de Chère (overlooking Lake Annecy), Samoens and the Désert de Platé, west of Chamonix. They are overlain by lacustrine limestones with *Lymnaea* and *Bulimus* (*B. cylindricus*, a large sinistral gastropod of late Lutetian age) and by beds with *Cerithium diaboli* (Platé area).

In the **Late Lutetian, Bartonian and Priabonian,** the "Aiguilles d'Arves flysch" accumulated in the more internal part of the Dauphiné zone. This flysch was derived from parts of the external zone, such as the Grandes-Rousses massif, as it was being uplifted. This flysch rests discordantly on earlier formations, including the Jurassic and even the Palaeozoic basement, which had been deformed by the Arvinchean phase, probably at the end of the Cretaceous or the beginning of the Tertiary. Further south, the Aiguilles d'Arves flysch is replaced by the Champsaur sandstone in the valley of the upper Drac and by the massive Annot sandstone in the valley of the upper Var (Fig. 3.58).

In Savoy and in Switzerland, the sandstones of Champsaur and Annot are represented by the Taveyannaz sandstones of the Diablerets massif. They overlie the *Cerithium diaboli* beds which often mark the base of the Eocene transgressions. The Taveyannaz sandstone is a diachronous lagoonal facies, similar to the Glauconie Grossière of the Paris Basin (see p. 67). The sandstones of Taveyannaz, Champsaur and Annot sometimes contain detritus of volcanic origin which gives rise to coloured patches in the grey beds near the base of the sequence. For this reason, some horizons are considered to be greywackés. The weathering of these sandstones leads to the formation of escarpments with masses of chaotic boulders, such as the "Grandes Tours" of the Lake of Allos (Basses-Alpes).

In Switzerland, the Taveyannaz sandstone, like the Diablerets Massif itself, is not in place, but is transported. It forms part of the Helvetic nappes, which originally stretched across the Aar-Gothard massif.

The Helvetic nappes rest on the ultra-Helvetic nappes, which were, however, more internal but became detached and slipped under gravity at an earlier date. The ultra-Helvetic Eocene is of the same age as the Aiguilles d'Arves flysch and, like it, rests on a folded sub-stratum which, in places, is eroded away to expose the underlying basement. Towards the interior of the chain, the ultra-Helvetic Eocene is overlain by flysch showing locally a continuous passage from the Maastrichtian to the Cuisian. This is the Schlierenflysch of the Freiburg pre-Alps, which came originally from the Sion-Val Ferret trench (or Valais zone) along the internal border of the St. Gotthard massif.

Further south, in the Maritime Alps and in Provence, the Annot sandstone is the highest member of the Palaeogene sequences. In the

---

1. This section has been reviewed and revised (in the French text) by MM. Debelmas and Pairis of the Institut Dolomieu at Grenoble.

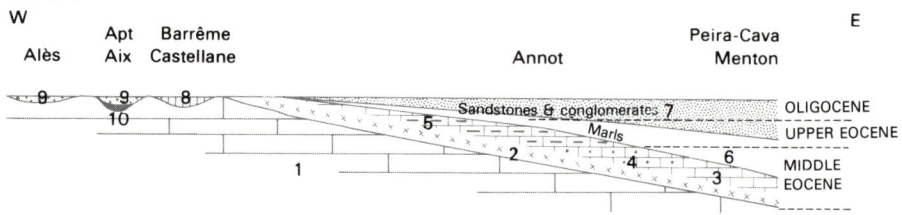

W   Apt   Barrême   Peira-Cava   E
Alès   Aix   Castellane   Annot   Menton

3.57 The Palaeogene transgression in the Maritime Alps and in Provence (after Blondeau, Bodelle and Campredon). 1 – Mesozoic beds; 2 – basal beds with *Microcodium* and *Cerithium diaboli*; 3 – Zone A (middle Lutetian), limestone with large nummulites (*N. millecaput, N. perforatus, Assilina exponens*); 4 – Zone B (upper Lutetian) limestone with *N. brongniarti, N. striatus, N. praefabianii*; 5 – Zone C (Bartonian and Priabonian) limestone with small nummulites (*N. garnieri, N. fabianii, N. incrassatus*); 6 – marls with *Globorotalia* (diachronous pelagic facies); 7 – Annot sandstone (also diachronous) overstepping the Eocene; 8 – Oligocene circum-Alpine channel through which the sea retreated (see Fig. 3.60); 9 – lagoonal-marine and lagoonal beds of the Rhone basin; 10 – continental Eocene deposits (Aix basin).

late Lutetian, the sea reached Nice from the region of Liguria (between France and Italy) where it deposited sandy calcareous marls containing *Nummulites perforatus, N. millecaput, N. striatus, N. brongniarti, Alveolina elongata* and *Truncorotaloides rohri*. These are exposed in the well-known outcrop at Cap de la Mortola, between Menton and Ventimiglia (Fig. 3.50).

Further west, in the Eocene synclines of the Var, which are oriented east to west, occur silty limestones, belonging to the transition zone to the upper Eocene, and comparable with the marly limestone of the Villa Marbella (Biarritz section, p. 99). They contain *Nummulites striatus* and *N. chavannesi*, but do not yet include *Globigerapsis semiinvoluta*. This is the Auversian of earlier authors (Fig. 3.57).

3.58 Annot sandstone (upper Priabonian and Oligocene) overlying Priabonian marls with Globorotalia (photo: Pomerol).

Finally, in the Annot syncline, west of Puget-Théniers, the Scaffarels section shows what has been called the "Priabonian trilogy". The sequences rest on upper Cretaceous with *Globotruncana* and commence with late Lutetian bioclastic sandy limestones with nummulites (*N. perforatus, N. striatus* etc.) and discocyclinids. These limestones pass up into early Priabonian beds with the appearance of *N. fabianii* and *Globigerapsis semiinvoluta*. The second division is represented by marls with *N. fabianii* and *Globorotalia cerroazulensis* of the late Priabonian. The third stage comprises the Annot sandstones (Fig. 3.58), now know to commence at the Eocene-Oligocene boundary, but which are essentially Oligocene (BESSON). Hence the bottom part of this famous "Priabonian trilogy" is upper Lutetian and the top is Oligocene. The study of marine palaeocurrents has shown that the transport of detritus into Provence and the Maritime Alps was *not* from the north as the present configuration of the chain seems to suggest, but from the south where the Maures, Esterel, Corsica and the western Mediterranean land mass were exposed to erosion. Of these areas, only the latter has been submerged (in the Pliocene).

At the end of the Eocene and the beginning of the Oligocene, major uplift accompanied by gentle folding (the Provencal phase) led to the emergence of almost the whole of the western Alps. Beginning in the late Lutetian, the **nappe of the Helminthoid** flysch (of late Cretaceous age) was pushed westwards by the uplift of the

The Palaeogene in France

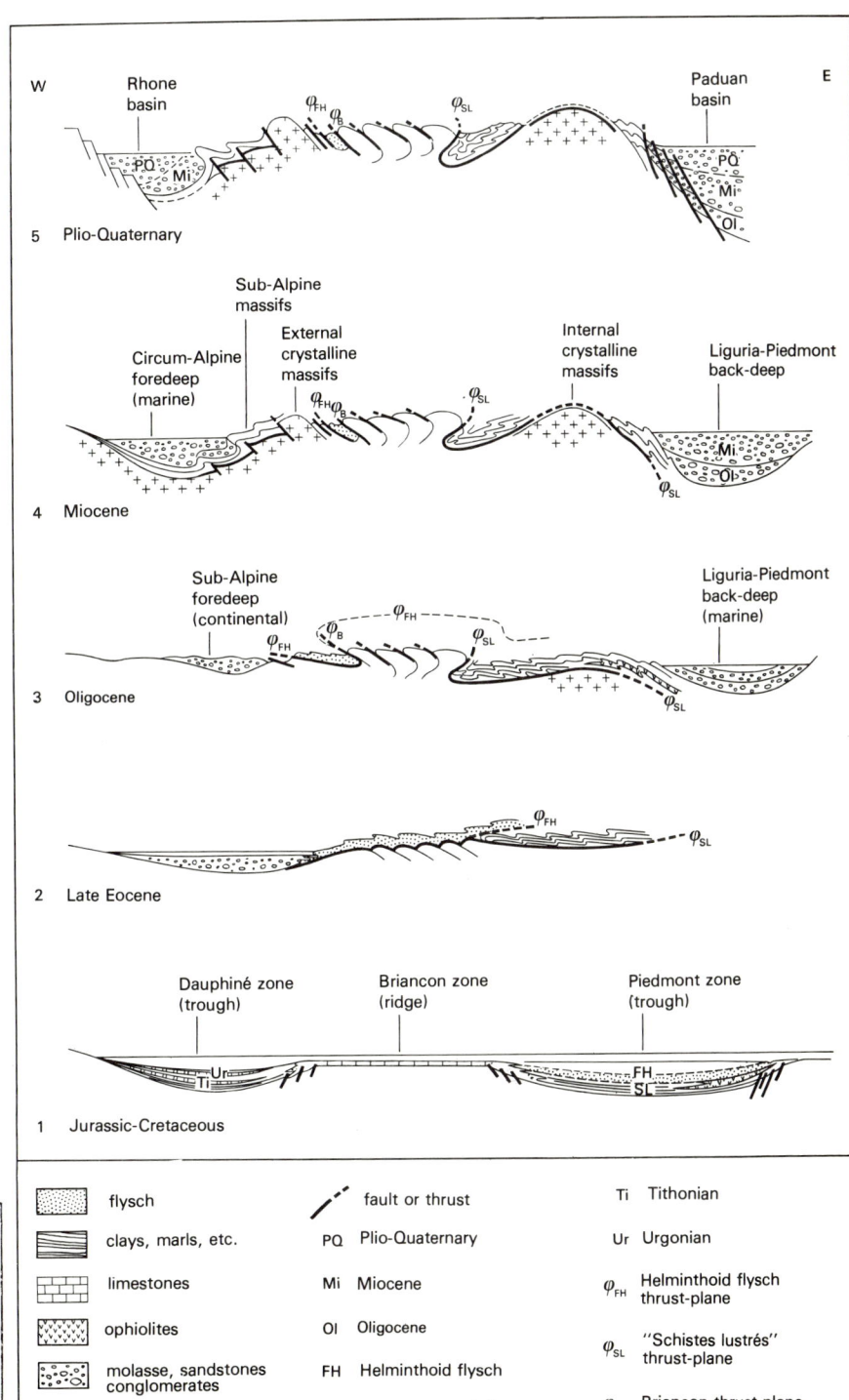

3.59 **Palaeogeographic and tectonic evolution of the western French-Italian Alps** (after Aubouin, by permission of the Encyclopedia Universalis).

107

Piedmont area to override the Brianconnais zone. This resulted in the westward displacement of the Palaeogene Sea during the course of the Priabonian. Overlying the nappe of the Helminthoid flysch is the **nappe of the "schistes lustrés"** (of lower Cretaceous and Jurassic age), which reached its final position in the early Oligocene (Fig. 3.59).

Thus in the Alps, during the Palaeogene, there were two episodes of nappe formation (separated by an interval of about 3Ma). The first was composed of marine sediments (black flysch and Aiguilles d'Arves flysch) while the second consisted of continental sediments (the Helminthoid flysch and the schistes lustrés) showing low grade metamorphism. The much older rocks of this second nappe now rest on the younger rocks of the first nappe and pebbles of the schistes lustrés from it are included in the Oligocene continental molasse, as are weathered clays from the Priabonian rocks of the first nappe.

The marine Oligocene with molluscs and small nummulites (*N. vascus* and *N. intermedius*) is known in a series of small basins which marked an arm of the circum-Alpine sea stretching from Provence to Switzerland through Castellane, Barrême, Dévoluy and Les Déserts near Chambéry. This was a forerunner of the future molasse trench which became well-developed in the Miocene (Fig. 3.60). To the west lay the lakes and lagoons of the Rhone valley. One of these, the Aix basin, will be referred to later (p. 112). The Oligocene trench expanded into the Swiss plain where the first molasse deposits were formed on its southern border. These included the Nagelfluh conglomerates of the Rigi and Mont-Pelerin at the foot of the Bernese Alps, which derive their name from the scattered pebbles which have been likened to the heads of nails (Nagel) driven into a wall (Fluh). The Oligocene trench was linked northwards, by way of the Délémont basin, to the Alsace plain (see p. 110) thus providing a connection between the North Sea and the Mediterranean region. The trench probably extended eastwards through the Vienna basin in Austria (Fig. 4.18, p. 126). In the Ligurian-Piedmontese basin, a back-deep was formed in which molasse was deposited discordantly on the "schistes lustrés" of the Piedmont Zone (Fig. 3.59 and 4.18).

In Switzerland, as in the Dauphiné area, the marine molasse was altered in late Oligocene times into a **red molasse** containing calcareous pebbles (gompholite) and overlain in places by lacustrine limestones with *Helix ramondi* which are attributed to the Chattian.

The tectonic history of the western Alps during the Oligocene may, according to DEBELMAS (1970), be summarised as follows:

1) *paroxysmal folding of the internal zones* of which at least the uppermost layers slipped *en bloc* or were imbricated to form nappes inclined to the west. In particular the nappe of Piedment shales, which had not been metamorphosed, was emplaced on the Briancon zone;

2) *metamorphism* of parts of this pile of nappes and its Palaeozoic substratum (i.e. the internal crystalline massifs, according to ELLENBERGER (1958)). The nappes covering the external zone escaped this metamorphism, as they had been emplaced in the late Eocene;

3) *reversed thrusting* of the internal part of the pile of nappes towards the east;

4) *uplift of the internal zones.*

This orogenic phase caused the sea to retreat from the area of the mountain chain to form a series of discontinuous laguno-marine basins in Switzerland and in the Rhine graben. As has already been emphasized, this succession of basins was the forerunner of the circum-Alpine trench and was progressively invaded by the Miocene sea. The sediments deposited in this sea were essentially molasse derived from the newly uplifted mountain chain.

IV. – THE CONTINENTAL LAKES AND BASINS

In the Paris Basin abundant detritus was initially supplied from the Massif Central, but this supply ceased at the end of the lower Eocene. A regime of lakes was then established on the periphery of the Palaeogene shorelines. At the same time siderolitic (iron-rich) deposits accumulated on the Mesozoic limestone plateaux around the Massif Central. It should be noted however that certain deposits, reputedly lacustrine in origin (Ducy limestone, St. Ouen limestone), are in reality lagoonal (see p. 73).

In the **Lutetian,** two lakes were formed south of the Paris gulf. The Lac de Provins lay to the south east, and the Lac de Morancez to the south west. Stone from this latter deposit was used in the building of Chartres cathedral. Other lakes existed further inland, as for example, that in which the Argenton-sur-Creuse limestone (Fig. 3.50) with mammalian fauna (*Lophiodon, Propalaeotherium*) was formed.

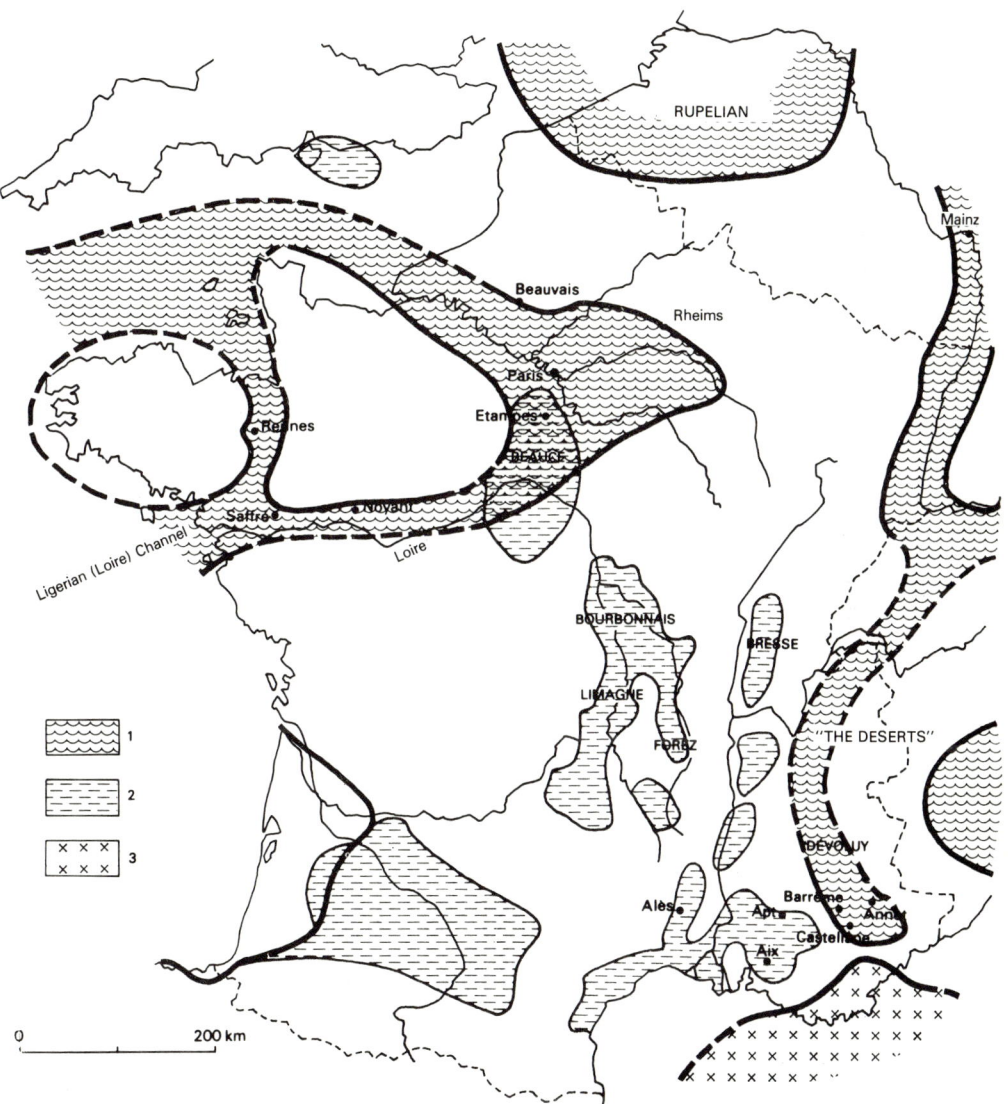

3.60 **Palaeogeography of France in the Oligocene.** 1 – marine formations; 2 – lagoonal or lacustrine formations; 3 – the Corsican-Sardinian massif. The sea in the Paris Basin retreated through the Loire channel, while that in the circum-Alpine trench retreated across Switzerland. The continental or lagoonal basins of the Massif Central and the Rhone Valley were temporarily in communication with the Alpine Sea. See also Figs. 3.61 and 3.63.

In the **Bartonian,** weathering of the lower Eocene sands and ferruginous clays was followed by the formation of continental sandstones around the edge of the Armorican massif which contain imprints of palm leaves (*Sabalites* sandstone). In the Sarthe area, these sandstones are overlain by limestones, once thought to be lacustrine, like those of Ducy and St. Ouen. These beds sometimes contain abundant foraminiferids (*Rosalina bractifera*) associated with laguno-marine ostracods. It is probable that these stretches of water were connected, at least temporarily, with the Bartonian gulf and so cannot be regarded as lakes. It is necessary to go to Touraine to find undoubted lacustrine deposits: the limestones of Loir, Anjou, Briare and Berry. These are contemporaneous with the lacustrine limestones of Champigny and Château-Landon, which are Ludian in age.

In the **early Oligocene** (Sannoisian) the lake in which the Touraine limestone was formed was contemporaneous with that of Brie and it may have had a connection with the Limagne. Sedimentation continued during the late Stampian with the formation of the *Calcaire d'Etampes* and into the Aquitanian with the deposition of the Calcaire de Beauce. The latter was locally separated from the Étampes limestone by the Gâtinais molasse, a detrital intercalation derived from the Massif Central. The Beauce limestone passes southwards into the limestones of Pithiviers, Orléans, Nevers and the Limagne.

In Alsace, continental Eocene beds are terminated by bituminous shales, lignites, the Bouxwiller lacustrine limestone and the *Melania* limestone containing a "Montmartre fauna" of Ludian age. They are overlain by the salt-bearing beds of the Mulhouse region and the oil-bearing beds of Pechelbronn, which mark the influx of the sea from the south at the end of the Eocene (Lattorfian)[1].

In the Stampian s.s. (Rupelian) the "Grey series" (from bottom to top) comprises the following (Table VII):

1) Marls with foraminiferids, grey, slightly pyritised.

2) The Fish shales (*Amphisile*), black, bituminous and strongly pyritised.

3) Meletta beds (with fish scales of *Clupea* (*Meletta*) *longimana*), shaly, sandy and micaceous marls.

4) *Cyrena* marls, grey and micaceous.

These indicate the influence both of the North Sea and the Alpine Sea, sometimes simultaneously and sometimes separately. The marls with foraminiferids are similar to those of the lower Rupelian beds of the Mainz basin and the circum-Alpine sea; the Fish and Meletta beds are known from the canton of Lucerne. Nevertheless, in the Oligocene the link between the North Sea and the Mainz basin was permanent, while that with the circum-Alpine sea was intermittent.

The Rhine graben acted as a strait linking the North Sea with the Mediterranean before tilting closed its southern end. This event diverted its drainage northwards into a gulf in the region of Cassel (see Fig. 4.18, p. 126) which persisted until the Chattian. In Alsace, the *Cyrena* beds of the upper Stampian show lagoonal features, while the Chattian limestones with *Helix ramondi* are wholly lacustrine. The Aquitanian sea did not extend south of the region of Haguenau.

Thus, during the Palaeogene, the Rhine graben deepened progressively as its margins (the Vosges and the Black Forest) were uplifted. It was repeatedly inundated by the waters of the Mediterranean and North Seas and was filled with an enormous volume of detrital and chemical sediments (sapropelic marls, clays and sands: halite, sylvite and gypsum). The climate appears to have been tropical in the late Eocene becoming sub-tropical to temperate, but more arid in the early Oligocene. This climatic sequence has already been referred to in connection with the evaporite sequence of the Paris Basin.

The **Massif Central** was uplifted and eroded continuously during the Eocene, but during the Oligocene was subjected to tension resulting from the Pyrenean-Provencal phase of movements. This produced a series of parallel faults which controlled the formation of a series of basins separated by low ridges, the erosion of which supplied sediments to the basins (Fig. 3.61). In the Limagne area, the most important basins are those at Brioude, Issoire and Clermont-Ferrand which now form

---

1. This paragraph was reviewed and completed (in the French edition) by C. Sittler of the Geological Institute of Strasbourg (Ch.P.).

Table VII
The Oligocene succession in Alsace
(after C. Sittler, in Géologie de la France, Paris: Doin)

		PECHELBRONN BASIN (type succession)	POTASH BASIN (saline deep)	MULHOUSE-SUNDGAU HORST (calcareous shelf)
QUATERNARY		Rhine and Vosges alluvium 0–250m	Plio-Quaternary	Loess 0–30m
PLIO-CENE		clays, sands and gravels 0–100m	Plio-Quaternary	Sundgau gravels 0–20m
MIOCENE — Aquit.		Upper Miocene — Hydrobia Beds with Corbicula 300m, Cerithium	Freshwater Beds — carbonates 250m / detrital 330m	Pontian: Vosges sands with Dinotherium
Chattian		Niederroedern Beds 600 m		Lacustrine marls and limestones 200m / Alsace molasse 90m — 10–30m / 150–300m
Rupelian	Pechelbronn Beds — Mottled series	100m / 300m ; 5–35m / 10–30m	Cyrena marls, Meletta Beds 100m / 300m ; Fish shales 2–17m ; Marls with foraminiferids 5–12m	Wolfersdorf marine sands 2–10m / 7–15m ; Wolschwiller marine sands
Lattorfian	Pechelbronn Beds — Mottled series (upper) / Grey series	Gypsum horizon with freshwater levels 90m ; Salt horizon with anhydrite nodules 80m ; Upper bituminous horizon: marls 60m with oilsands	Salt-bearing Zone (upper): Gypsum horizon with freshwater levels 40m ; Salt horizon 400m ; Upper bituminous horizon: marls with rock salt and potash 100m	Haustein upper { mottled clays 40m, sandy clays } ; lower { limestones, grey limestones 50m, conglomerates }
Lattorfian	Grey series (middle / lower)	Fossiliferous horizon — Marls with Hydrobia, bryozoa, Mytilus 80m ; Lower bituminous zone: marls mottled 150m	Salt-bearing Zone (middle): Fossiliferous horizon Marls with Hydrobia, bryozoa, Mytilus 80m ; (lower) Lower bituminous zone: marls grey, with salt 200m	Flaggy limestones and fossiliferous marls 4–40m ; Green marls with gypsum 80m
Lattorfian	Red beds	Red beds: marls with anhydrite	Conglomerates	
Lutet.	Dolomitic horizon	Green dolomitic marls with Lymnaea 250m ; Calcareous marls with anhydrite and salt	Green dolomitic marls with Lymnaea ; Calcareous marls with anhydrite and salt 700m	Green marls and limestones with Melania 300m
MESOZOIC: Lias, middle and upper Jurassic		Basal clay or limestones and lacustrine marls or iron ore deposits 0–100m		

the valley of the Allier. The Loire valley crosses the basins of Roanne, Montbrison and Le Puy, while those of Montlucon, St. Flour, Ambert and Alès are part of the Rhône basin.

In the Limagne basins, the Tertiary beds may reach a thickness of 3000m. The sequence begins with arkoses followed by marls with *Striatella* (small, highly ornate gastropods) attributed to the lower Oligocene (Sannoisian). This fauna has South European affinities which suggests that these lagoons were at some time in communication with those of the Rhône basin, the Cevennes not yet having been uplifted. This link is more probable than a connection with the Paris Basin, where the Stampian transgression had barely begun. Furthermore, the fauna comes from the south, since it shows a marked similarity to that of the basins of Provence (REY, 1966).

The *Striatella* marls are overlain by marly limestones with *Potamides lamarcki*. The upper part becomes lacustrine once more and in the region of Clermont-Ferrand is represented by the larval cases of the caddis-fly (*Phryganea*) and the land mollusc *Helix ramondi*. Near

Vichy, these marly limestones are succeeded by lacustrine limestones of Aquitanian age which have yielded a large fauna of birds and mammals at Saint-Gérand-le-Puy, reminiscent of that of the limestones of the Beauce and Orléans areas to the north and the Agenais to the south. This fauna flourished here in a lake which has been compared by MILNE-EDWARDS (1867) to certain lakes in central Africa.

The **basin of Alès** is similar to those of the Limagne, with which it was originally connected, and leads us to the **Rhône valley** where small basins, filled by debris from the Massif Central to the west and the earliest Alpine folds to the east (Arvinchian phase), occur between Dijon and Provence. The sediments of these basins (at Royans, Crest and Mormoiron) are essentially late Eocene and Oligocene in age. In the Aix region a large lake, elongated east to west, a relic of the continental environment of the late Cretaceous, persisted throughout the Palaeogene. To its north was a basin encompassing Apt, Manosque and Forcalquier, while to the south lay the basin of Marseilles, which only came into existence in the Oligocene (Fig. 3.62). In the Aix-en-Provence lake the sediments show a lacustrine sequence beginning in the Maastrichtian, continuing into the Danian = Rognacian[1] (which consists of red marls and sandstones with reptile remains) and succeeded by the white limestone of Rognac, which stands up in relief. In the Montian, locally called the Vitrollian[2], are the red clays of Vitrolles and the *Physa montensis* limestones which pass upwards into the Langesse limestones, of Thanetian age, to be followed by lacustrine Lutetian limestones.

The Eocene formations were folded in the Pyrenean-Provencal phase of movement and are overlain unconformably by Oligocene beds commencing with a conglomerate which is followed by a gypsiferous limestome extremely rich in molluscs, insects, centipedes, spiders, fish, amphibians, mammals and tropical vegetation. Short-lived marine invasions from the Alpine sea can be recognised at several levels in the Oligocene of Provence. They occur in the

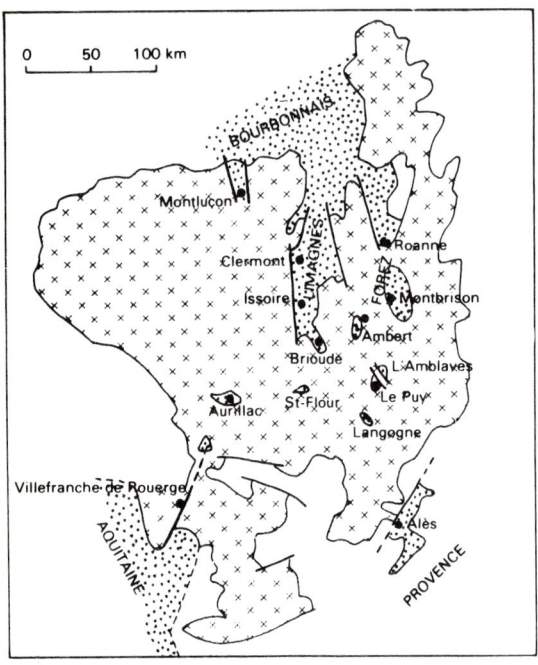

3.61 **The principal Oligocene basins of the Massif Central** (see Fig. 3.62).

1. Rognacian, from Rognac, a village north east of the Étang de Berre (Villot, 1883). Plaziat has shown recently that the Rognacian is more probably of late Maastrichtian age, while the Begudian is likely to be early Maastrichtian.

2. From Vitrolles, a locality east of the Étang de Berre (Matheron, 1878).

middle Stampian (in the small basins of Manosque, Haut-Var, Luynes and Estaque) and in the upper Stampian at Aix-en-Provence, where the gypsum is of marine rather than continental origin.

In the Languedoc, the tip of the marine gulf of Aquitaine penetrated between the Montagne Noire and the Corbières during the Montian and the Ilerdian. This gulf was flanked by a series of lagoons in which were deposited red clays and sandstones, followed by limestones which continued into the Lutetian. Continental Oligocene beds occur locally in small basins and are of interest on account of their mammal faunas and of a flora which is particularly well developed in the basin of Armessan.

In France, during the Oligocene, the area covered by lagoons and lakes was considerable in an area of low relief, except in the south, where the Pyrenean orogeny had reached its most intense phase, and the south east, where the Alps and Provence were being uplifted.

The study of the palaeogeography and stratigraphy of these lagoonal and lacustrine deposits has led to a better understanding of the vegetation and the tropical conditions in which it and the vertebrates (in particular, the mammals) evolved. The transport of their remains by floods into shallow marine deposits has allowed precise correlations to be made between the marine and continental formations.

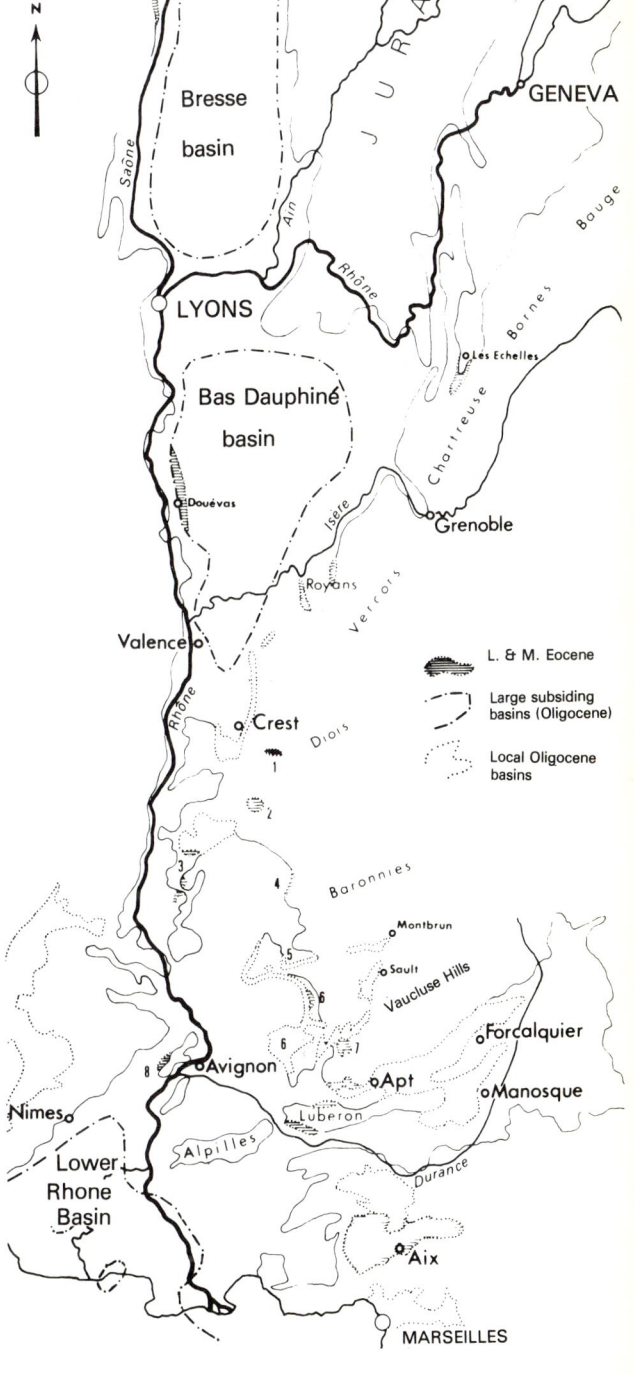

3.62 **The principal Palaeogene basins of the Rhone Valley** (after Demarcq (1970)).
1 – Saou; 2 – Dieulefit; 3 – Tricastin; 4 – Nyons; 5 – Malaucène; 6 – Mormoiron-Pernes; 7 – Muers; 8 – Les Angles.

113

Table VIII
Stratigraphic correlation between the Paris Basin, the Belgian Basin and the London/Hampshire Basin in the Palaeogene

The stage names in parenthesis are now little used
(see also the horizons corresponding to the stages of the Paris Basin in Table VI, p. 58)

ENGLAND	PARIS BASIN		BELGIUM	
			Formation	Stage
emergence / Upper Hamstead Beds	STAMPIAN	OLIGOCENE	Voort Sands / Boom Clay / Berg Sands	Chattian? / Rupelian
Lower Hamstead Beds	(Sannoisian)		Vieux-Joncs sands / Henis clay	upper Tongrian
Bembridge Beds / Osborne Beds / Middle-Upper Headon Beds	BARTONIAN (s.l.) — Ludian = Priabonian		Neerrepen sands / Grimmertingen sands	lower Tongrian or Lattorfian
Lower Headon Beds / Barton Beds	Marinesian / Auversian		Asse sands / Asse clay	Assian
Upper Bracklesham Beds (XVI, XVII & Huntingbridge Bed)	upper	EOCENE	Wemmel sands	Wemmelian
Middle (IX to XV) Bracklesham (VI to VIII) Beds	LUTETIAN — middle		Lede sands	Ledian
	lower		Brussels sands / Aeltre sands	Bruxellian
Lower Bracklesham Beds (I to V)	YPRESIAN — Cuisian		Mont Panisel sands / Ypres clay	(Paniselian) / Ypresian
"Lower Bagshot Beds" / London Clay	Sparnacian	PALAEOCENE	A. scutellaria sands	Landenian
Woolwich and Reading Beds			Tuffeau de Lincent / Gelinden marls	(Heersian)
Thanet Beds	Thanetian		Calcaire de Mons / Tuffeau de Ciply	Montian / Danian
	Dano-Montian			Maastrichtian
Coniacian – Campanian	Campanian		Campanian	

The Palaeogene in the rest of Europe

4

I. – THE NORTHERN BASINS

The Palaeogene formations of Belgium and England will be discussed first and this discussion will be followed by a consideration of Denmark, Germany and Poland.

1. Belgium (Table VIII)

The Belgian Palaeogene has close affinities with that of the Paris Basin. In the south, in the Mons Basin, the first transgression of the Palaeogene sea gave rise to sediments deposited unconformably on a Maastrichtian sub-stratum (the Saint-Symphorien limestone). The Tertiary beds commence with the **Tuffeau de Ciply** (Tuffeau = sandy limestone), which passes upwards through intermediate beds into the **Calcaire de Mons,** designated as the type for the Montian stage by DEWALQUE in 1868. This sequence is followed by a lacustrine limestone containing *Physa* and the whole constitutes the **Dano-Montian,** of early Palaeocene age. In Limburg, to the west of Maastricht, there is a similar series of beds (Fig. 4.1).

The **Landenian** (late Palaeocene) of Belgium was defined as a stage by DUMONT in 1839 from the locality of Landen, south east of Louvain (Fig. 4.1), where the sequence is more complete than in the Paris Basin. The Gelinden marls containing *Arctica morrisi* are sometimes referred to the Heersian stage (from Heers, east of Landen). This unit is followed by the Tuffeau de Lincent, which is the equivalent of the Tuffeau de la Fère of the Laonnois. As the transgression spread southeastwards, it deposited sands with *Arctica scutellaria,*

equivalent to the Sables de Bracheux. This period marked the maximum extension of the Landenian sea over the Ardennes massif and in Belgium marked the maximum transgression during the whole of the Cenozoic (Fig. 4.2).

The top of the Landenian is continental and contains many fish, plants and mammals. The pollen in these beds resembles that of the basal beds of the Sparnacian of the Paris Basin. As a consequence of this, the boundary between the Palaeocene and the Eocene is taken slightly lower in the Paris Basin than in Belgium (Table VIII).

The sea again transgressed this area in the **Ypresian** (DUMONT, 1849) at a time when Sparnacian lagoons still occupied the Paris Basin. This sea deposited more than 100m of glauconitic sands and clays (Ypres Clay) which contain *Globorotalia aquiensis*, which occurs in the Ilerdian of the Mediterranean. The Ypres sands and clays are overlain by the Mons-en-Pévèle sands and then by those of Mont-Panisel (the type locality of the Paniselian, created in 1851 by DUMONT). The Mont-Panisel sands contain *Nummulites planulatus* and *Alveolina oblonga*, characteristic of the Cuisian. The top of the Ypresian is represented by the Oedelem sands, east of Bruges and the Aalter or Aeltre sands, west of Gand, with *Venericardia planicosta* and *Turritella solanderi*. These were collectively called the Den Hoorn formation by NOLF and form a transition to the middle Eocene (Bruxellian, DUMONT, 1839).

In **early Lutetian** time, the **Bruxellian** sea, with *Nummulites laevigatus* and *N. variolarius*,

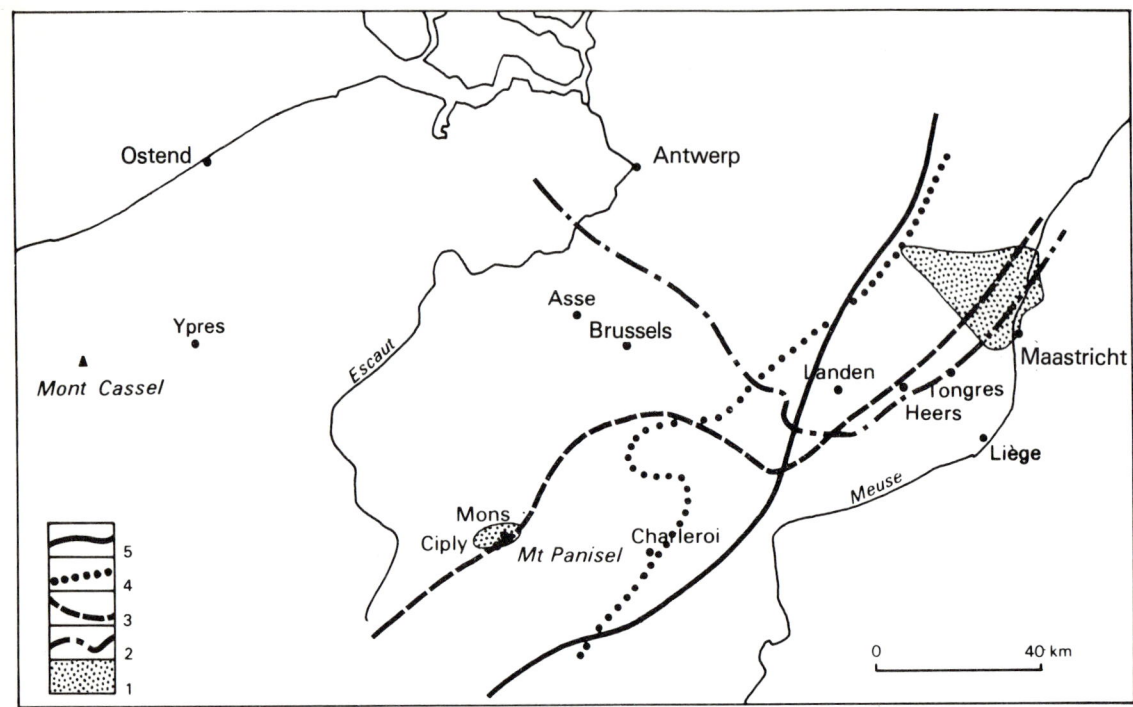

4.1 The eastern limits of the principal Lower Palaeogene formations (Dano-Montian to Bruxellian) in Belgium, before the uplift of the Artois dome (after Gulinck). Limits of: 1 – Dano-Montian; 2 – Lower Landenian (Heersian); 3 – Middle and Upper Landenian; 4 – Lower Ypresian (Ypres clay); 5 – Bruxellian.

was the last to be in communication with that of the Paris Basin (Fig. 4.3). The Paris Basin in Lutetian times was essentially one of limestone deposition, whereas the Belgian basin received sandy limestone and sands derived largely from the Ardennes-Rhine massif. The final separation between the two basins resulted from the uplift of the Artois dome in mid-Lutetian times. An earlier uplift of this dome in the Ypresian had effected only a temporary separation (see Fig. 4.3).

The early Lutetian (Bruxellian) beds are followed by the sandy, or locally, calcareous beds of the middle Lutetian or **Ledian** (from Lede, MOURLON, 1887). These beds contain *Orbitolites complanatus, Ditrupa strangulata* and *N. variolarius*, as in the middle Lutetian of the Paris Basin, where however, *N. variolarius* is less common. The late Lutetian, known locally as the **Wemmelian** (VINCENT and RUTOT,

1878) consists of the Wemmel sands, which contain an evolved form of *N. variolarius* now called *N. orbignyi* (= *N. wemmelensis*). Finally the clays and sands of Asse (the **Assian** or Asschian of RUTOT, 1882) are of Bartonian (*s.l.*) age.

All the units above the Bruxellian, that is the Ledian, Wemmelian and Asschian, were deposited in a sea open to the north. They are often glauconitic and contain planktonic foraminiferids of a northern aspect and heavy minerals derived from the north (Fig. 4.5).

In the east of Belgium, the Grimmertingen sands, rich in molluscs, and the overlying Neerrepen sands form the lower part of the Belgian **Tongrian** stage (from Tongres, north of Liège). Though some authors place this stage at the base of the Oligocene, it is here correlated with the Ludian of the Paris Basin and with the

4.2 Submarine slumping in fine sands of the littoral Landenian at Linsmeau, near Landen (photo: Gulinck). The contortion of the beds is due to their mobility (thixotropy) when saturated with water.

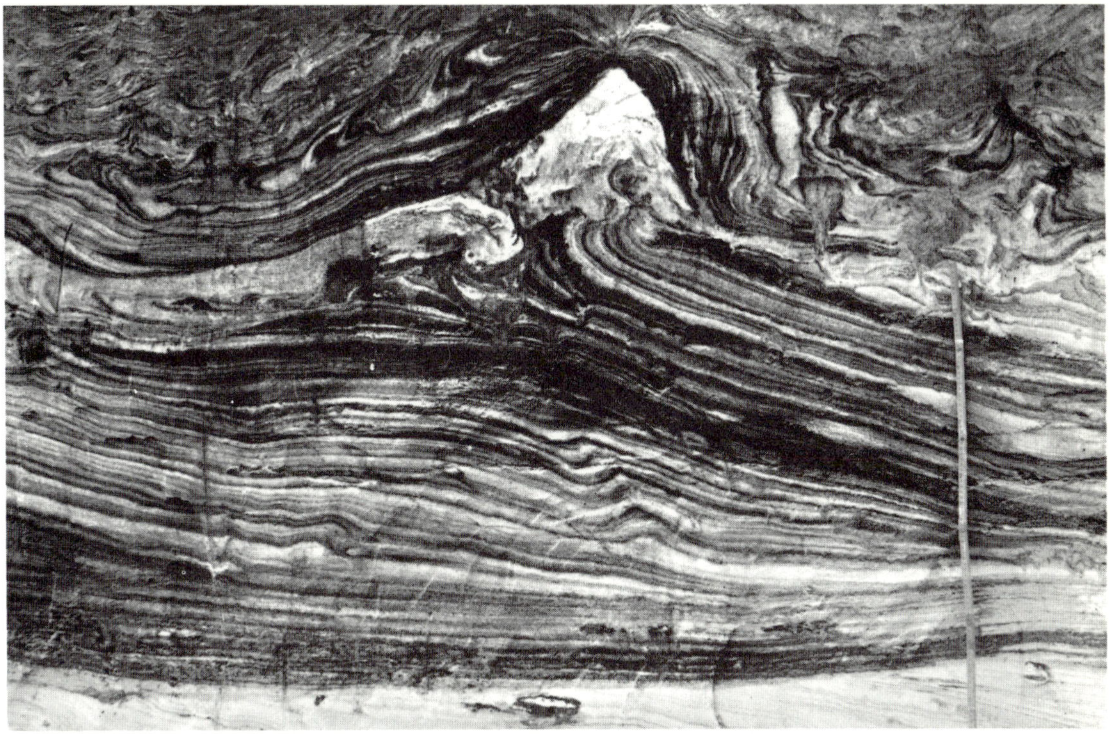

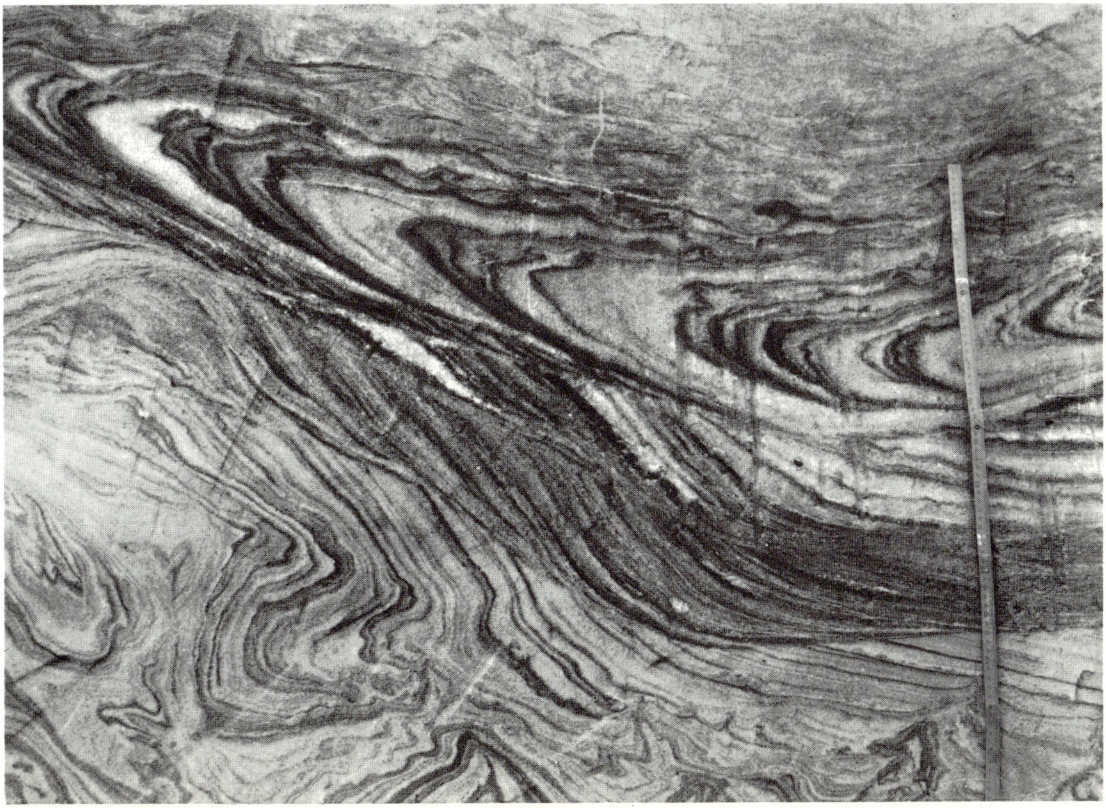

The Palaeogene in the rest of Europe

N BELGIAN BASIN		S PARIS BASIN (Northern part)		
RUPELIAN	Boom Clay / Berg Sands	Fontainebleau Sands		STAMPIAN
U. TONGRIAN	Vieux-Joncs Sands / Henis Clay	Sannois limestone / Green Clay	SANNOISIAN	
L. TONGRIAN	Neerrepen Sands / Grimmertingen Sands	Gypsiferous marls / Gypsum	LUDIAN	BARTONIAN
ASSIAN	Asse Sands	Cresnes-Marines Sands	MARINESIAN	
	Asse Clay	Auvers-Beauchamp Sands	AUVERSIAN	
WEMMELIAN	Wemmel Sands	Shelly Sands / BIARRITZIAN / Shelly Sands	UPPER	LUTETIAN
LEDIAN	Lede Sands	Limestone with *Miliola* and *N. variolarius*	MIDDLE	
BRUXELLIAN	Brussels Sands	Calcaire grossier	LOWER	
YPRESIAN	Mont-Panisel Sands	Cuise Sands	CUISIAN	YPRESIAN
	Ypres Clay	Clay with lignite	SPARNACIAN	
LANDENIAN	Sands & Lignites / Sands with *Arctica scutellaria*	Mortemer limestone / Bracheux Sands		THANETIAN
	Gelinden marls			DANO-MONTIAN
DANO-MONTIAN				

(ARTOIS dome shown in centre of diagram)

4.3 Diagrammatic representation of the Palaeogene formations on either side of the Artois dome. The beds of the Belgian basin are almost all marine (some brief lagoonal or continental episodes have been omitted from the diagram), whereas lagoonal and continental formations are well-developed in the northern part of the Paris basin. Note the role of the Artois dome in early Ypresian times when it acted as a barrier between the Sparnacian to the south and the lower Ypresian in the north. From the middle Lutetian onwards it formed a permanent barrier until the end of the Stampian (see also Table VIII, p. 114). Stippled: dominantly marine formations; broken lines: mainly lagoonal formations; blank: continental formations.

4.4 Alternations of sands and calcareous sandstones in the Ledian at Balegem quarry (photo: Pomerol).

Lattorfian of north Germany. It is therefore regarded as **Priabonian** (latest **Bartonian** *s.l.*) and as forming the top of the Eocene (Table VIII).

Overlying the Neerrepen sands are fossil soils and lacustrine beds with a fauna of terrestrial vertebrates (the *Hoogbutsel horizon*), referred to the upper Tongrian. This fauna is comparable with that of Ronzon (see p. 36) and is of Sannoisian, that is, of lowermost Oligocene age.

In the eastern part of Belgium, continental or lacustrine beds succeed the Hoogbutsel level, but borings in the Antwerp area show a continuous marine sequence from the Eocene to the Oligocene. The main unit of the Oligocene consists of the Berg sands containing *Nucula* overlain by the Boom clay with septarian nodules (Fig. 4.7), the latter unit being recognizable as far as Poland. These beds,

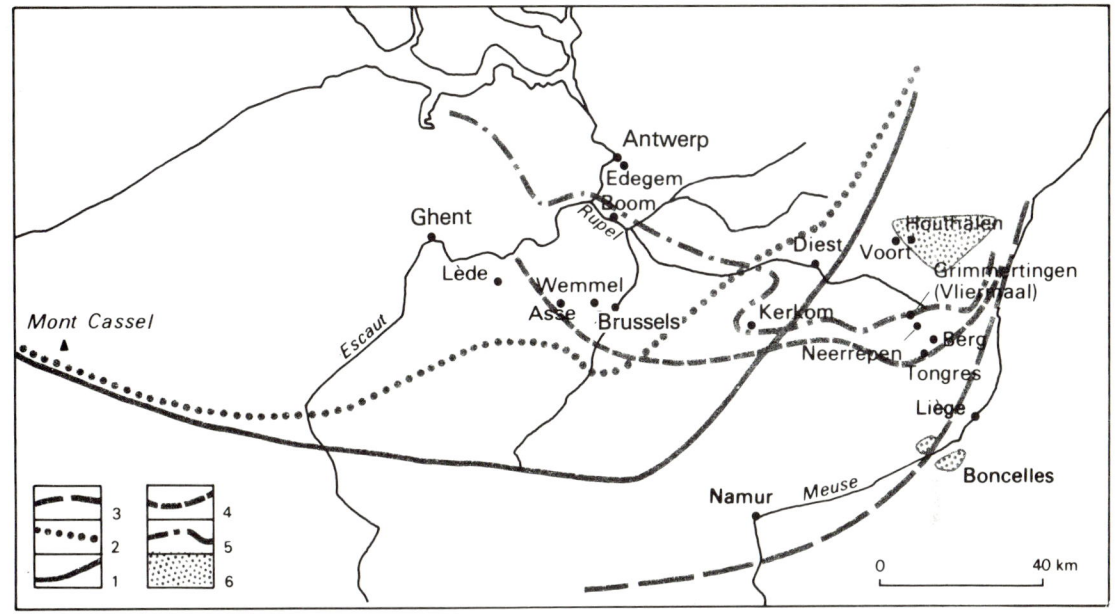

4.5 The southern limits of the principal formations of the late Palaeogene (Ledian to Chattian) in Belgium, separated from the Paris Basin by the Artois dome (after Gulinck). 1 – Ledian; 2 – Assian; 3 – Tongrian; 4 – lower Rupelian (Berg sands); 5 – upper Rupelian (Boom clay); 6 – Chattian (sands of Voort and of Boncelles).

contemporaneous with the Stampian of the Paris Basin, are called the **Rupelian** in Belgium (from Rupel, a tributary of the River Scheldt, DUMONT, 1849).

Finally, the **Chattian** appears to be represented in the eastern Campine by the Voort sands, and near Antwerp by the Edegem sands. The latter were, until recently, thought to be Miocene and placed in the Anversian. However, radiometric dating (30Ma) and the planktonic foraminiferid assemblage suggest that these beds are older than Miocene, the lower limit of which is placed at 23M. A regression marks the Oligocene-Miocene boundary, following which the Houthalen sands were deposited (see p. 165).

Thus in the Palaeogene, from the beginning of the Landenian until the end of the early Lutetian, with the exception of a brief interval in the Sparnacian (early Ypresian), the Belgian basin linked a northern ocean with the Paris Basin. Then, in the middle Lutetian, there occurred a crucial change in the palaeogeography of this area. The uplift of the Artois dome between the two basins severed

this link which was never again re-established. The Belgian basin thus developed wholly northern characteristics, though to the south and east, in Hainaut, Brabant and southern Flanders, there was a tendency towards emergence. At the end of the Oligocene, the sea was restricted to northern Flanders and the Campine. Except for brief regressive episodes, it remained there for the whole of the Neogene, until the Quaternary.

2. England (Table VIII)

In England, Palaeogene formations occur only in the two synclinal regions of the **London Basin and the Hampshire Basin,** now separated by the Weald anticline (Fig. 4.8). The Dano-Montian is absent and the earliest Tertiary deposits are the Thanet Beds.

Thanet Beds

These are slightly clayey, fine grained, glauconitic sands, with a maximum thickness of 40m, transgressive on the Chalk. Molluscs, including *Cucullaea crassatina, Arctica morrisi*

4.6 The junction between the Neerrepen sands (lower Tongrian) and the Hénis clay (upper Tongrian) near Tongres. Palaeosoils at the top of the Neerrepen sands indicate that the Tongrian spans two sedimentary cycles. The beds below the soil are of late Eocene age, whereas those above are early Oligocene, and are correlated with the Argile verte de Romainville in the Paris Basin (photo: Pomerol).

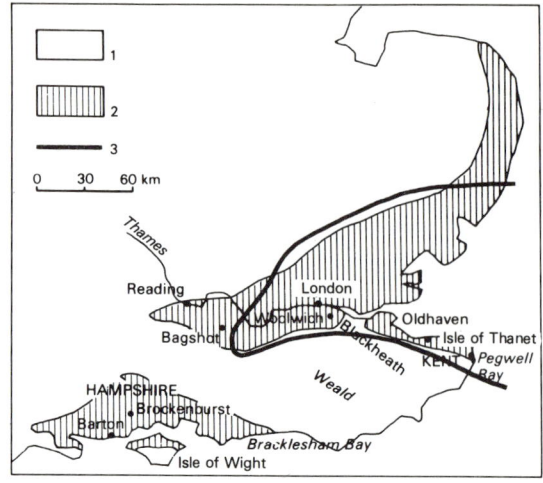

4.8 Main outcrops of the Palaeogene formations in England. 1 – pre-Cenozoic beds; 2 – Palaeogene beds; 3 – Limit of the Thanetian gulf.

4.9 The Chalk-Thanet Sand junction, Pegwell Bay, Kent (see Fig. 4.8). The sands are partially cemented. A thin band of flint occurs at the boundary between the chalk and the sand (photo: Pomerol).

4.7 A septarian nodule in the Boom clay, near Antwerp (Rupelian = Stampian) (photo: Pomerol).

and *Pholadomya oblitterata* are sometimes abundant, but badly preserved. The Thanet Beds are limited to the London Basin and are best seen in the local cliffs of the Isle of Thanet on the southern side of the Thames estuary (Fig. 4.9). They correspond approximately to the Landenian of Belgium and the Thanetian of the Paris Basin, but their fauna suggests somewhat colder waters. However, it is possible that the highest beds of the Sables de Bracheux (Châlons-sur-Vesle sands), which have faunistic similarities with the English Woolwich beds, are younger than the highest beds of the English Thanetian.

Woolwich and Reading Beds

The **Woolwich Beds** are laguno-marine sediments of the London basin containing *Ostrea bellovacina* and consist of alternating layers of sand and clay of variable thickness. They pass westwards into the **Reading Beds** beyond which lay a continental area from which the sediments were probably derived. In the Hampshire Basin the Reading Beds form the earliest unit of the Palaeogene and are well displayed in the Isle of Wight in the magnificent sections at Alum Bay in the west and at White-cliff Bay in the east (Fig. 4.10). Here the Chalk and the Palaeogene beds have been tilted into a near-vertical position by Miocene tectonism. According to CURRY (1967), the Reading and Woolwich Beds (the latter known also near Dieppe) can probably be correlated with the top of the Thanetian and the base of the Sparnacian of the Paris Basin and with the top of the continental Landenian of Belgium.

In the London Basin, the Woolwich and Reading Beds are locally overlain by the **Oldhaven and Blackheath Beds** which are sands and gravels of estuarine facies, whose fauna is similar to that of the Sinceny sands at the top of the Sparnacian of the Paris Basin (see p. 65).

London Clay

The Oldhaven and Blackheath Beds are overlain by the London Clay in the eastern part of the London Basin. In the western part of the basin and in the Hampshire Basin, the London Clay rests on the Reading Beds. The London Clay is a thick (180m) dark grey-blue clay which also occurs in the English Channel and in discontinuous outcrops along the French coast from Dieppe to Boulogne. Its fauna includes molluscs (*Astarte, Arctica*) and foraminiferids, some of which indicate a deep water facies (150 to 300m) occurring in a sea closed to the west. This was relatively cold and of Baltic type. Waters flowing from the nearby continental landmass carried a great profusion of vegetable matter. Recent work on the dinoflagellates of the London Clay indicates that it is younger than the type Sparnacian and is equivalent to the lower part of the Cuisian before the arrival of *Nummulites planulatus* (Sables de Cuise inférieurs and Varengeville Formation).

Bracklesham and Bagshot Beds

The appearance of *N. planulatus* is associated with the opening of the English Channel and the development of a neritic facies in a tidal sea, similar to the present day North Sea. *N. planulatus*, which is a useful marker,

4.10 **Panorama of Whitecliff Bay, Isle of Wight.** Note the steep northerly dip of the Palaeogene beds. The contact between the Chalk and the Tertiary Beds is nearly vertical and the dip remains high as far as the Headon Beds but then decreases rapidly. The Bembridge Beds are almost horizontal.

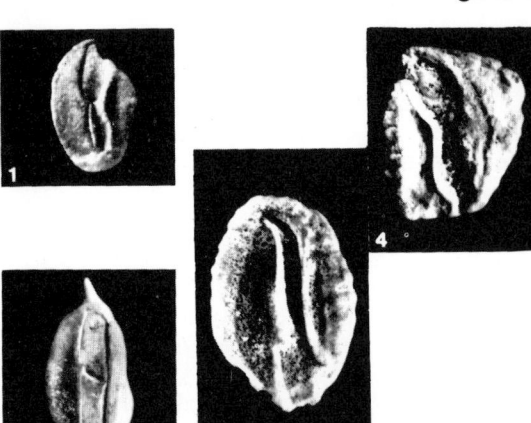

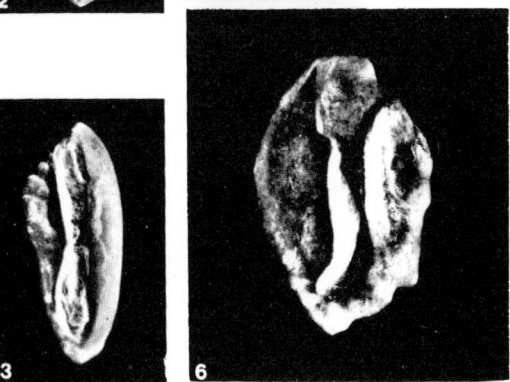

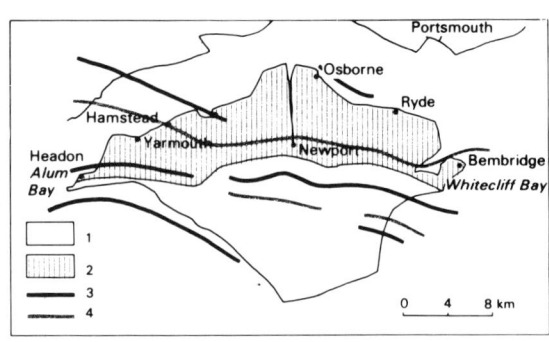

4.12 The Palaeogene of the Isle of Wight and its type-localities. 1 – Mesozoic; 2 – Palaeogene; 3 – anticlines; 4 – synclines (after White (1921)).

4.11 English Palaeogene otoliths (photo: Stinton). 1 – *Apogon bellovacinus* (Upper Bracklesham Beds) (3.5mm); 2 – *Trisopterus dimidiatus* (Upper Bracklesham Beds) (4.5mm); 3 – *Palaeogadus serratus* (London Clay) (8mm); 4 – *Beryx lerichei* (Lower Barton Beds) (4.5mm); 5 – *Xenistius notus* (Upper Bracklesham Beds) (5mm); 6 – *Pagrus bognoriensis* (London Clay) (10mm):

occurs in the sandy facies of the upper part of the Ypres clay and in the argillaceous sands of the Bracklesham Beds of Hampshire and the Isle of Wight but has not been found in the London Basin. It should be noted that the Bagshot Sands of Whitecliff Bay and Alum Bay in the Isle of White are barren sands with pipeclays, probably of continental origin, occurring between the London Clay and the Bracklesham Beds. They are therefore older than the Bagshot Beds of the London Basin.

The Bracklesham Beds have a rich fauna of molluscs with more than 500 species, many of which indicate a warmer sea coming through the English Channel from the west. These beds can be divided into three parts:

1) the *lower Bracklesham Beds* with *N. planulatus*, corresponding to the Cuisian, the upper Ypres clays and the Paniselian of Belgium;

2) the *middle Bracklesham Beds* with *N. laevigatus* and early forms of *N. variolarius*, which are correlated with the lower Lutetian and Bruxellian;

3) the *upper Bracklesham Beds* with *N. variolarius, Orbitolites complanatus, Fabiana cassis, Alveolina elongata, Linderina brugesi* and *Discorinopsis kerfornei* (middle and upper Lutetian and Ledian-Wemmelian). This unit yields many species of otoliths (Fig. 4.11).

Barton, Headon and Bembridge Beds (Fig. 4.12)

The type locality of Barton was selected by MAYER-EYMAR in 1857 for his Bartonian stage but the Barton Beds are also well exposed in the coastal sections of the Isle of Wight. They contain nummulites which have evolved from *N. variolarius*, beginning with *N. prestwichianus* followed by *N. rectus* which has also been found in the eastern part of the English Channel near Fécamp (AUFFRET, 1973). Radiometric dating of the glauconitic sand containing *N. prestwichianus* has indicated an age of 43Ma which is accepted for the age of the base of the Bartonian in the type locality. These beds can be correlated with the upper part of the Asse clay in Belgium and with the Auversian-Marinesian of the Paris Basin, which was separated from the basins of England and Belgium, but in which *N. variolarius* was abundant.

The Barton Beds are followed by the Headon Beds, which are marls with sandy intercalations. These are correlated with the *Pholadomya ludensis* marls and with the base of the Tongrian. The succeeding Bembridge Beds are lagoonal or lacustrine marls and limestones containing a fauna similar to that of the Montmartre gypsum. English authors traditionally place the Eocene-Oligocene boundary at the base of the Headon Beds. In France, however, the Montmartre gypsum is included in the Ludian, and French geologists would therefore place the Bembridge Beds in the Upper Eocene.

Hamstead Beds

The base of the Hamstead Beds is continental like the upper part of the Bembridge Beds which they overlie. The Oligocene transgression is represented by marls with *Ostrea longirostris* which occur in the upper part of the Hamstead Beds. These may be correlated with the Oyster marls of the lower Stampian (Sannoisian facies) of the Paris Basin, and with the upper part of the Belgian Tongrian (Table VIII).

No Palaeogene beds younger than the Hamstead Beds are known from the Hampshire Basin; the sequences in the London Basin end at the top of the Bagshot Beds, of late Lutetian age.

This early termination of the English Palaeogene sequences contrasts with the continuing deposition in the nearby Paris and Belgian basins. The colder, deeper waters of the English Basin compared with the Paris Basin and the differences in sedimentary facies suggest that the concept of a single Anglo-French-Belgian Basin may have been exaggerated. In the Cuisian and Lutetian, the Atlantic was linked by way of the English Channel to the North Sea (cf. Fig. 3.13) and the concept of a single Anglo-French-Belgian basin is acceptable. This union and subsequent separation of the basins can be traced readily from the distribution of the nummulites. *N. planulatus* occurs in all three areas, as do *N. laevigatus* and *N. variolarius* in the Lutetian. *N. variolarius* continued to exist in the

Bartonian of the Paris Basin, but was replaced by *N. prestwichianus* and then by *N. rectus* in England and by *N. orbignyi* in Belgium, suggesting the separation of the three basins.

Denmark-Germany-Poland

At the end of the Cretaceous, the northern seas were everywhere in retreat and persisted only as a long, broad strait stretching from the eastern part of the North Sea, across Denmark and central Poland to the Crimea (Fig. 4.14). It was in this area that the **Danian limestone at Fakse** (Fig. 4.13) was formed. (Fakse, south of Copenhagen is the type locality of the Danian, DESOR, 1846). It has already been noted that the Cerithium and Bryozoan limestones of the Danian contain Tertiary species of *Globigerina* (*G. daubjergensis, G. eugubina*), *Globorotalia* (*G. triloculinoides*) *and nannoplankton* (*Markalius astroporus*) characteristic of the earliest beds of the Tertiary era, whilst

4.13 **Reef breccia and limestone banks in the Danian at Fakse, Denmark** (photo: Pomerol) (see also Fig. 2.1, p. 28 and Fig. 9.5, p. 205).

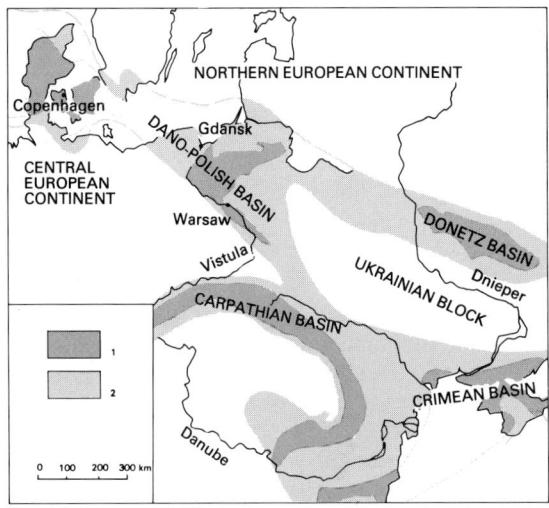

4.14 The Dano-Montian in eastern and northern Europe (after Požaryska (1971)). 1 – areas where sediments have been preserved; 2 – areas where sediments have largely been destroyed by erosion.

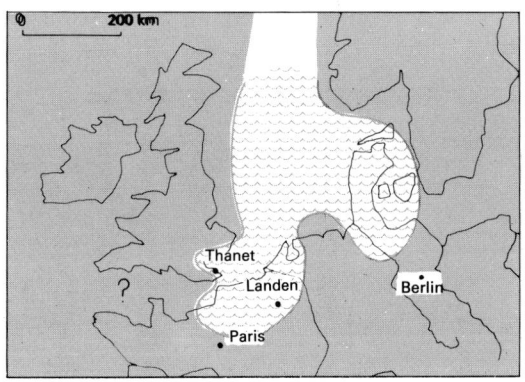

4.15 The Thanetian (Landenian) sea in northern Europe.

Globotruncana, characteristic of the upper Cretaceous, is absent. The lower part of the Montian of Mons (Tuffeau de Ciply) is equivalent to the middle Danian (Coral limestone) and its upper part (the Calcaire de Mons) corresponds to the upper Danian (Bryozoan limestone). For these reasons, it seems preferable to use the term Dano-Montian to designate the first sedimentary cycle of the Palaeocene.

In the **late Palaeocene and early Eocene,** the sea retreated from Poland and occupied a gulf or embayment extending from the North Sea basin across Denmark and part of North Germany whilst another embayment extended across Holland, Belgium, northern France and part of south east England. In Denmark, the **Selandian** sandy glauconitic marls, which rest on the Danian, are the lateral equivalent of part of the Thanetian of the Paris Basin (Fig. 4.15). The eastern extension of the sediments of the Danish embayment is mostly covered by more recent deposits, but they occur at the surface between Hanover and Osnabrück, where the Lehrte greensands have a fauna similar to that of the Wemmelian (late Lutetian). In Germany, a small inlet of the sea extended southwestwards and, in the middle Eocene, reached the Elbe valley in Saxony. The localities of Helmstedt and Latdorf, near Magdeburg are in this area (Fig. 4.16 and 4.17).

In Brandenburg and near Lublin (Poland) patches of glauconitic sand of **Lattorfian age** (from Lattorf or Latdorf, MAYER-EYMAR, 1893) **overlie the upper Lutetian.** The Latdorf sands contain a fauna which has close affinities with that of the Grimmertingen sands (lower Tongrian of Belgium) and with that of the Brockenhurst Beds (Fig. 4.8). The latter beds occur in the New Forest of Hampshire and form the lowest part of the middle Headon Beds of the Isle of Wight. The molluscan fauna of the Lattorf beds also has strong affinities with that of the upper Eocene of the Ukraine at Mandrikovka. The presence of *Nummulites germanicus*, a southern species, indicates a late Eocene age. For this and other reasons (see p. 26) *the Latdorf beds are here placed* (as CAVELIER (1972) has proposed) *at the top of the Eocene and not at the base of the Oligocene* as BEYRICH had intended when he created this epoch (Fig. 4.17); an intention which some geologists still support.

The Oligocene sea (s.s.) spread further north in Poland and Russia (Fig. 4.18), while a depression occupied by the late Eocene lagoons gradually deepened and extended southwards. This became the **Rhine Graben** some 50-100km wide, crossing western Germany, into which the Oligocene sea flooded. The area covered includes Hanover and Cassel (which the sea had reached in the Lattorfian), Mainz, Strasbourg and Alsace, where there was a link with the marginal lagoons of the Alpine sea. Near the

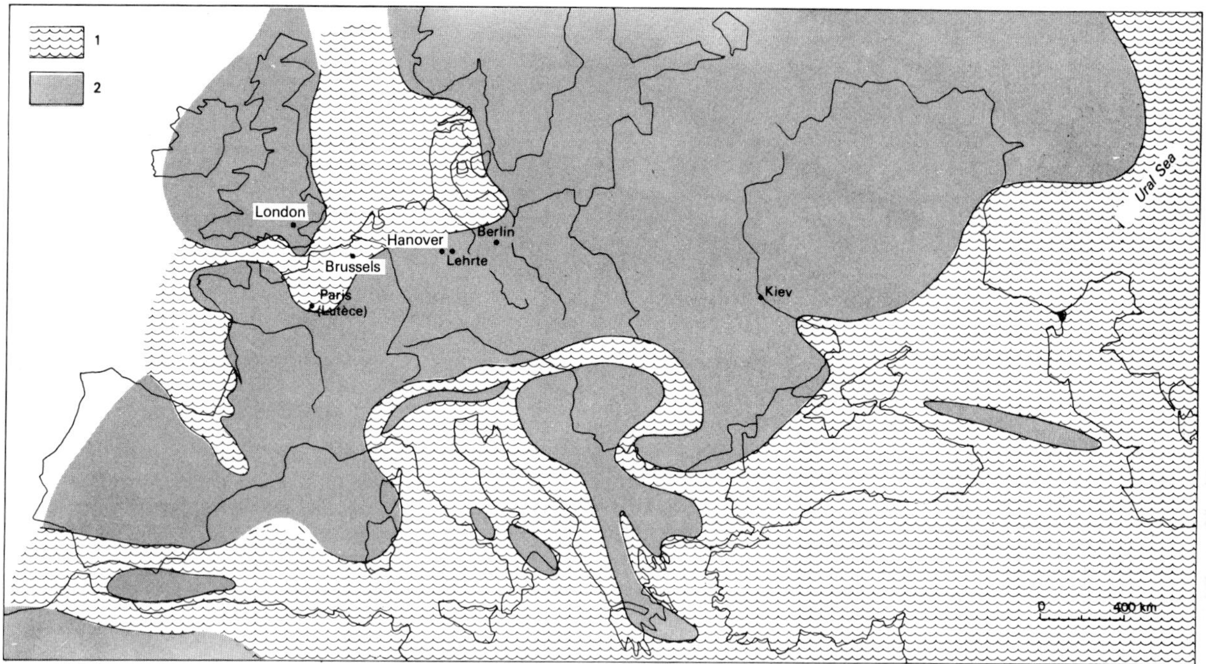

4.16 Europe in the middle Eocene (early Lutetian). The Paris Basin was still linked with the North Sea. The Ural Sea joined the Sea of Tethys with the northern ocean and separated Europe from Asia. 1 – marine areas; 2 – continental areas.

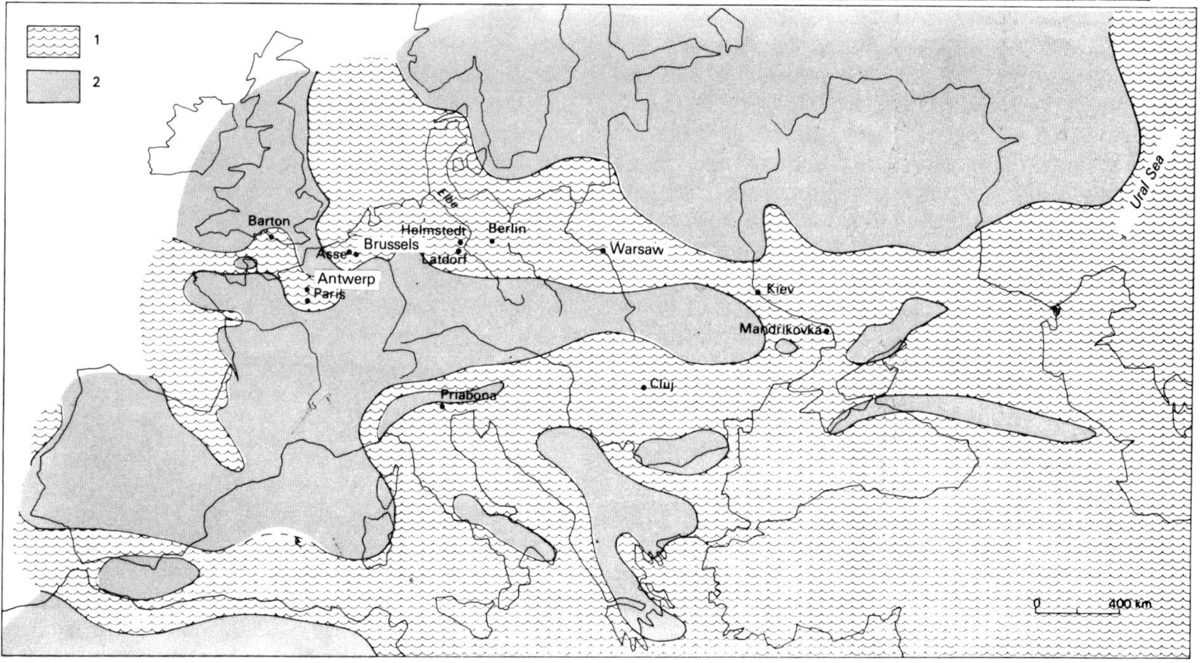

4.17 Europe in the late Eocene. The Ural Sea was still in existence and linked with the North Sea by a channel across northern Europe. The Paris and Hampshire Basins had been separated from the North Sea in the late Lutetian. However it is possible that a strait was present in Priabonian times, linking Hampshire with the North Sea, which would explain the similarity between the mollusc faunas of Lattorf, and the Barton and Middle Headon Beds.

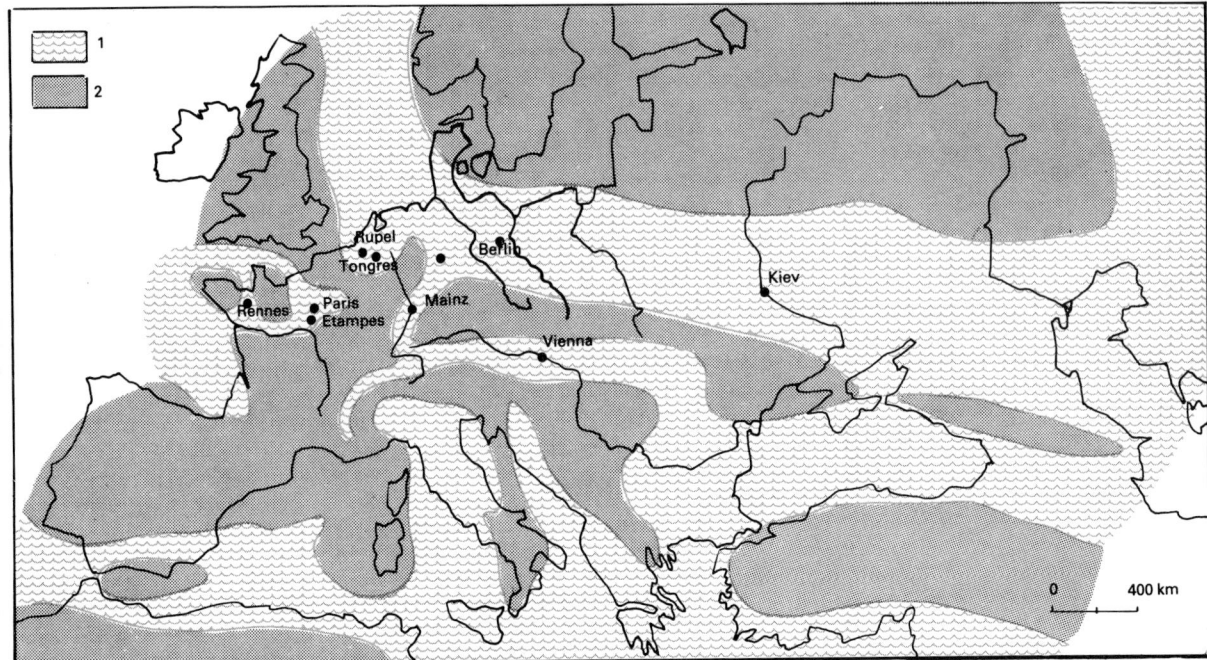

4.18 **Europe in the Oligocene.** The Ural Sea was still in existence but receded at the end of the Oligocene. The North Sea was connected to the Tethys by a major link across northern Europe and temporarily by a smaller channel through the Rhine valley, Switzerland and the Vienna basin. In France, the circum-Alpine channel had regressed northwards (see p.109) . 1 – marine area; 2 – continental area.

margins, the sediments were sandy (e.g., the "Meeressand" of the Mainz Basin) but in the axis of the graben, thick beds of grey clays were deposited, known as the *Septarienton* (septarian clay), which are the lateral equivalent of the Boom clay with septaria of the Antwerp basin (Rupelian = Stampian) (p. 114) and of the foraminiferal marls and fish beds of Alsace (p. 110).

In the Mainz Basin, the Septarian clay is covered by lagoonal beds with *Corbicula* overlain by others with *Cerithium*, the equivalents of the *Meletta* beds of Alsace. These mark the beginning of a regression, well marked in the Chattian, from the south of the graben. At the same time the sea deposited glauconitic sands in the gulf of Cassel (the type locality of the **Chattian**[1] to the north). These sands are considered by some to belong to the late Oligocene, but they have been dated at 30Ma and ROTH (1970) has found nanno-

plankton of middle Oligocene age in them. Throughout the Rhine graben, the Oligocene terminates with the deposition of lagoonal and lacustrine beds which include the *Corbicula* and *Hydrobia* limestones of the Mainz Basin. On the Polish coast of the Baltic, the sea throws up pebbles of amber, lumps of fossil resin probably derived from the Eocene. The amber often contains remarkably well-preserved insects, fish scales and leaves.

The Palaeogene seas of northern Europe were relatively permanent in Denmark and in north Germany. They extended to the east in the Dano-Montian and to the south in the late Eocene and especially in the Oligocene. A direct link between the North Sea and the Mediterranean area was established in the late Eocene (Lattorfian) across Poland and the Russian platform, as is shown by the distribution of *Nummulites germanicus*. A second connection, though a rather precarious one, was established by way of the Rhine Graben in the Stampian, but this was broken following the general regression at the end of the Oligocene (Fig. 4.18).

---

1. The Chatten, an ancient tribe of the Cassel region. The term Chattian was proposed by FUCHS, 1894, (as Chattische Stufe) and used by HAUG (1911).

II. – SOUTHERN EUROPE

In the Palaeogene, *the Mediterranean area became a region of rapidly changing palaeogeography where stratigraphy and tectonism are closely associated*. In this area, above all others, the distribution of land, sea and relief is far removed from that of the present day. Thus, in place of the western Mediterranean there was a continent or more probably, an archipelago, called the Tyrrhenian province, made up of at least two major units.

The first unit is a northern one, made up of Catalonia, the eastern Pyrenees, Languedoc, Maures, Esterel and the greater part of Corsica and Sardinia. This craton split up to form a northwestern basin in the Oligocene by the rotation of the Corsican-Sardinian massif towards the south east.

The second unit lay to the south and comprised the Betic-Rif massif (the Straits of Gibraltar did not open until the Pliocene), the Alboran sea, the Kabylie, the Mediterranean in the region of the Maghreb, the north of Sicily, the Tyrrhenian sea and the Island of Elba (Fig. 4.16). By contrast, Italy was almost entirely submerged, as were the Dinarides and the greater part of Greece, Libya, Egypt and the Middle East, to form part of the Mesogean (Tethyan) sea which spread across Turkey and Iran to reach the southern part of central and eastern Asia. The Indian peninsula was at this time still far to the south and was not yet attached to Asia.

The connection between the Tethys and the Atlantic was by way of two straits. One in the north (via the Betic Cordillera), and the other in the south (via the Rif Cordillera), flanked the Betic-Rif massif (see p. 179). The Atlantic lay over part of southern Portugal and penetrated a gulf in Aquitaine and Aragon covering the western Pyrenees. This gulf reached the Languedoc and Catalonia but did not link with the Tethyan sea. The Mistral borehole in the Gulf of Lions has shown that the Miocene rests directly on a NW-SE ridge of Palaeozoic rocks joining the south eastern part of the Pyrenees to the Maures massif. From the early Eocene (Ilerdian) this gulf was split into two approximately equal parts as a result of the uplift of the Pyrenees. In 1960, HOTTINGER and SCHAUB chose the southern part of this gulf as the stratotype of the Ilerdian (see p. 128).

The connection between the Tethyan and North Seas which had been established by way of the Ural sea late in the Cretaceous continued until the end of the Oligocene. Another connection was present in late Eocene and early Oligocene times and this linked the two by way of Poland and the Russian platform. This is the route towards the north west which was followed by *Nummulites germanicus* and the Priabonian molluscan faunas of the green sandy clays of Mandrikovka in the Ukraine. A third, rather short-lived link occurred in the Oligocene through the circum-Alpine depression, Switzerland, Alsace and the Rhine Graben (Fig. 4.18).

Two types of sea characterised the Mediterranean region in the Palaeogene. One was a shelf sea and the other occupied the deep trench or geosyncline on the edges of elongated chains (cordilleras) during their uplift.

The shelf seas, which are typically rich in nummulites, can be recognised in a number of regions such as, for example, around Verona and Vicenza in northern Italy, where the stratotype of the Priabonian has been established. The Pannonian basin in Hungary (in the Transdanubian Mountains, north west of Budapest and the Bakony Mountains, north east of Lake Balaton) and the Cluj basin (in Transylvania) (Fig. 4.17) are also of this type. The Cluj basin is the type area of the Napocian (BOMBITA, 1962), proposed for the early part of the late Eocene (see p. 131). Shelf seas covered basins in Moldavia, the Ukraine and the Crimea on the Russian platform and also occurred on the northern edge of the African craton, where they are recorded in the Libyan basin and in Egypt. Nummulitic limestones (*N. gizehensis*) were used in Egypt for the construction of the Pyramids and the Sphinx.

The seas of trench-type are characterised by the presence of rhythmically bedded sequences of sands and clays, known as *flysch*, derived from the rising mountain chains. Deposits of this type are found in the External zone of the

Western Alps and the Carpathians, the zones of Parnassos and Gavrovo and the Ionian zone in Greece and Albania, in the Dinarides and the Apennines and the Betic and Rif cordilleras.

Stratigraphy of the neritic basins of Spain, Italy and central and eastern Europe will be discussed here, but that of north Africa will be dealt with later (chapter 8, p. 179).

1. The south-Pyrenean gulf

The Tertiary deposits which today occupy the Ebro basin between the Cantabrian mountains and the Pyrenees to the north, the Iberian chain to the south and the coastal chain of Catalonia to the east were originally part of a vast Aquitaine-Aragon gulf extending from the Atlantic in the north east, and formed the landward extension of the Aturian graben (see p. 95).

This gulf was bounded to the north by the Mesozoic landmass of Aquitaine (the southern edge of the Massif Central) and to the south by the land mass of Asturias and Ebro, where the continental shelf was narrower than that on the north side. The median trench of the gulf (the site of the subsequent mountain chain) was filled initially with Maastrichtian *Globotruncana* flysch deposits which pass upwards without interruption into Danian limestones with thin-walled globigerines (*Globigerina daubjergensis*). In the Ebro region, these limestones are dolomitic and pass laterally into

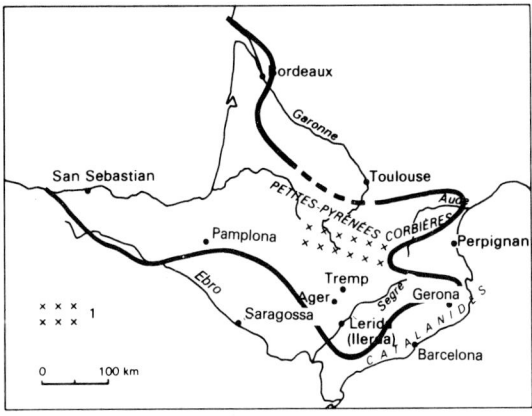

4.19 **The Pyrenean Gulf at the maximum extent of the Ilerdian transgression.** 1 – zone of uplift (after Plaziat and Kromm, original).

lacustrine limestones resembling those of the Ariège (middle Garumnian = Rognacian).

Sedimentation continued into the Montian with essentially calcareous beds containing algae and discocyclinids (*D. seunesi*) which are followed in the Thanetian (or Landenian, as MANGIN (1959) prefers) by detrital deposits (Fig. 3.48, p. 99). The axial trench, which had been largely filled before the Thanetian, reappeared and once more accumulated flysch sediments. On the southern border these were replaced by limestones containing *Alveolina primaeva* and *Globorotalia velascoensis*.

At the beginning of the Eocene, the transgression spread eastwards into Catalonia and the detrital sediments became thicker (fig. 4.19). It is this series which HOTTINGER and SCHAUB (1960) have proposed as the stratotype of the **Ilerdian** to characterise south European formations deposited in the time interval between the deposition of the Bracheux sands or Thanet sands and the deposition of the Cuise sands. The type section has been chosen in the *Ager formation* in the Tremp basin (Lerida province = Ilerda, in Latin) on the road from Tremp to Pont de Montañana. It is defined by five alveolinid biozones which from bottom to top are: *Alveolina cucumiformis, A. ellipsoidalis, A. moussoulensis, A. corbarica* and *A. trempina* (see Table III, p. 24). However, the position of the Ager formation in the general stratigraphic scale is not clear since it is intercalated between continental formations. To overcome this problem, a parastratotype at Campo in the Val d'Esera in Huesca province has been proposed (Colloque sur l'Éocene, 1968, **69**, p. 467). In this section the Ilerdian beds rest on the *Alveolina primaeva* zone of the Thanetian and are overlain by the zone of *Alveolina oblonga* and *Nummulites planulatus* of the Cuisian. Beds of Ilerdian age have now been recognized all round the Mediterranean and particularly in France in the Petites Pyrenées and the Corbières.

The Cuisian limestones and siltstones are succeeded by Lutetian beds, with large nummulites and alveolines and a varied lithology, sometimes calcareous, especially at the margins, sometimes flyschoid in the median trench. In the late Lutetian a major palaeogeographic inversion occurred: the axial zone was uplifted while the Ebro continent slowly

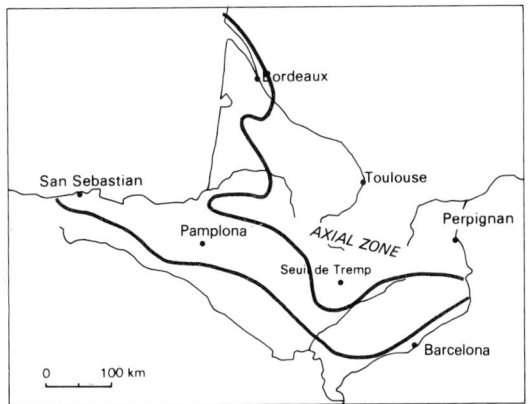

4.20 **The Pyrenean Gulf in the late Lutetian.** Note the regression in the Aquitaine Basin and the existence of a shelf in the region of Tremp (after Plaziat, 1973).

subsided (Fig. 4.20). As a result the deposition of flysch was displaced southwards in the Auversian, the transition stage between the middle and late Eocene. The Bartonian is marly and is similar to that of the cliffs of Biarritz (see p. 100). During this stage the Montserrat conglomerates were deposited on the south-eastern margin of the gulf in Catalonia. They are analogous to those of Palassou though they are younger and are derived from the Catalan cordillera.

In the Oligocene there was a general regression. In Navarra, the basal Stampian marine sandstones are succeeded by evaporite and molasse. Deposits rich in mammalian remains occur in Catalonia and the province of Lerida. Furthermore, potash deposits of similar age to those of Alsace are found in Lerida.

2. Northern Italy and Istria

Northern Italy was divided into two palaeogeographic regions in the Palaeogene. The one in the east, in the hills of Verona and Vicenza, in Istria and in the Trieste peninsula was an area of neritic sedimentation rich in large foraminiferids and planktonic organisms. Volcanic tuffs and basalts are interbedded, making stratigraphical analysis difficult. The neritic facies is also found in the extreme west of the region at Mortola, near the Italian frontier, referred to earlier (p. 106).

The other major region is essentially of

pelagic sediments (scaglia rossa, scaglia cinerea, flysch) and occupies Liguria and Lombardy. This region is the northern part of the Apennines and had a highly complex history during the Palaeogene. It is not considered further here.[1]

a) **Vicenza, Verona and Istria**

In Istria, as in Aquitaine, there is a continuous passage from the Maastrichian to the Danian. The Tertiary era begins with the "Lower limestones with foraminiferids" attributed to the Danian. This is followed by the Cosina beds of marly limestone and the "Upper limestones with foraminiferids", which probably correspond to the Ilerdian. STACHE (1872) gave the name *"Liburnische stufe"* **(Liburnian)** (from Liburnia, a region in the Quarnaro gulf, south of Trieste) to this Danian-Ilerdian series. BIGNOT (1970) has shown that this term, in fact, spans the sequence from the Senonian to the upper Lutetian. It is, therefore, preferable not to use it.

---

1. For details consult Azzaroli & Cita, Geologia stratigrafica. La Goliardica, Milan.

4.21 **Map showing Italian type-localities** (Palaeogene-Neogene-Quaternary). Spilecco and Monte Bolca are about 20km west of Priabona.

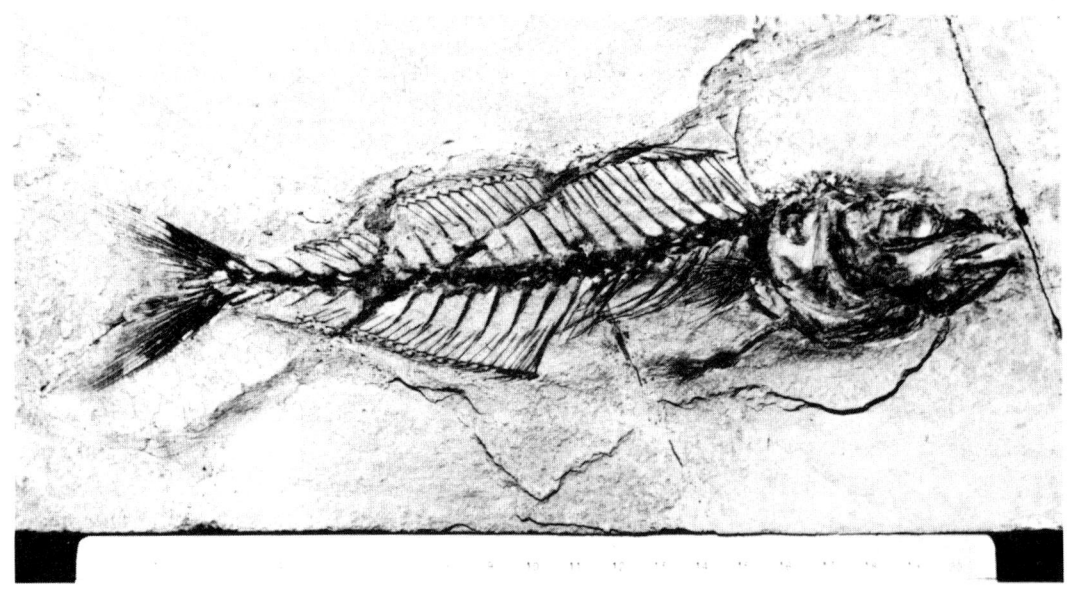

4.22 **Seriola prisca,** Lower Eocene, Monte Bolca (x ½) (photo: Blot).

4.23 **Mene rhombea,** Lower Eocene, Monte Bolca (x ¼) (photo: Blot).

Near Verona, the Palaeogene begins with the Spilecco beds (near Bolca) for which FABIANI (1912) created the stage name, **Spileccian** (Fig. 4.21). These are fossiliferous limestones with echinoids and rhychonellids which rest on the scaglia rossa of the upper Cretaceous and are overlain by Lutetian limestones. Since the series is incomplete, only the presence of *Globorotalia velascoensis* and of *Discoaster multiradiatus* indicate the presence of the Ilerdian stage.

The **Monte Postale** series (also near Bolca) begins in the Ypresian with limestones containing *N. atacicus*. These beds are therefore Ilerdian in age, rather than Lutetian as is generally supposed. The limestones are followed by thin-bedded, marly limestones which have yielded a remarkable fish fauna (Figs. 4.22 to 4.25). These are, in turn, overlain by plant beds and then by limestones with molluscs and large nummulites of mid-Lutetian age (*N. millecaput, N. gizehensis*). Limestones with *N. brongniarti* associated with volcanic tuffs outcrop about 15km to the south at Ronca and are overlain by a regressive sequence of lignites with freshwater molluscs. These units were referred by BOUSSAC to the Auversian, but are now considered to be late Lutetian.

130

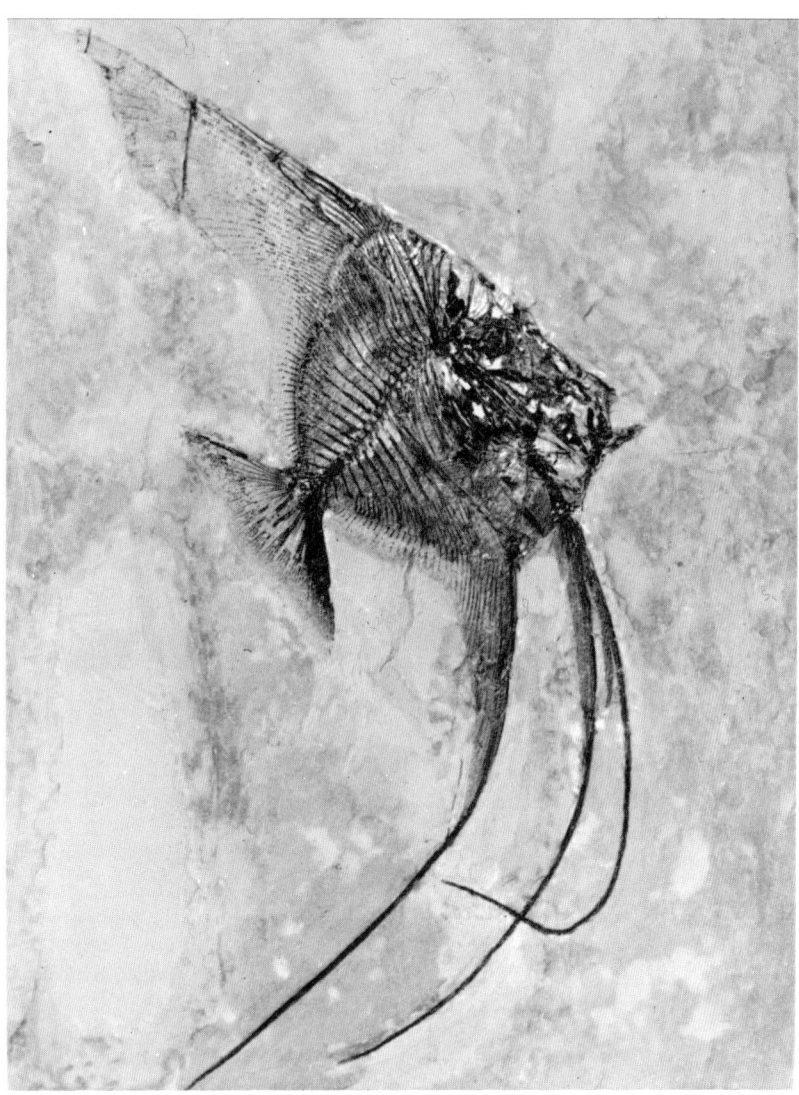

4.24 **Ceratoichthys pinnati-formis,** Lower Eocene, Monte Bolca (x ¼) (photo: Blot).

The upper Eocene outcrops around Priabona, also near Vicenza, where MUNIER-CHALMAS and DE LAPPARENT defined the **Priabonian** stage in 1893. In the Venetian dialect "priabona" means "good stone". However, the stone to which the term is applied overlies the Priabonian marls and is, in fact, of Oligocene age.

In their definition of the Priabonian MUNIER-CHALMAS and DE LAPPARENT subdivided the beds into three formations, which from bottom to top, are:

1) Granella limestone with *Cerithium diaboli*;
2) Limestones with discocyclinids – the

Priabona group;
3) Brendola marls with bryozoans.

It should be noted that in the neritic facies, whilst *Nummulites fabianii* is abundant, planktonic micro-organisms are rare. At the same time bivalves (*Anomia, Pecten, Spondylus*), gastropods (*Pleurotomaria, Turritella*) and echinoids (*Clypeaster, Schizaster*) are abundant.

The section is incomplete at the base (which is why Bombita defined the Napocian (see p. 127)) and at the top it has not been possible to identify the boundary with the Oligocene. For these reasons, an international committee of

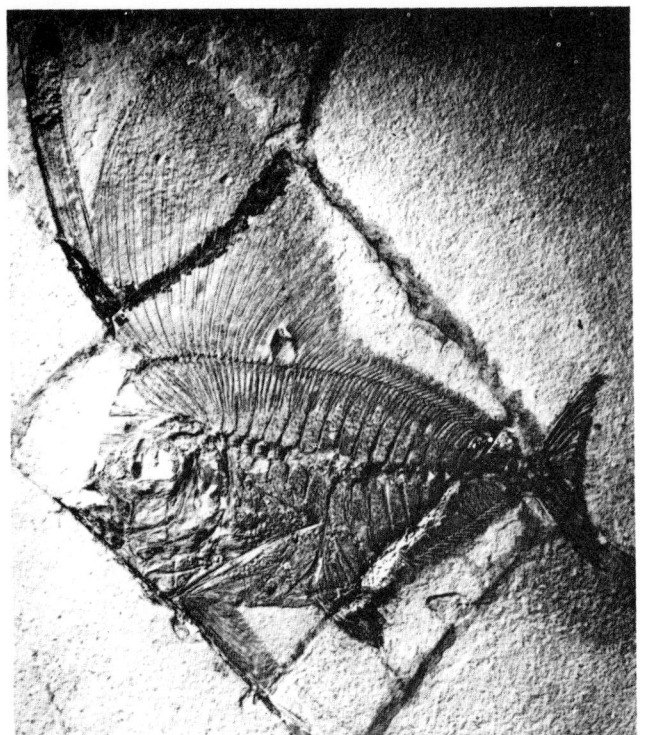

4.25 **Exelia velifer,** Lower and Middle Eocene, Monte Bolca (x ½) (photo: Blot). These remarkably well-preserved fishes occur in a calcareous dolomitic mud laid down in a littoral lagoon. The fish were killed by poisonous gases emitted during a volcanic eruption.

geologists, under the chairmanship of M. B. CITA has examined the sections in the Priabona area in order to complete the sequence of the stratotype before giving the Priabonian stage the status of full equivalence with other stages of the upper Eocene. The principal parastratotypes are represented by sections at Brendola, Mossano and Possagno, which are described in the "Colloque sur l'Éocene", 1968.

The Oligocene limestones (with *Nummulites intermedius* and *N. vascus*) forming the hills of Montecchio Maggiore and Castelgomberto, are well-known for their coral reefs and belong to a single sedimentary cycle. It is not, however, possible to distinguish the Lattorfian-Tongrian-Sannoisian or Chattian stages and the series ends with the Aquitanian (early Miocene) Schio beds which contain large lepidocyclinids (*L. elephantina*), molluscs and echinoids (*Scutella, Clypeaster*), marking the base of the Miocene.

b) Liguria and western Lombardy

North of Milan, *the Paderno d'Adda section* shows a sequence of beds from the Cretaceous to the Cuisian. The facies is pelagic in the lower part (the scaglia rossa) and is flyschoid in the upper. The planktonic foraminiferal zones range from that of *Globorotalia pseudo-bulloides* (Danian) to *Hantkenina aragonensis* (Cuisian). The nannoplankton zones cover the same interval, which thus includes the sequence Danian-Thanetian-Ilerdian-Cuisian.

East of Genoa, the Oligocene transgression has been studied in minute detail by LORENZ (1962) who demonstrated the presence of lower and middle Stampian beds with *Nummulites intermedius, N. vascus* and *N. bouillei;* upper Stampian beds with nummulites and lepidocyclinids (*Nephrolepidina* and *Eulepidina*); a zone transitional to the Miocene with *Miogypsinoides*; and finally, true Aquitanian beds with *Miogypsina*. The Chattian cannot be identified in this section.

In the hills of Turin an east-west anticline exposes a Priabonian limestone overlain by an Oligocene-Miocene sequence which begins with a conglomerate overlain by two marl beds with pteropods, separated by a sandy horizon. The **Bormidian** stage was created in 1865 by PARETO to include this series where it outcrops in the Bormida valley in Liguria. However, these deposits, although linked with the Oligocene transgression, are heterochronous (LORENZ, 1965), ranging from early Oligocene to early Miocene in age. Thus the stage extends across a series boundary and so should not be adopted.

To summarize, northern Italy shows an almost complete sequence of Palaeogene beds, with passage beds from the Cretaceous and into the Miocene. The precise identification of the Lattorfian-Sannoisian and of the Chattian stages however, is not possible. Rigorous correlation within this area is made difficult by the lack of continuous exposures. The area has one advantage over the Paris Basin in so far as the presence of planktonic organisms simplifies correlation over long distances. This has enabled Italian geologists to set up a stratotype for the late Eocene, the Priabonian, which is much better defined than the Bartonian of the northern basins.

3. The Russian platform

During the Palaeogene, the greater part of eastern Europe was above sea-level. Detritus from this platform was carried southwards and deposited in the Ukraine, in the neighbourhood of the Black Sea and in the Caspian syncline (Fig. 4.26).

The palaeogeography of the Russian platform is closely linked to the Alpine orogeny. Following the regression at the end of the Cretaceous, the sea returned in the Palaeocene and reached its maximum extension in the late Eocene, when communication was established between the North Sea and the Ural Sea across northern Europe (Fig. 4.17). During the Alpine movements, the Oligocene sea was in regression. The new invasion which ushered in the Neogene developed in the area of the lagoons and lakes around the Paratethys.

In the **Palaeocene** (Fig. 4.26), limestones and marls were formed where the platform was most deeply submerged, while sedimentation was more detrital or siliceous (opoka = cherty sandstone) near the northern shoreline. In the Dano-Montian, communication had been established across the Ukraine between the Russian platform, Poland and the North Sea (Fig. 4.14, p. 124), but this link was broken in the Thanetian (Fig. 4.15).

During the **Eocene** the sea transgressed towards the north and the north-west. In the late Eocene it reached Poland and north Germany (the Lattorfian sea with *Nummulites germanicus*). The sediments deposited were essentially marly except in the northern margins where there are sandy shelf-deposits (Mandrikovka, near Dnepropetrovsk) (Fig. 4.27).

The **Oligocene** sea was less extensive in the south and south-east of the basin, where marine silty beds were formed, while lacustrine sediments were deposited in the Ukraine.

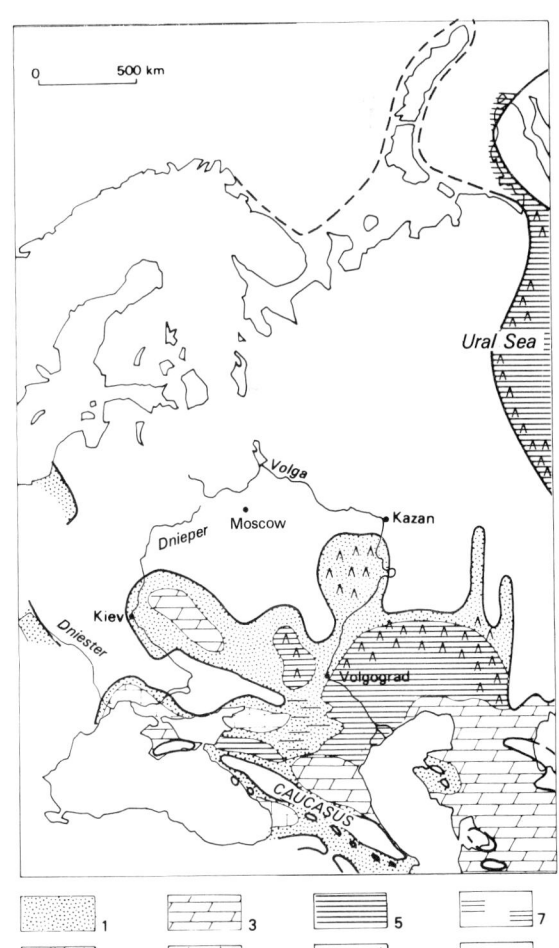

0   500 km

4.26 **The Russian platform in the late Palaeocene** (after Nemkov, original). 1 – sand; 2 – clayey sand; 3 – marl; 4 – limestone; 5 – clay; 6 – friable sandstone; 7 – diatomite; 8 – gypsum and anhydrite.

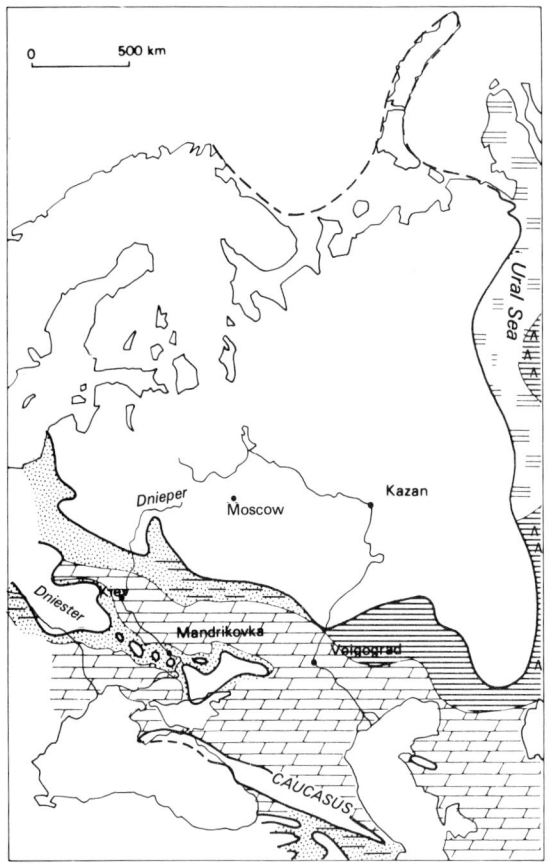

4.27 The Russian platform in the late Eocene (after Nemkov, original) for key, see Fig. 4.26).

The thickness of Palaeogene beds reaches 1000m in the Caspian synclinal basin, but is less than 300m in the Ukraine. Except in the Dano-Montian and in the late Eocene, the western part of the Russian platform was continental. The southern border of the platform was submerged by the sea in the Miocene (Fig. 4.28) but the seas advanced further and became brackish in the Pliocene. As will be shown later (p. 174), the southern part of Russia eastwards to the Aral Sea became part of the Paratethyan region during the Neogene.

The southern side of the Crimean peninsula exposes a long sequence of Palaeogene formations. These are best seen in a series of south-facing cuestas on the northern flank of an anticline, the southern limb of which sank into the Black Sea as a result of a series of Pliocene

faults with a total throw of 2000m. The coastal chain of hills near Yalta is close to the axis of this anticline (Fig. 4.29).

Following a brief continental episode at the end of the Maastrichtian, the Danian began with the deposition of glauconitic sands overlain by 30 to 40m of limestones and marly limestones. These pass imperceptibly into the Montian (**Inkermanian**) stage with *Globigerina daubjergensis*, a compact saccharoidal limestone about 20 to 30m in thickness. The Dano-Montian forms a limestone plateau limited to the south by a cuesta and by cliffs along the valley sides.

The Thanetian (**Kachian** stage, Table IX) is represented by 30m of grey-blue glauconitic marl which increases in thickness to about 300m in the centre of the Crimea. These marls contain molluscs similar to those of western

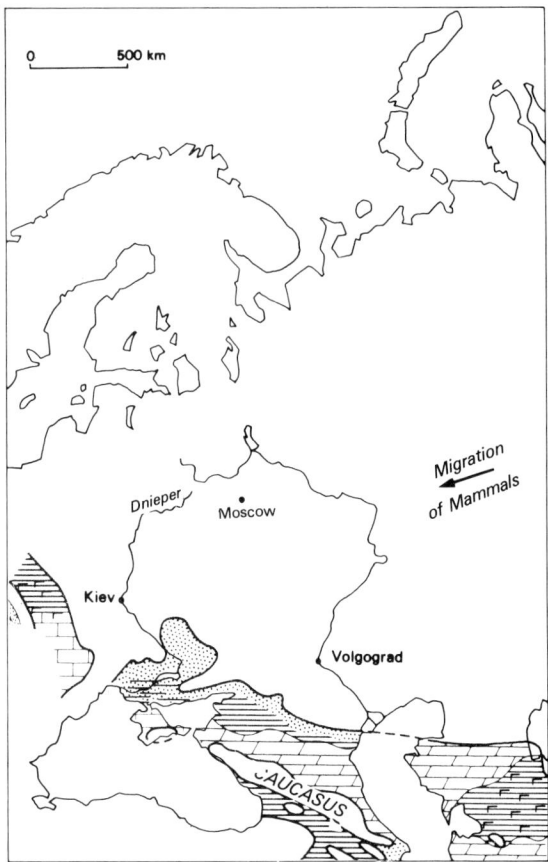

4.28 The Russian platform in the Miocene (after Nemkov, original) (for key, see Fig. 4.26).

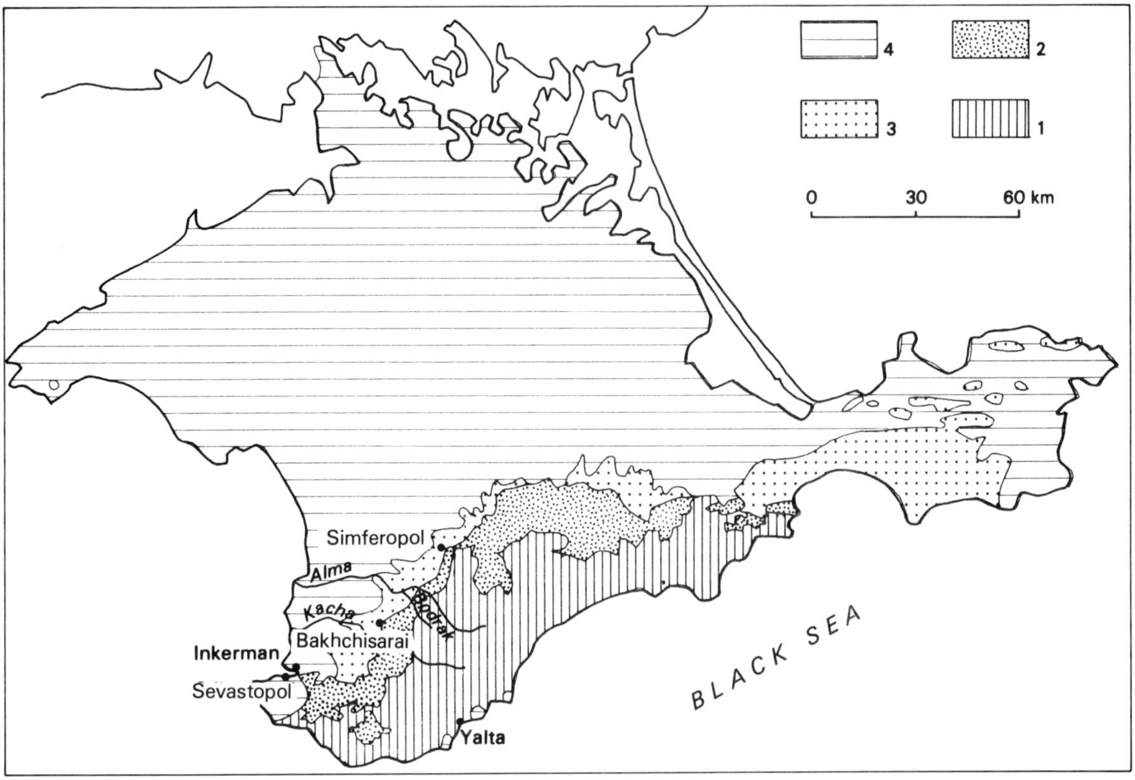

4.29 **Geological map of the Crimea.** 1 – Triassic and Jurassic; 2 – Cretaceous; 3 – Palaeogene; 4 – Neogene.

4.30 **Bodrakian chalk (Upper Lutetian) in the Alma Valley, Crimea** (photo: Pomerol).

Table IX
Correlation between Palaeogene stages of Western Europe and the Crimea

WESTERN EUROPE		CRIMEA
Oligocene		Maikop Series
	Priabonian	Almian
	Bartonian, Middle and Late Lutetian	Bodrakian
Eocene	Early Lutetian / Late Cuisian	Simferopolian
	Early Cuisian / Ilerdian	Bakhchisaraian
Palaeocene	Thanetian / Montian / Danian	Kachian / Inkermanian / Danian
	Maastrichtian	Maastrichtian

Russian authors have traditionally correlated this stage with the Lutetian and the Auversian of the Paris Basin, but recent studies have indicated that the Simferopolian is older than had been supposed. The Simferopolian stage is composed of limestones (45 to 60m thick in the Crimean mountains) whose outcrops are bounded by cuestas giving the area an appearance reminiscent of the limestone plateaux of the Causses of southern France. The area is one in which karstic features are well developed and the many caves gave shelter to prehistoric men. The Simferopolian contains abundant assilinids and large nummulites and is similar to those facies of northern Italy in which *Assilina exponens* and *Nummulites distans* occur.

In the middle Eocene, it seems likely that there were two Tethyan faunal regions; one to the north and the other to the south. The Crimea belonged to the northern region while the Caucasus together with North Africa belonged to the southern area. However, the facies with large nummulites seems to have commenced earlier in the Crimea (in the middle and late Cuisian) than in Italy where it is typical of the middle Eocene.

In the Crimea, the middle and late Lutetian and the Bartonian are represented by the **Bodrakian** which consists of a chalky limestone (15m) followed by bituminous marls (40m). The latest Eocene is represented by the **Almian,** a sequence of soft, greenish marls and argillaceous limestones, rich in foraminiferids, about 100m thick. This stage transgressed towards the north-west and as a result established in the Lattorfian a direct link with the North Sea.

The Oligocene contains an impoverished fauna corresponding to a phase of regression and is represented by the **Maikop** series. This is a varied sequence of marl, sand and clay beds which is overlain by the formations of the Neogene transgression.

Europe such as *Arctica morrisi*. The Palaeocene ended with the uplift of the area.

The lower Eocene (**Bakhchisaraian**) is everywhere transgressive and slightly discordant on the Palaeocene. It is argillaceous at the base and becomes more calcareous and marly at the top. Nummulites (including *N. planulatus*) and assilinids appear for the first time and, together with the micro-fauna (*Globorotalia subbotinae* etc.), enable the Bakhchisaraian to be correlated with the early and middle part of the Ypresian of western Europe, that is, with the late Ilerdian and early Cuisian.

The late Cuisian and early Lutetian appear to be represented by the **Simferopolian** stage.

General characteristics of the Neogene

5

The Neogene, the second period of the Cenozoic, was erected by HOERNES (1853). His justification for this division was the appearance of new forms of life which have continued to evolve to the present day. It thus encompasses the Miocene and Pliocene epochs and should by original definition include the Quaternary. With a duration of only about 21Ma between the end of the Oligocene (23Ma) and the beginning of the Quaternary (2Ma), the Neogene is the shortest of all the geological periods if, as is customary, the Quaternary is excluded.

I. – LIMITS AND SUB-DIVISIONS

The Neogene is sub-divided into two epochs: the Miocene and the Pliocene, both defined by LYELL in 1833.

1. The Miocene (Table X)

The lower limit of the Miocene coincides with the upper limit of the Oligocene, discussed earlier (p. 28). The molluscan fauna which is associated with the Miocene transgression resembles that of the present day. On the basis of the foraminiferids, it has generally been accepted that the *Oligocene ended with the extinction of the nummulites*, while the *Miocene began with the appearance of Miogypsina and Globigerinoides*. This implies that the boundary occurs 1 or 2Ma before the deposition of the lowest beds of the type Aquitanian, which are later than the first appearance of *Globigerinoides*. As we have seen, in Aquitaine the Escornebéou beds are still Oligocene, although they contain the first *Globigerinoides* (p. 102). Thus a precise boundary is difficult to trace and even if one supposes that the evolution of a faunistic assemblage is very rapid, there is often a zone of transition between two horizons whose stratigraphic position is unanimously accepted.

On the other hand, it is difficult to apply a radiometric date to a boundary which cannot be fixed with precision stratigraphically whilst the question of the Chattian has not been resolved.

The *Miocene-Pliocene boundary* is marked in terms of planktonic foraminiferids by the rapid increase in abundance of *Sphaeroidinellopsis* and by the appearance of *Globorotalia margaritae*. In southern Europe it is essentially a change in palaeogeography which marks the boundary between the two epochs. The Miocene ended with the *Messinian salinity crisis* and its associated evaporite deposits, which extend over the greater part of the dried-up Mediterranean basin. This was followed by a general transgression accompanied by folding which completely changed the distribution of land and sea and marked the beginning of the Pliocene which is dated at about 5Ma.

The Miocene is traditionally subdivided into three parts which Dutch geologists (DROOGER and MARKS, 1971) and those of Milan (CITA, *et al.*, 1971) have suggested should be called super-stages. However, the division of the Miocene into two groups, separated by a boundary based on the first appearance of the very characteristic planktonic foraminiferid

Table X
Correlation between the principal biozones and stages of the Miocene
(see also Table XII, p. 164)

Epochs	Western European stages	Planktonic foraminiferids	Nannoplankton	North American continental stages	Californian marine stages
PLIOCENE — 5.3 MA —	Tabianian or Zanclean	Globorotalia margaritae	Ceratolithus tricorniculatus	Hemphillian	Repettian
UPPER MIOCENE	Messinian Tortonian	Globorotalia menardii	Discoaster quinqueramus Discoaster calcaris	Clarendonian	Delmontian Mohnian
MIDDLE MIOCENE	Serravallian ("Helvetian") Langhian	Globigerina nepenthes Globorotalia fohsi Orbulina suturalis	Discoaster kugleri Discoaster exilis Sphenolithus heteromorphus	Barstovian Hemingfordian	Luisian Relizian
EARLY MIOCENE — 23 MA —	Burdigalian Aquitanian	Globigerinita dissimilis Globigerinoides primordius	Discoaster druggi Triquetrorhabd- ulus carinatus	Arikareean	Saucesian
OLIGOCENE	Chattian	Globorotalia kugleri	Sphenolithus ciperoensis	Orellan	Zemorrian

Orbulina, can also be justified (ANGLADA, *pers comm.*). The three-fold division is adopted here.

The **early Miocene** includes two classical stages which were first defined in south-west France; the **Aquitanian** and the **Burdigalian** (see p. 160). VIGNEAUX *et al.* (1954) proposed that these stages should be combined in the Girondian super-stage, based on borehole information in an offshore region. However, this inaccessible stratotype and the uncertainty of its boundaries has led many authors to reject this suggestion and to regard the Girondian as synonymous with the early Miocene. This is not, however, strictly true for, as has been shown, the sequences in Aquitaine contain both *Miogypsinoides* and infra-Aquitanian *Globigerinoides* which might be held to justify the use locally of the term Chattian or Bormidian (see p. 132).

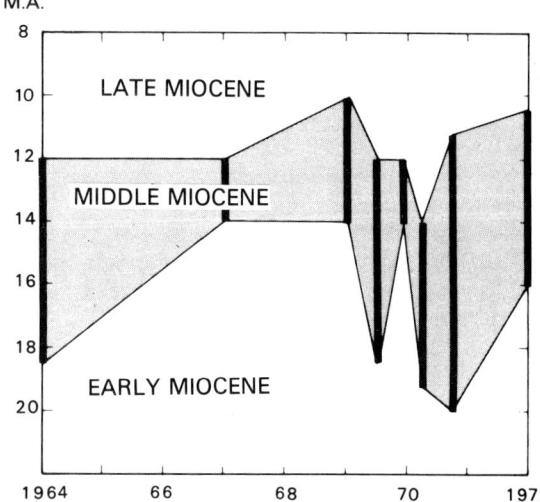

M.A.

5.1 **Limits of the middle Miocene as proposed at International conferences between 1964 and 1972** (after *Geotimes*).

The **middle Miocene** begins with the appearance of *Praeorbulina* and corresponds only very approximately to the classical **Helvetian** stage (see p. 162). Because of the uncertainty of the limits of the Helvetian stage, it has been suggested that the **Langhian-Serravallian** succession in Piedmont, which is rich in planktonic organisms, should be substituted for it. However, the boundaries of the middle Miocene are still undecided (Fig. 5.1).

The **late Miocene** is marked by the incoming of *Globigerina nepenthes* and *Globorotalia menardii*. The latter is very abundant and is easily identified. At the same time, the land vertebrate *Hipparion*, which originated in North America, arrived in western Europe by way of Asia.

The late Miocene is sub-divided into two stages: the **Tortonian** below and the **Messinian** above. Spanish geologists (e.g. PERCONIG, 1966) would substitute the name **Andalusian** for the upper stage. This is based on a section in the gulf of Guadalquivir, which is the only part of southern Europe where Miocene beds remain marine right up to the Mio-Pliocene boundary.

2. The Pliocene (Table XI)

As with the Oligocene, this period is little more important than the average stage. Its duration of about 3Ma (from 5Ma to 2Ma) is less than that of most of the stages of the Mesozoic or Palaeogene.

The subdivisions of the Pliocene are based on Italian sequences and are difficult to recognize elsewhere. The lowest horizons contain *Sphaeroidinellopsis*, and *Globorotalia margaritae*, which replaces *G. menardii*. In the Mediterranean region this boundary is especially easy to recognize because the Messinian evaporites are abruptly succeeded by thick marls with a shallow marine fauna. In contrast, the upper boundary of the Pliocene is still a matter of controversy since it is also the lower boundary of a somewhat singular period, the Quaternary. It can be defined by one or more fixed events, for example, a *reversal of the earth's magnetic field*, a *radiometric age*, the *base of a characteristic marine horizon*, or it may be defined by events originally thought to be fixed, but which subsequent research has

5.2 **Megacardita (Cardita) jouanneti** (x 1) Helvetian, middle Miocene, Salles, Gironde (photo: Fay).

139

Table XI
Subdivisions of the Pliocene

T: Age in millions of years. M: Magnetic Stratigraphy
The Olduvai event (a period of normal polarity in the Matuyama inversion at 1.9Ma) corresponds very closely the Plio-Pleistocene boundary
The Mammoth event (a period of inverse polarity in the Gauss normal at 3Ma) coincides with the extinction of *Sphaeroidinellopsis* (see also Table XIII, p. 226)

T	M	(polarity)	Epochs	Stages: marine	Stages: continental	Planktonic foraminiferids	Nannoplankton
2	Olduvai (1.9)	Matuyama inv.	EARLY PLEISTO-CENE	Calabrian	Late Villafranchian	*Globorotalia truncatulinoides*	*Pseudoemilia lacunosa* end of Discoasters
2,5	Mammoth	Gauss norm.	PLIOCENE	Plaisancian or Astian	Early Villafranchian	*Globorotalia inflata* *Globorotalia crassaformis* *Globigerinoides obliquus extremus* extinction of *Sphaeroidinellopsis*	*Discoaster brouweri*
3							*Discoaster surculus*
3,3		Gilbert inv.		Tabianian or Zanclean	Ruscinian	*Globorotalia margaritae* *Sphaeroidinellopsis* at the base in the Mediterranean region	*Ceratolithus rugosus* *Ceratolithus tricorniculatus*
5,3		Period V (normal)	LATE MIOCENE	Messinian or Andalusian	Turolian or Pikermian	*Globorotalia menardii*	*Discoaster quinqueramus*

shown to fluctuate, such as the *appearance of Man* or the *beginning of glaciations.*.

Unfortunately, the last two events were used to justify the creation of a "period" of a mere two million years duration which is less than half the length of most stages. Twenty years ago, the Pleistocene period was thought to have existed for only half-a-million years, but the discovery of hominid remains of considerably greater age has pushed the Plio-Pleistocene boundary further and further back in time. It is now believed that the first hominids appeared about four million years ago and that the first glaciations, albeit polar ones, commenced nine million years ago. Some authors believe that the Antarctic was glaciated 50 million years ago. Is it necessary therefore, to push the Plio-Pleistocene boundary still further back in time? Many geologists now accept that the appearance of Man and the occurrence of glaciations are not satisfactory markers for the beginning of the Pleistocene and have agreed to place this boundary at the *beginning of the marine Calabrian stage* (cf. Table XIII, p. 226). This stage began with a transgression which is characterised in the Mediterranean by the appearance of a "cold fauna" with *Arctica islandica* and *Hyalinea* (*Anomalina*) *balthica* (p. 224). This boundary corresponds to a radiometric age of about 2Ma which is also the age of the Olduvai gorge occupations and a brief return to normal magnetic polarity in the long Matuyama period of reversal. Palaeontologically, it corresponds with the *extinction of the discoasters* (nannoplankton) and the appearance of *Hyalinea balthica* (benthonic foraminifer) and *Globorotalia truncatulinoides* (cf. Table XI).

II. – PALAEONTOLOGY

1. Introduction

As in the Palaeogene, *the mammals and the micro-organisms are the principal stratigraphic markers.* Nevertheless, the other groups are of considerable importance particularly in neritic, lagoonal and lacustrine environments. In the Neogene, the climate and palaeogeography were less uniform than in previous periods and palaeoecology becomes an important aspect of the palaeontology.

This is well-illustrated, for example, in the associations of molluscs (*Ostrea crassissima*, Fig. 6.3, *Murex turonensis*, Fig. 6.4, *Cardita jouanneti*, Fig. 5.2) which were abundant in the epicontinental seas of Touraine and the Rhone Basin. The Pectinacea, including the genera *Pecten, Flabellipecten, Amussium* and *Chlamys* (Fig. 5.3), were used by DEMARCQ (1970) to determine the detailed stratigraphical succession of the Rhone Valley. DAVID *et al.* (1967) have successfully used bryozoans (branched cyclostomes in particular) as palaeogeographic indicators of the nature of the substrate, water depth, salinity and water temperature.

In shallow seas molluscs are found in association with regular (*Echinus*) and irregular

5.3 **Chlamys northamptoni** (x ½) Burdigalian, lower Miocene, Rhone Valley (photo: Demarcq).

echinoids (*Echinolampas, Clypeaster*, Fig. 5.4, *Scutella*, Fig. 5.5). A nautilid cephalopod, *Aturia aturi* (with a tightly coiled shell) is found in a muddy facies (Schlier) of the Miocene.

The lagoons of eastern and central Europe were the domain of Cerithiidae, *Congeria* and *Limnocardium*. In the Pliocene, the lagoons were replaced by lakes in which the *Dreissenidae*, the *Unionidae* and especially the *Viviparidae* evolved (Fig. 7.11, p. 175).

Amongst the lake basins, that of Oeningen (near Lake Constance in Switzerland) includes a late Miocene molasse which has yielded a very

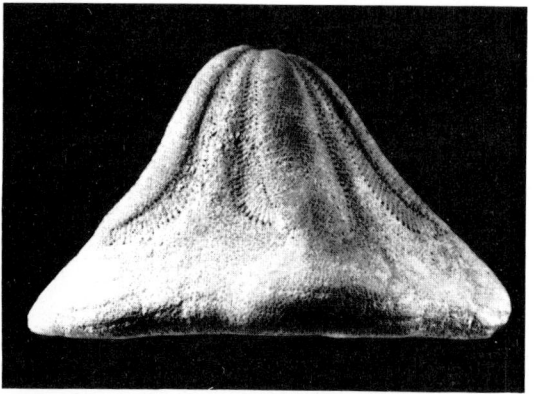

5.4 **Clypeaster portentosus,** side view (x ½) Burdigalian, lower Miocene, Malta (photo: Fay).

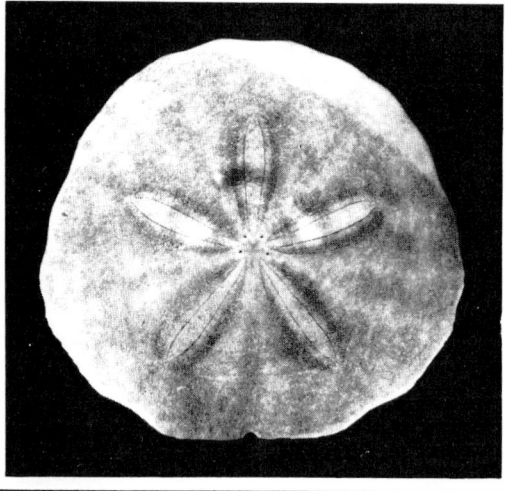

5.5 **Scutella leognanensis,** dorsal view (x ³/₅) Burdigalian, lower Miocene, Léognan, Gironde (photo: Fay).

rich flora of at least 475 species. These include *Sequoia, Taxodium*, cypress and many deciduous trees, both temperate (poplar, willow, birch, oak) and tropical (laurel, banana, avocado, camphor, palms). The climate was probably sub-tropical and humid with a mean temperature of 18 to 20°C, resembling that of Madeira and the Azores at the present time. The fauna was no less rich. BERTIN (1939) has described it thus: "In the prairies of grasses and legumes were herds of mastodons, dinotheriums and anchitheriums. The oak forests were frequented by pigs and anthracotheriums. Tapir and rhinoceros wallowed in the marshlands. Squirrels and monkeys climbed the trees. Beavers built their dams in the still waters. A rich fauna of insects supplied the table of the insect-eaters. For the hyenas, civets and cats (*Machairodus*), the woods and prairies provided an abundance of prey.

The aquatic fauna is no less interesting with crocodiles, turtles, salamanders and a great variety of fishes, including pike, perch, roach, tench, eels and several tropical forms."

In the Pliocene, the temperature fell slightly over most of the world. In France, the Meximieux (Ain) flora indicates a reduction in temperature of 4-5°C in comparison to that of Oeningen.

2. Mammals (see Table V, p. 36)

The mammals, with the exception of Man, reached their zenith in the Neogene, when the land was covered by prairies. Then followed a slow decline which commenced at the end of the Pliocene and was accentuated in the Quaternary.

The diversification and the increase in size of the proboscidians were remarkable. The earliest Palaeogene forms were *Moeritherium*, followed by *Palaeomastodon*. These were succeeded in the Miocene by mastodons with four tusks (*Tetrabelodon*, Fig. 5.6) and by *Dinotherium* (Fig. 5.7) which was 5m to the withers and had tusks in the lower jaw which curved backwards. This animal lived on in Africa until the Pleistocene.

In the Pliocene, large mastodons with two tusks and a well developed trunk appeared. The

5.6 **Mastodon:** Complete skeleton, about 2.5m high, of *Tetrabelodon* (*Mastodon*) *angustidens* from the Miocene of Sansan (Gers), France. A proboscidian with four well-developed tusks. The true Mastodons of the Pliocene possessed only two upper tusks (Fig. 5.8). They survived until late Villafranchian (Quaternary) times and are the ancestors of the elephants (photo: Serrette).

lower incisors (the lower tusks of the Miocene mastodons) disappeared with the reduction in size of the lower jaw. The teeth were still bunodont (i.e. with conical tubercles) and later became lophodont (in which the tubercles become ridge-like). The selenodont type (with crescentic ridges) and the hypsodont (high crowned) type appeared in *Stegodon* in the late Pliocene, which heralded the coming of the elephant in the Villafranchian (Figs. 5.8, 12.12 and 12.13). The *evolution of the horse* (Fig. 5.9) showed a gradual increase in size while the lateral digits of the foot slowly atrophied. Simultaneously, the teeth became hypsodont. The early Miocene horse (*Merychippus*) and its descendant *Hipparion* (Figs. 5.10) appeared in America and migrated to Europe about 13 million years ago (GABUNIA, 1968). *Pliohippus*,

a small animal about the size of a pony and having a single functional digit on each foot, appeared in the Pliocene and was followed by *Equus* in the Pleistocene. In Europe, where the sequence of the *Equidae* is discontinuous, *Hipparion* was preceded in the early Miocene by *Anchitherium* which was descended from the American Oligocene horse *Miohippus*.

The Vallesian and Turolian (ex-Pikermian) stages of the late Miocene were based essentially on the evolutionary stages of the horse (see p. 169 and Table XII, p. 164).

From the Miocene onwards, the carnivores are represented either by present day genera (dog, cat), by intermediate types (*Amphicyon*) or by extremely specialised "end of series" forms which are now extinct. Among the latter, the Mio-Pliocene Felidae, such as *Machairodus*

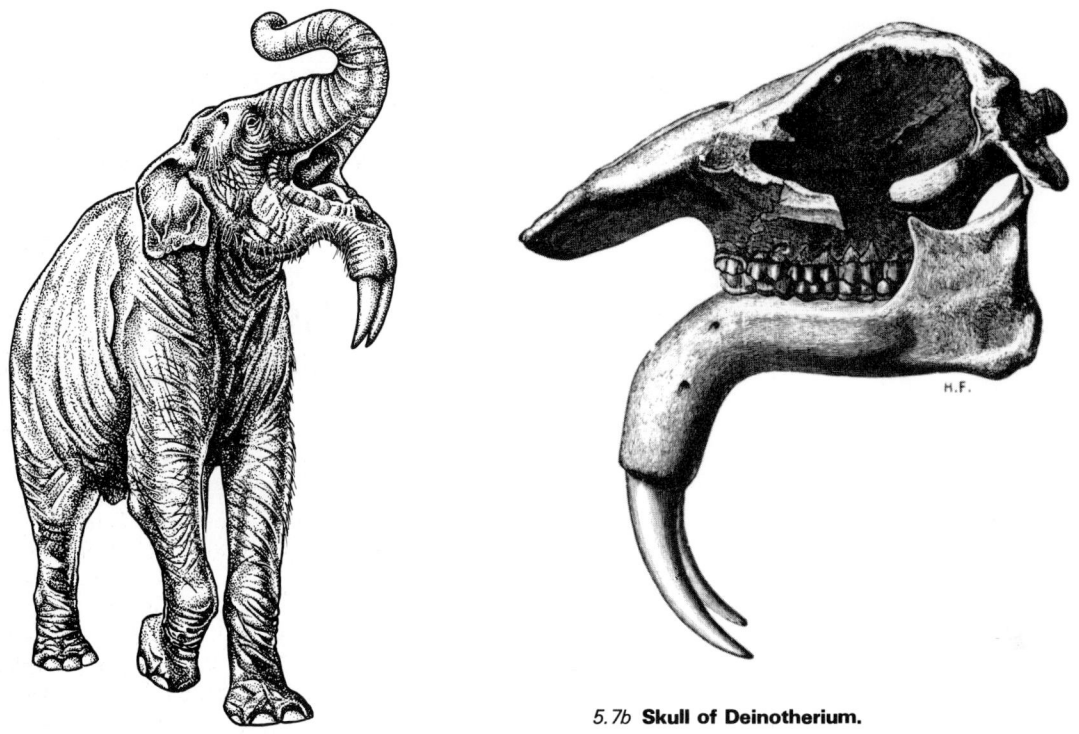

5.7b Skull of Deinotherium.

5.7 **Deinotherium giganteum** (from *deinus*, terrible and *ther*, savage animal). This was an aberrant proboscidian which lived from the Miocene to the Pleistocene. Like *Mastodon* it became extinct in the late Villafranchian. It is thought that its tusks (see Fig. 5.76) were used for digging. The dimensions of the animal were: height – 5m, length – 6.5m) (after Augusta and Burian – by permission of *Encyclopaedia Universalis*).

in Eurasia and *Smilodon* (Fig. 5.11) in North America had enormous upper canine teeth.

The primitive artiodactyls were succeeded by ruminants either with hollow horns (antelopes, oxen) or with solid horns (deer, giraffe), which were accompanied by hippopotami and pigs. The first rhinocerids appeared in the Oligo-Miocene (Fig. 5.12), but a true rhinoceros, characterised by a single horn, (*Rhinoceros megarhinus*) occurred in the Pliocene. Monkeys with tails (*Mesopithecus*) and monkeys without tails (*Pliopithecus, Dryopithecus*) also appeared for the first time in the Miocene.

At the end of the Pliocene and the beginning of the Quaternary many large mammals became extinct, though the genera *Leptobos, Equus* and *Elephas*, which had appeared in the lower Villafranchian at the end of the Pliocene, are still living.

The late Miocene saw, too, the beginning of the ascent of Man, with the appearance of *Oreopithecus*, to be followed in the Pliocene by *Australopithecus*. These early ancestors of man lived about 4 million years ago and were discovered near Lake Rudolph in Kenya and in the Omo valley in Ethiopia (Fig. 12.18, p. 242).

3. Micro-organisms (Figs. 5.13 and 5.14)

The comments made about the micro-organisms of the Palaeogene (see pp. 40 to 54) are equally valid for the Neogene, but certain groups, such as the nummulites and the alveolinids were restricted to the Palaeogene, while the miogypsinids replaced the lepidocyclinids (Fig. 2.2, p. 30).

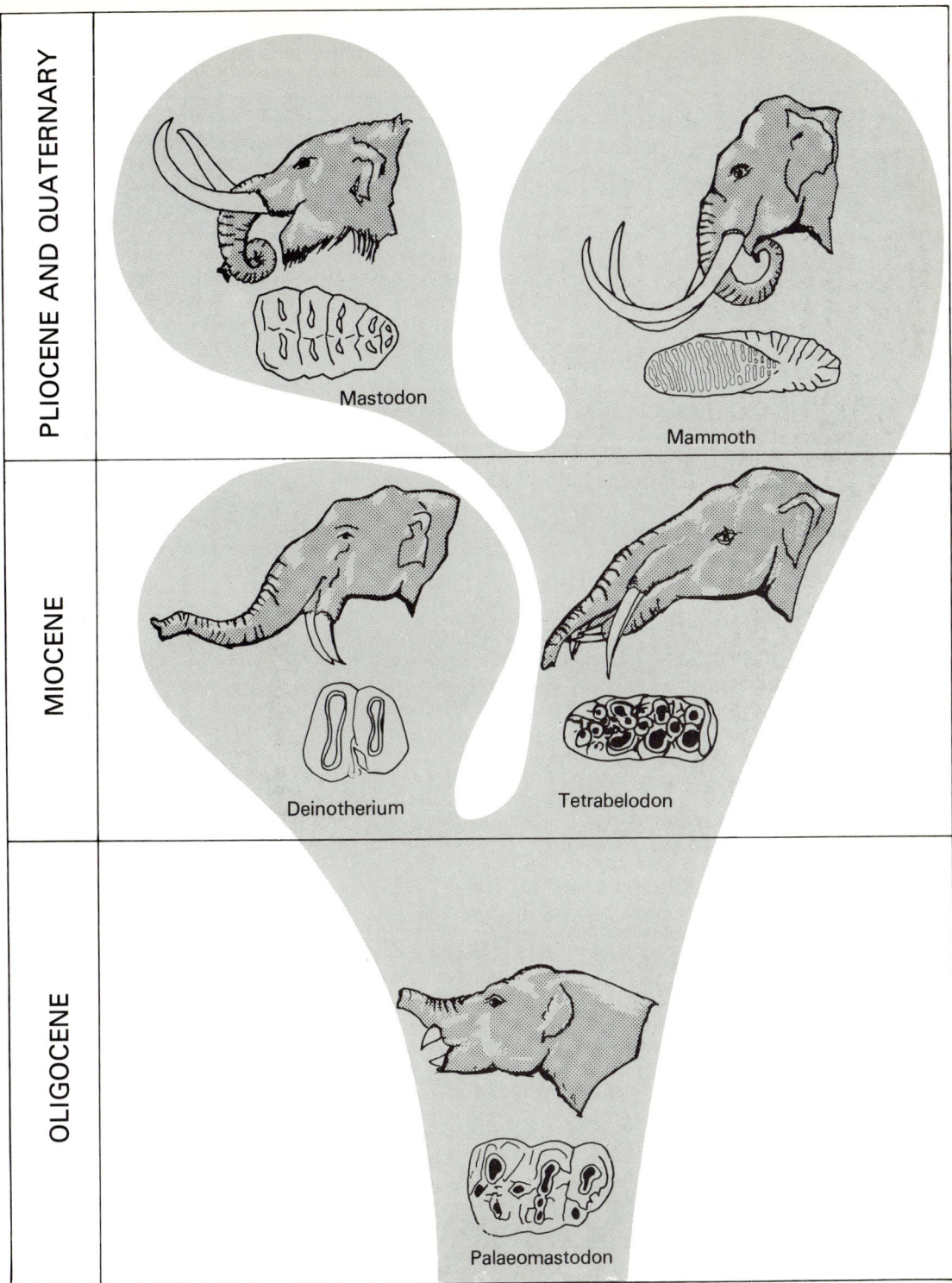

PLIOCENE AND QUATERNARY

MIOCENE

OLIGOCENE

Mastodon

Mammoth

Deinotherium

Tetrabelodon

Palaeomastodon

5.8 **The evolution of the proboscidians.** The tetrabelodonts are mastodons with four tusks, the true mastodons having only two tusks (upper incisors). The earliest proboscidian (not shown here) was *Moeritherium* of Eocene age from the Fayum of Egypt (after Kummel, History of the Earth (1961), modified).

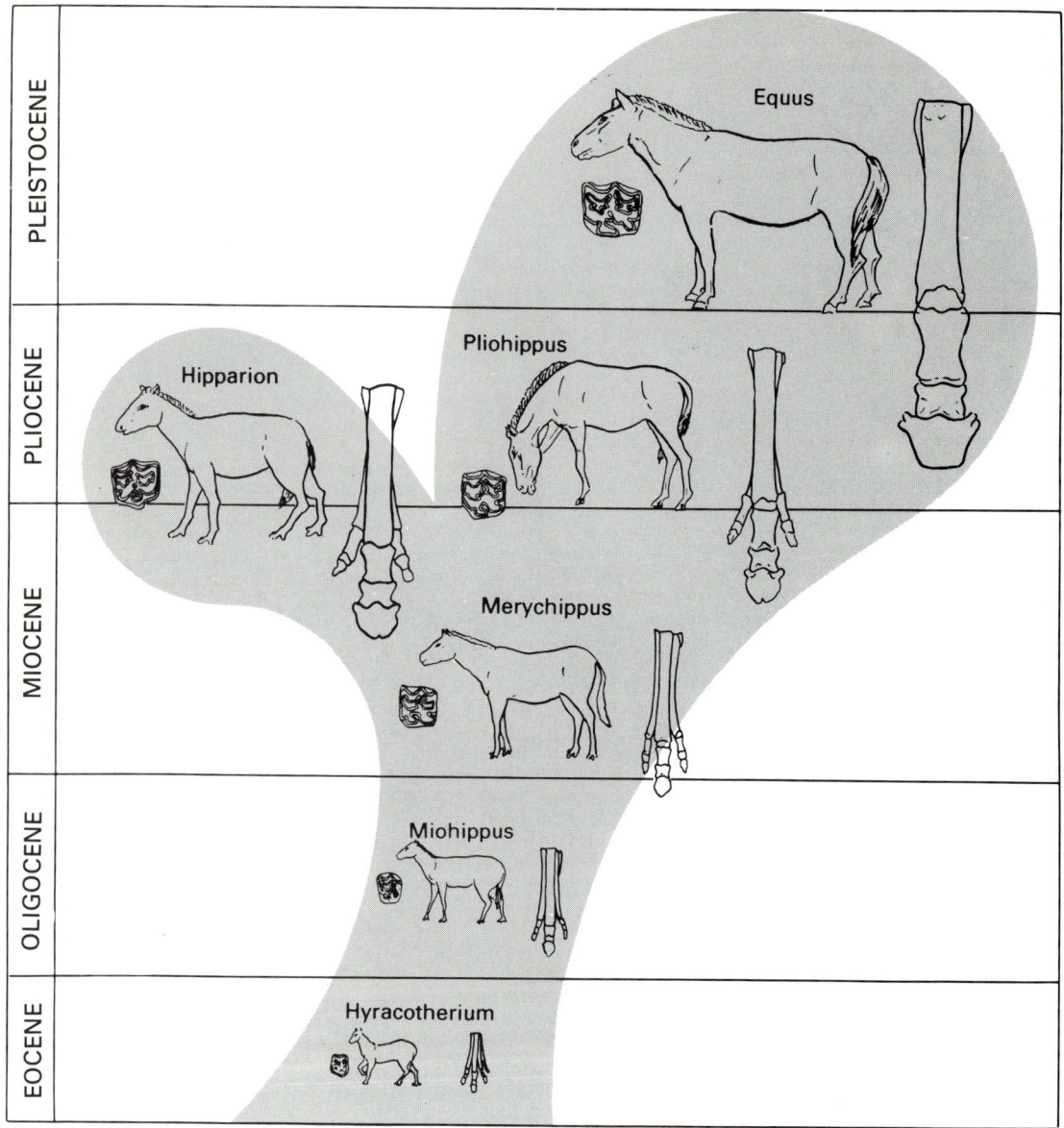

5.9 **Evolution of the horse.** The teeth, limbs and animals have been drawn to the same scales. Note the increase in size, the reduction and disappearance of the lateral digits and the increasing complexity of the grinding surfaces of the molars.

III. – PALAEOGEOGRAPHY

With the disappearance of the Ural Sea at the end of the Oligocene and in the Miocene, the collision of India and Asia and the crustal shortening of 400 to 500km which resulted in the uplift of the Himalayas, the *Eurasian continent took on a familiar appearance* in the

Neogene. To the south the *confusing palaeogeography of the Tethyan region still posed a number of problems.*

It is certain that the circum-Alpine trough, filled with molasse, reached its maximum extent at the same time as the tectonic events attained their paroxysmal phase. The Alpine arcs in central and eastern Europe enclosed a "Paratethys" which from the late Miocene, became progressively less saline, a tendency which reached its maximum extent in the Pliocene.

At about this time, Tethys itself began to change. About 18 millions years ago, at the end of the early Miocene, *the Middle East region began to close,* separating the Tethyan Province from that of the Indo-Pacific, *linking Africa with Asia and Europe and allowing African proboscidians and bovids to enter Eurasia. The foraminiferids of these provinces then evolved*

independently. Several million years later, the western end of Tethys was closed and became separated from the Caribbean province. This closure, linking Africa to Europe in the region of Gibraltar, occurred in the middle/late Miocene, following an orogeny which affected the whole of the Berberides (see p. 193) and the land bridge thus formed allowed the invasion of Europe by the pre-hominids.

Thus, for the first time in the Earth's history, *Africa and Europe were united at their two extremities in the late Miocene.* As the "evaporite belt" had been displaced southwards since the Palaeogene (see p. 56) the enclosed sea now became a gigantic evaporating basin, not unlike the Dead Sea, with a strongly negative hydraulic balance. Continuous evaporation caused its level to fall by as much as 2000m. As a consequence of eustatic changes probably associated with the first glacial and

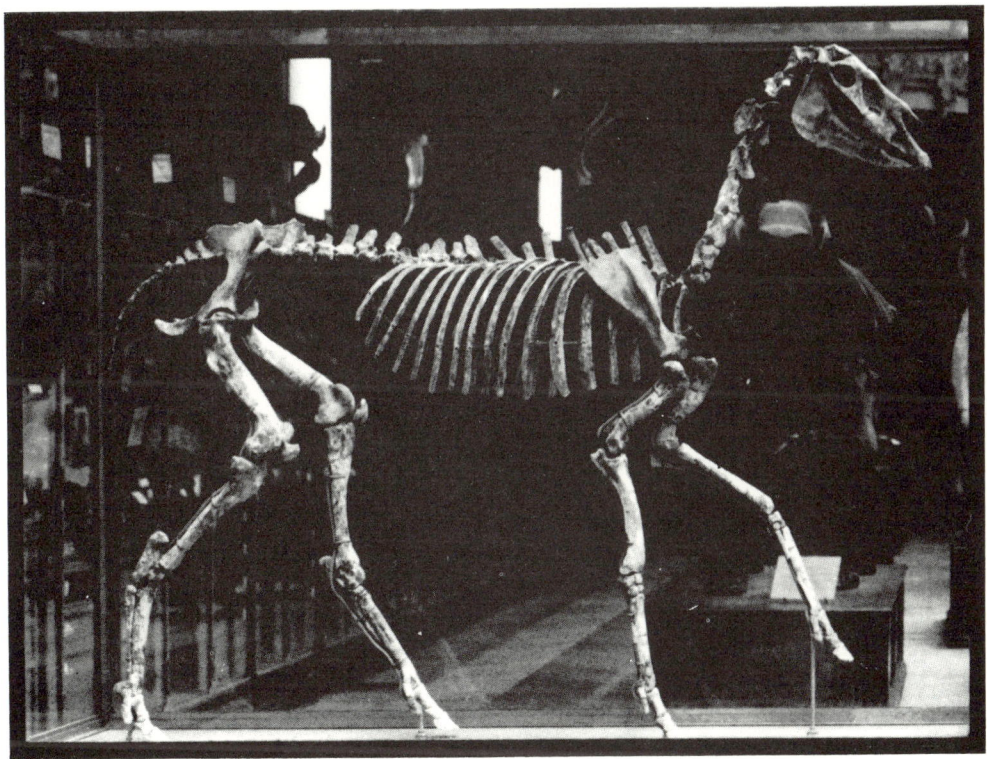

5.10 **Hipparion gracile, Turolian = Pikermian (ex upper Pontian) upper Miocene, from Pikermi, Greece.** This animal survived in Asia until the late Pliocene (early Villafranchian). Only the middle digit touched the ground, the two lateral digits being much reduced in length. This is an animal which diverged from the line of the true horses as is shown by the complex folding of the enamel of the teeth. *Hipparion* was smaller than the horse (1.4m to the shoulder) (photo: Serrette).

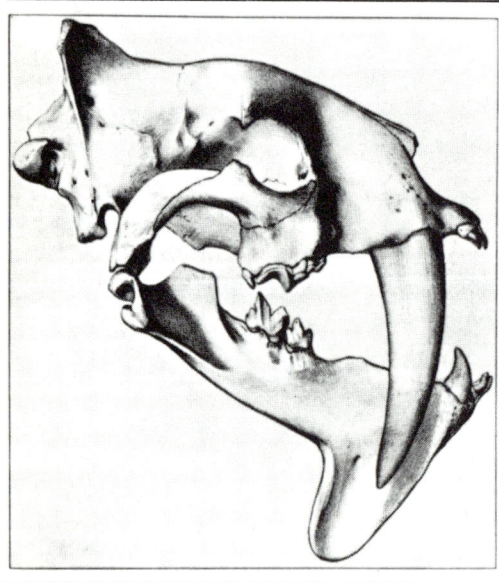

5.11a **Smilodon from the Pliocene and Pleistocene of the Americas.** Note the sabre-like canine teeth. *Smilodon* is, with *Machairodus*, the most highly evolved member of the cat family. Smaller forms, such as *Eusmilius* (Fig 5-11b), were present in the Oligocene. This form was about the size of a panther and lived in Europe and North America (photo: Serrette).

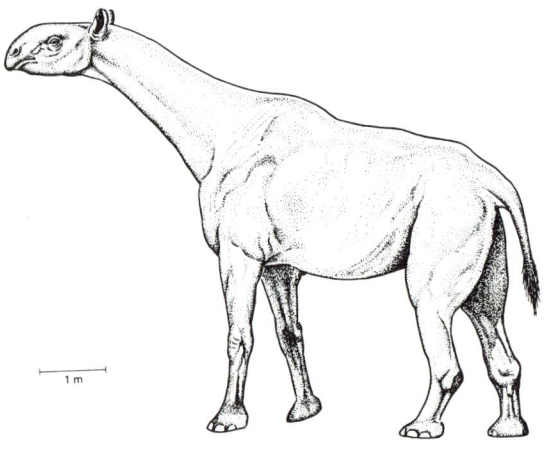

5.11b **Skull of Eusmilius dakotensis** (after Scott).

5.12 **Baluchitherium = Indricotherium.** A giant rhinoceros (4m to the shoulder, neck 1.3m long), known from the Oligocene and Miocene of Asia (after Augusta and Burian, reproduced, by permission, from the *Encyclopaedia Jniversalis*).

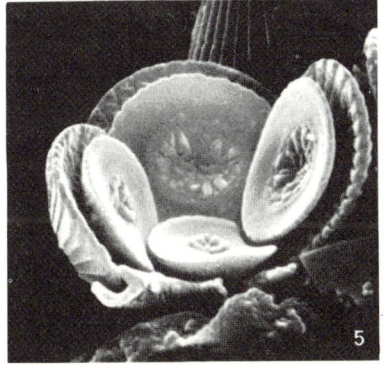

5.13 **Miocene nannoplankton from North Africa** (electron micrographs by Clocchiati). 1 – *Coccolithus pelagicus:* coccosphere showing arrangement of the coccoliths (x 6000); 2 – *Discoaster challengeri*, distal view (x 6300); 3 – *Cyclococcolithus leptoporus*, oblique distal view of a coccolith (x 10500); 4 – *Helicosphaera carteri*, proximal face of a placolith (x 6000); 5 – *Coccolithus pelagicus*, fragment of a coccosphere. The broken coccolith shows two discs connected by a tube (x 6000).

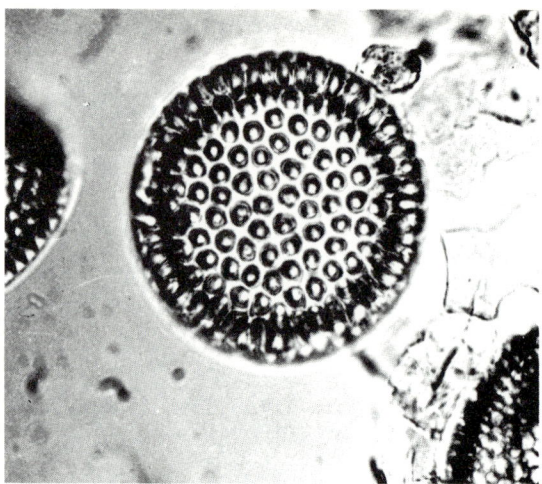

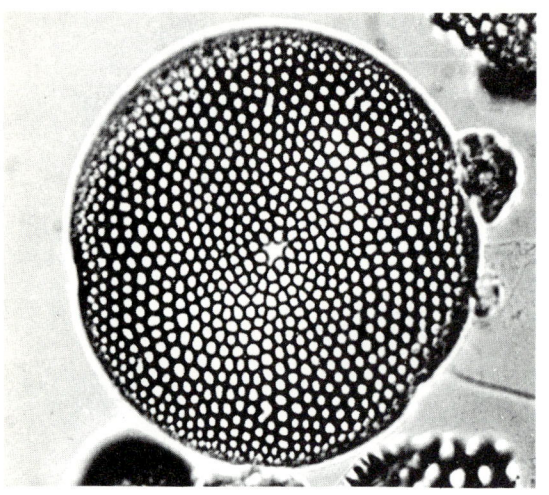

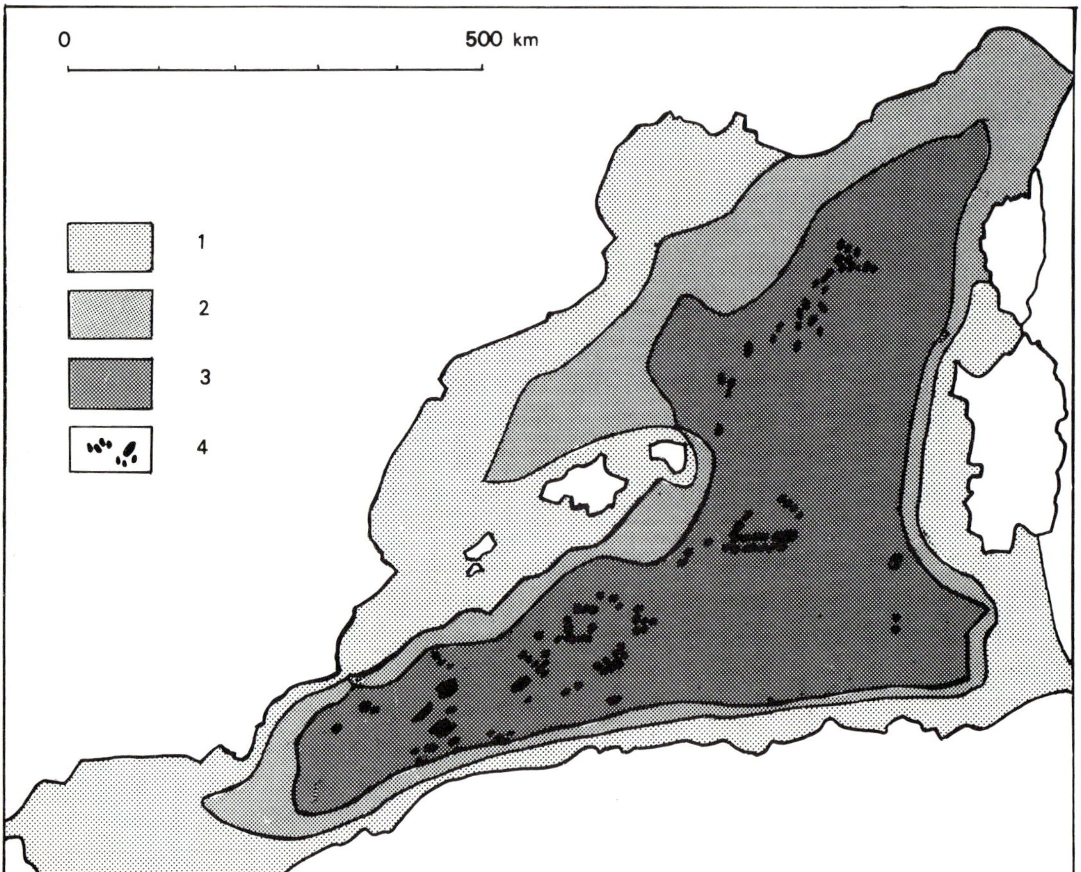

0   500 km

1

2

3

4

5.14 **Burdigalian marine diatoms from Uzès (Gard)** (photo; Ehrlich). 1 – *Coscinodiscus obscurus*; 2 – *Coscinodiscus marginatus* (x 200).

←

5.16 **Birds of the Tertiary era.** left: *Diatryma*, Eocene, North America; right: *Phororhacus*, Oligocene-Miocene, Patagonia. These are large, primitive flightless birds. (After Augusta and Burian, by permission of *Encyclopaedia Universalis*.)

←

5.15 **The evaporitic basin of the western Mediterranean at the end of the Miocene (Messinian)** (after Hsü (1973)). Carbonates (1) were deposited around the periphery; next came the sulphates, anhydrite and gypsum (2) and finally halite (3), which was responsible for the formation of domes or diapirs (4).
Several basins of this type occurred in the Mediterranean during the periods when it dried out at the end of the Miocene, before the opening of the Straits of Gibraltar in the Pliocene.

interglacial stages of high latitude (Antarctica) the barrier was submerged from time to time, and oceanic waters flooded back to refill the basin. Clays with planktonic foraminiferids were thus intercalated in the several hundred metres thickness of the late Miocene evaporites. During the periods of drying out of the basin by evaporation, water courses cut deep canyons, which remain as submarine canyons at the present time, while the mouths of major rivers, such as the Nile, were the sites of the accumulation of very great thicknesses of sediment (Fig. 5.15).

This remarkable palaeogeography came to an abrupt end at the beginning of the Pliocene (5Ma). *The opening of the Straits of Gibraltar by faulting* led to the replacement of the water lost by evaporation and the shorelines of the Tethys began to assume their present day outline. At the same time, uplift raised mountains and continents to their present level. Minor marine incursions, less important than those of the Miocene, transformed the lower reaches of rivers into rias. The Red Sea separated Arabia and Africa without impeding the passage of cattle and horses during the Villafranchian. The East African Rift Valley commenced to open as a prelude to a new break up of Gondwanaland which will only become effective in several million years time (Fig. 12.35, p. 255).

It was also in the *Pliocene that North and South America were joined together* by the Isthmus of Panama, although the resultant migrations were somewhat restricted by the barrier of the Mexican desert. Horses, mastodons, tapirs and llamas moved towards the south, while *Megatherium*, opossums and armadillos moved northwards. *The Atlantic and Indian oceans each continued to expand at about 2 to 3cm per year. As in the Miocene, volcanism was active*, andesites being emplaced along the Tethyan and circum-Pacific belts, and basalts in the interiors of cratons (e.g. the Massif Central).

The climate still remained warmer than that of the present day, but it was more strongly differentiated than previously as a result of the first periods of polar cooling. There thus appeared in the late Miocene a "cold water" microfauna in both California and New Zealand giving support to the idea that the closing and opening of the Gulf of Guadalquivir, which, as has been shown, made communication between the Atlantic and Tethys possible, was the result of glacio-eustatic changes controlled by the growth or melting of Antarctic glaciers. In particular, the rise in world temperature which occurred during the Aquitanian was maintained throughout the Miocene (palms, cycads, laurel) but there is strong evidence by Pliocene times of cooling of the seas as is shown by the changing morphology of the foraminiferids. The first globigerinids with sinistral coiling occur at this time (see p. 224).

But the earliest hominids, which appeared about four million years ago in the heart of Africa, were not aware of this climatic cooling and indeed, their descendants in European latitudes remained unaffected by this change until the Quaternary. Thus the two criteria, the appearance of Man and the development of glaciation must be abandoned in the definition of the base of the Quaternary system.

The Neogene in France 6

After the great transgressions and regressions of the Palaeogene, the Neogene was a relatively quiet period, at least as far as the sedimentary basins were concerned. It is marked by *two brief marine incursions* into the synclinoria of the *English Channel and the Loire*, in the middle Miocene and the early Pliocene, which isolated the western part of the Armorican massif. It is also marked by *increased marine penetration into the Aquitaine basin*, which was bordered by lakes and lagoons.

The uplift of the Alpine chain at the end of the Oligocene was associated with the forma-tion of a submarine trench around the western and northern margins of the chain which in Miocene times filled up with molasse, built up from the detritus produced by the erosion of the mountains. In France, the Rhône valley area was part of this *peri-alpine molasse trough*. In the Pliocene the Molasse trough became a ria, a branch of which is known as the Lake of Bresse, north of Lyons (Fig. 6.1). In Provence, and in the Alpine chain, important vertical movements occurred during the Pliocene which gave to these regions the outline of their present day topography.

THE GULFS OF THE WESTERN CHANNEL, THE MIDDLE LOIRE AND THE VENDÉE.

At the end of the Aquitanian, Lake Beauce dried out completely leaving behind lacustrine beds which at Selles-sur-Cher (Fig. 3.34) yield a mammal fauna which can be correlated with that well-known at Saint-Gérand-le-Puy, near Vichy.

Uplift of the Massif Central during the Savic phase of the Alpine orogeny resulted in the flooding of the area of the Beauce Lake, and hence the Paris Basin itself, with a mass of debris in the **Burdigalian**. This debris was brought by powerful rivers and consists not only of sands, but also of large boulders, often of granite. The latter decomposed to form the **sands and marls of the Orléans and Blois districts** and the **sands and clays of Bourbon, Sologne and Lozère,** which can be traced as far as Étretat on the Channel coast (Fig. 6.1). These sands are characterised by an absence of augite, which shows that they were derived from the Massif Central before the extrusion of the basalts. It should be noted that the rivers which emplaced these sands were earlier than either the Loire or the Seine and flowed across the area of Beauce. In so doing, they abandoned the route through the Loire region which had been utilised by the Stampian sea.

In the middle Miocene (**Helvetian**) the sea penetrated once more into the lower valley of the Loire, depositing the **shelly sands of Anjou and of Touraine.** This sea extended northwards into the western part of the English Channel and southwestwards into the Atlantic Ocean. The passage of the transgression from west to east is clearly demonstrated by the occurrence of earliest Helvetian beds in the west, while in the east, the lowest beds laid down were of middle Helvetian age. GINSBURG (1971) has

The Neogene in France

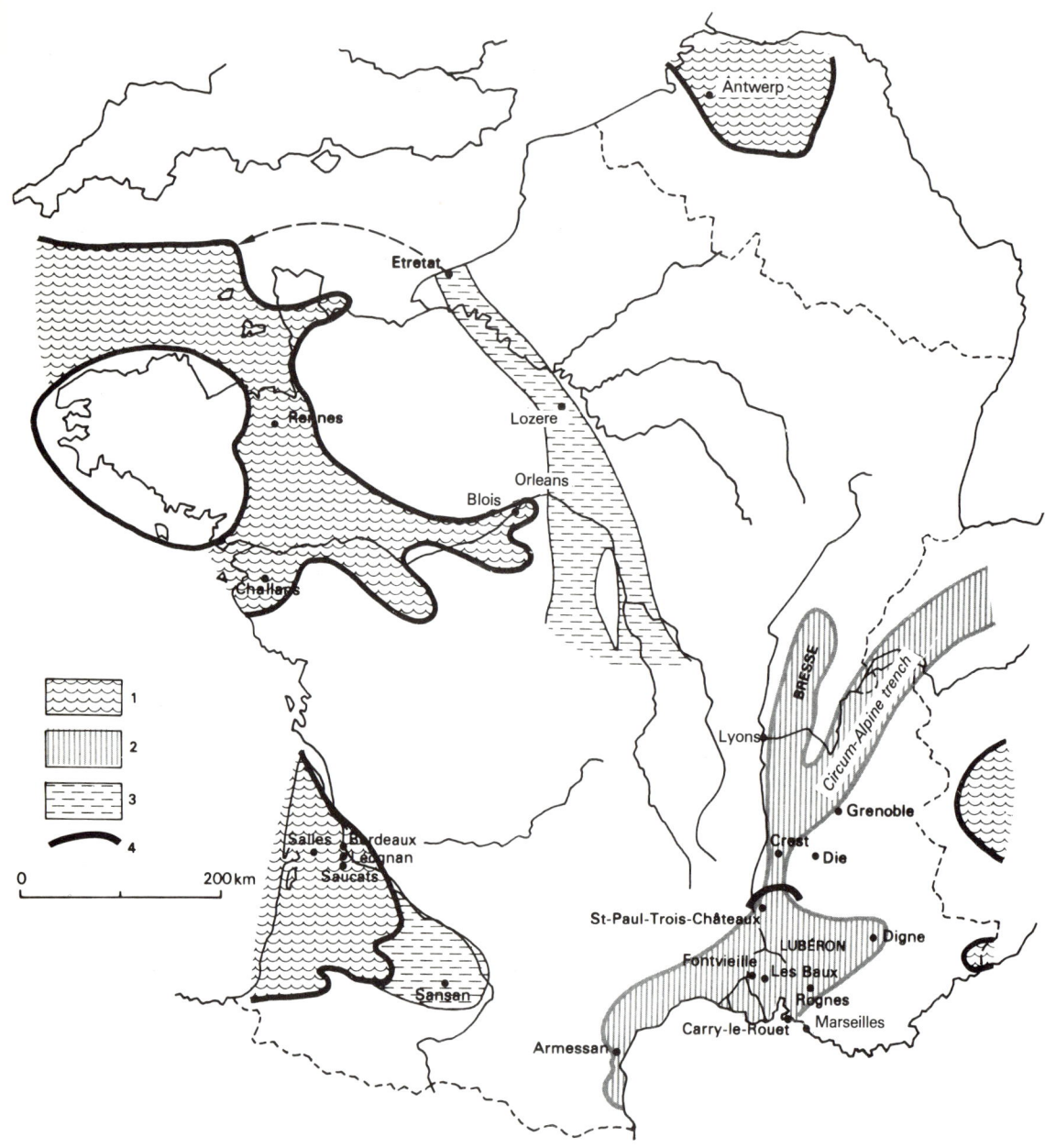

6.1 **The Miocene in France.** 1 – epicontinental seas; 2 – circum-Alpine molasse basins; 3 – continental formations (in the Paris Basin, trains of granite boulders deposited in the Burdigalian weather into clayey sands known as the Sologne and Lozère sands); 4 – northern limit of the early Burdigalian transgression in the Rhone Valley. The circum-Alpine trench was formed in the late Burdigalian and the Bresse trench in the Helvetian.

6.2a **Cross-stratification and solution pockets in the Helvetian shell sands of the Pontilevian facies at Thenay, 3km east of Pontlevoy (Loire-et-Cher)** (photo: Pomerol).

6.2b **Helvetian shell sands with Arca turonica (Pontilevian facies) at La Robardière,** 3km S.S.W. of Channay-sur-Lathan (Indre-et-Loire). The majority of the valves of *Arca* have the convex face turned upwards, as always happens in bivalve thanatocoenoses (photo: Pomerol).

shown, on the basis of mammalian remains occurring in the marginal beach sands, that the Anjou sands (in the west) are definitely older than the Touraine sands (to the east).

In the **Pontilevian facies** (from Pontlevoy, Loir-et-Cher) there are coarse shelly sands whose cross-bedding indicates the presence of strong tidal currents (Fig. 6.2a). The rich fauna, indicative of a tropical sea, includes many bivalves (*Ostrea crassissima,* Fig. 6.3,

Pecten scabrellus, P. albinus, Arca turonica, Fig. 6.2b, *Cardium turonicum, Cardita monilifera, Venus subrotunda, Pholas dujardini*), gastropods (*Calyptraea sinensis, Turritella triplicata, Voluta lamberti, Conus dujardini, Pleurotoma tuberculosa, Murex turonensis,* Fig. 6.4), corals, echinoids, calcareous algae and bryozoans. The latter are particularly abundant in the **Savignéan facies** (from Savigné-sur-Lathan, Indre-et-Loire).

6.3 **Ostrea crassissima;** a giant oyster, common in the Mediterranean Helvetian (note the scale).

Shark's teeth and fragments of turtles are common. A large variety of mammalian remains are also found including carnivores (*Amphicyon*), rodents, proboscidians (*Mastodon angustidens, Dinotherium cuvieri* (Figs. 5.7a and 5.7b), pigs, deer (which shed their antlers), horses (*Anchitherium*), rhinoceros and early primates (*Pliopithecus pireteani*). The proboscidians migrated westwards (in the opposite direction to the marine transgressions), driving before them an

6.4 **Murex (Truncilariopsis) turonensis,** Helvetian (x 1) (photo: Fay).

anthracothere (*Brachyodus onoideus*), which took refuge in Anjou, a region in which this species is present in younger beds than in the area around Orléans.

At the maximum of this transgression the gulf reached the region of Blois. Traditionally, it has been believed that this was the moment when the *upper part of the proto-Loire – proto-Seine river system was captured by the sea*, thus interrupting the transport of detritus from the Massif Central towards the Paris basin. However, recent sedimentological studies (TOURENQ, 1972; GERMANEAU, 1971) have shown:

1) that the heavy minerals of the shelly marls with abundant garnet were derived from Armorica rather than from the Massif Central: *the proto-Loire river*, which had discharged into the Channel (Fig. 6.1) in the Burdigalian, *had by the Helvetian lost its importance*,

2) that the augite, derived originally from the volcanics of the Massif Central from the late Miocene onwards, is found until the Quaternary in the alluvium of the Loing and the Seine, which implies at *least partial communication between* the basins of *the Loire and the Seine*[1]. This communication was probably established in the region between Gien

1. Transport by the wind across the watershed between the Loire and Seine basins is also a possibility.

and Sully-sur-Loire and survived until very recent (Würm) times[1]. The capture of the rivers flowing northwards from the Massif Central by the Helvetian sea is thus thought to be neither so important nor so complete as formerly supposed (Fig. 6.5).

In the late Miocene an epeirogenic elevation caused the sea to retreat. It returned in the

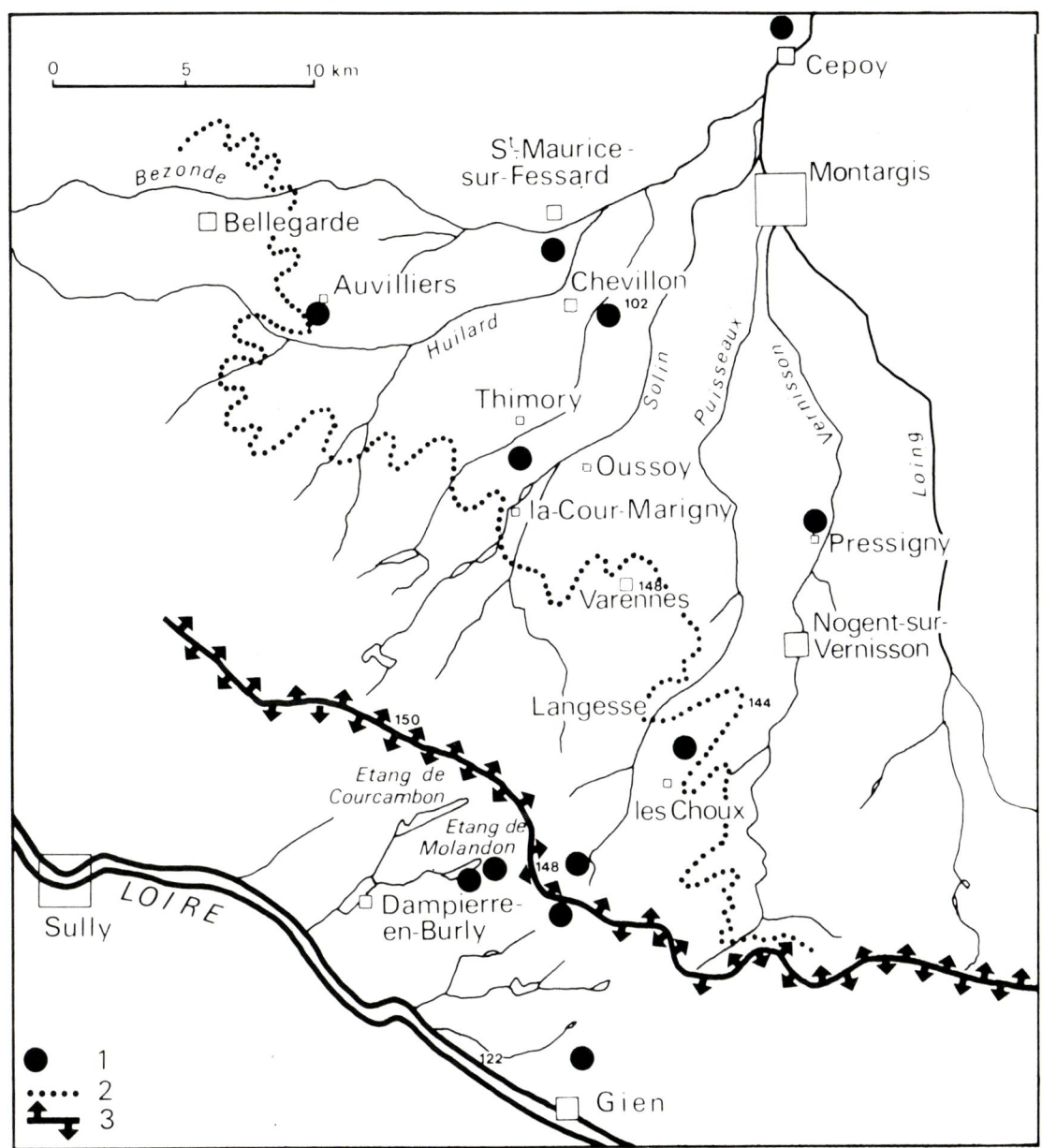

6.5 **The passage of detrital augites from the Loire Basin into that of the Seine across the present-day watershed.** 1 – localities from which augites have been collected; 2 – boundary between the Sologne and Orléanais sands (Burdigalian) to the south and the Gâtinais molasse (Stampian) to the north; 3 – present watershed. The fact that Plio-Quaternary augites are found in the molasse shows that certain stratigraphic attributions may be erroneous (after J. Tourenq).

The Neogene in France

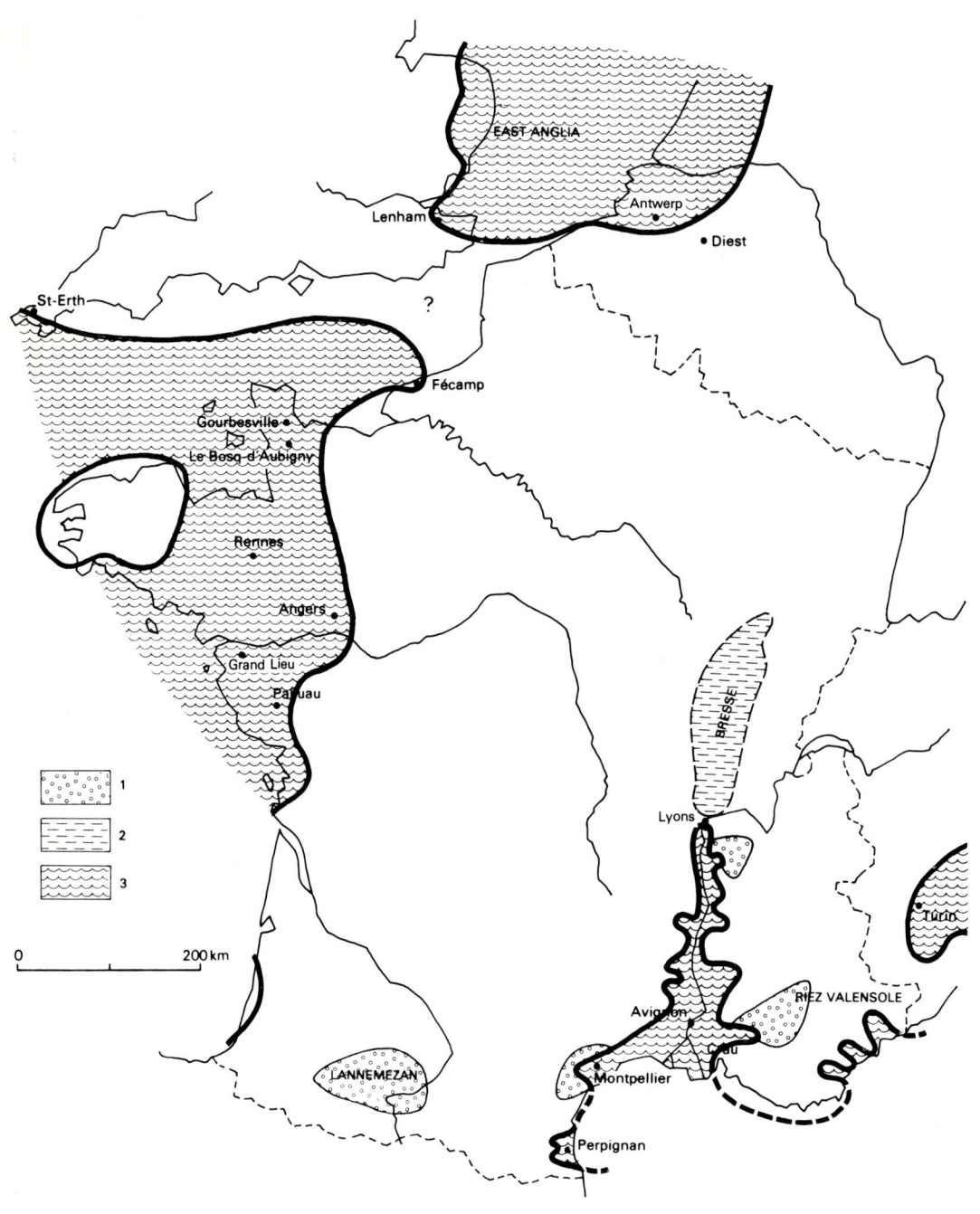

6.6 **The Pliocene in France.** The extension of the Pliocene over the Armorican massif is based on the distribution of red sands (after Estéoule-Choux, 1967). 1 – continental formations; 2 – lacustrine formations. 3 – marine formations.

Pliocene, though shallower than in the Miocene, and deposited the shelly marls of the lower Loire (VASSEUR, 1881), type of the **Redonian** stage of DOLLFUS (1900) (derived from Rhedonium, the Roman name of Rennes). Formerly thought to be of late Miocene age, these shelly marls are now placed in the **Pliocene** on the basis of a detailed study of several different faunal groups. While they contain undoubted Miocene species of molluscs, they also have species which are typically Pliocene including *Potamides basteroti*.

In the Brittany peninsula, isolated for the second time during the Neogene, are to be found *red sands* (white when pure) which overlie the shelly marls or occur laterally to them. These sands were formerly thought to be of continental origin. However they contain glauconite and, in associated clays, a marine Pliocene microfauna. This fact, and the occurrence of red sands and marls with molluscs and foraminiferids in the Morbihan, suggests that in the Pliocene the "Isle of Brittany" was penetrated by the Pliocene sea much more deeply than was once supposed (S. DURAND, 1960 and Fig. 6.6).

The Pliocene shelly sands are similar to the Coralline Crag of East Anglia (p. 168) and both show for the first time *evidences of northern influences*. It is therefore necessary to consider the question of the opening of the English Channel towards the north. During the Miocene, the western part of the Channel was certainly open, but there is no evidence that it extended much eastwards of the Cotentin peninsula (cf. Fig. 6.1). The recent discovery (1970) of an *outlier of marine Pliocene sediments near Fécamp* (see Fig. 6.6) at 60m above sea level and 120km further east than the most easterly deposits previously known (at Gourbesville and Bosq d'Aubigny, in the Cotentin peninsula) seemed to suggest that the English Channel might have been open in the Pliocene. However, the fauna at Fécamp is less northern in character than that at Gourbesville or even that at Palluau (Vendée). The idea must thus be abandoned although a passage to the north of the Weald, between the Hampshire and London basins has sometimes been proposed. The small deposits of Pliocene age at St.

Erth, near the north coast of Cornwall near Land's End, contribute nothing to the problem of the opening of the eastern end of the English Channel. Nevertheless, the greater part of the western Channel was submerged in the Pliocene and, according to LARSONNEUR (1972), the eastern part was already being formed, though it was not fully developed until the Quaternary (see Chapter 10, p. 210).

In the **coastal regions of the Vendée,** the Miocene transgression flooded into the river valleys from the west. Shell marls have been preserved in the **Challans basin** and, in very small, scattered patches, in the western Vendée. *The Pliocene (Redonian) transgression* invaded the southern coast of Brittany, the Loire estuary and the Vendée, forming rias and submerging the whole of the coastal region to a depth of about 65m. The regression which followed is marked by the presence of raised beaches with rolled pebbles. This transgression and regression is marked by two facies. The lower beds are clays, with marls and limestones at the base, formed when the sea was at its maximum depth, with sands and gravels above. When the latter are decalcified they form the "red sands" referred to earlier at Challans and Touvois. The principal occurrences of these deposits are at Palluau, Touvois, Falleron, Bourgneuf Bay, and the low ground of Grand-Lieu and the Marne. Towards the east, slight uplift has raised the Pliocene sediments to a height of 80m at Chaillé-les-Ormeaux (TERS, 1961) (Fig. 6.7).

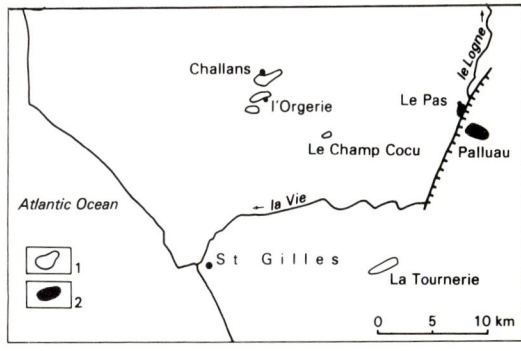

6.7 The Miocene (Helvetian) (1) and Pliocene (Redonian) (2) deposits of the Vendée (after Ters, 1961). Recent researches have shown a substantial area of Pliocene, particularly to the south and east of Grand-Lieu lake, north of the area shown (see Fig. 6.6).

II. – THE AQUITAINE BASIN

In contrast to the central part of the Paris Basin, which was finally abandoned by the sea at the end of the Oligocene, the Aquitaine basin was submerged by a new transgression in the early Miocene. This Atlantic transgression moved in from the south west towards the north east forming this time a single embayment reaching Agen in the east (Fig. 6.8).

In the Bas-Adour region, the lowest beds containing *Globigerinoides* and *Miogypsina* (at Escornebéou) have been attributed to the late Oligocene. However, deposits situated to the east of Marseilles at Carry-le-Rouet (Fig. 6.1) are certainly Miocene, but are earlier than the Aquitanian s.s.

Furthermore, the type Aquitanian-Burdigalian sequence has been established on the basis of a littoral or neritic facies poor in planktonic foraminiferids, which makes correlation difficult in the Mediterranean Miocene. These two "stages" are difficult to separate in borings and have only a local biostratigraphic value based on species of *Pecten* and *Miogypsina*. They probably relate to a single sedimentary cycle for which VIGNEAUX has proposed the name **Girondian,** a superstage unsatisfactorily defined on the basis of a boring at Soustons (Landes).

The Aquitanian sensu stricto[1] is represented by the shelly marls and **sandstones of Bazas** and the lower part of the section at Saucats containing the echinoid *Scutella subrotunda*. The lagoonal marls containing cerithiids and *Ostrea aginensis* between the Garonne and the Dordogne are at the same horizon. Further east, the dark foetid **Agenais limestone** of Laugnac has yielded a mammalian fauna comparable to that of St. Gérand-le-Puy (Limagne) and that of the Beauce limestone. A good correlation between the continental formations of the Paris basin and the marine formations of the Aquitaine basin is thus possible.

In the **Burdigalian**[2], the **upper shelly marls of Saucats,** the **Léognan molasse** and the shelly marls of **Le Coquillat** contain *Pecten burdigalensis, Cardium, Turritella, Melongena* and several species of *Miogypsina*, and, in the south only, the last lepidocyclinids. Further east the lower part of the Armagnac molasse is attributed to the Burdigalian, though this formation probably extends up to the Tortonian.

In the **middle Miocene,** the area of the gulf was reduced. Only the Helvetian is represented

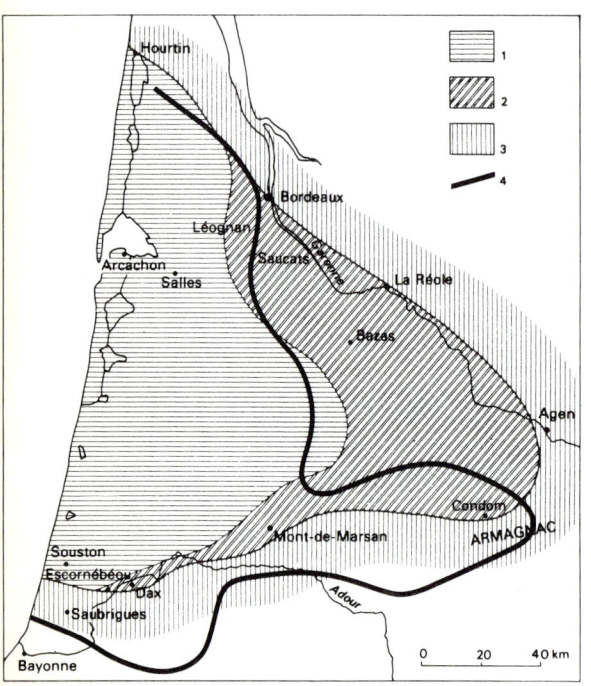

6.8 **The lower Miocene in Aquitaine.** 1 – marine formations; 2 – shallow-water formations of alternating marine and continental facies; 3 – continental formations; 4 – limit of the sea in the middle Miocene (after Alvinerie, 1969) (see also Fig. 3.42, p. 91).

1. Stage established by Mayer-Eymar (1857) in the stream section of the River Saint-Jean-d'Étampes between the Moulin de Bernachon and the Moulin de l'Église in the region of La Brède and Saucats.

2. Stage created by Depéret (1892) on the basis of the rich fauna of the shelly marls of Bordeaux and of Léognan. Depéret noted the position of the Saucats shelly marls above the Aquitanian.

in the Gironde, where the **Salles shelly marl** with *Cardita jouanneti* (Fig. 5.2) and *Pecten latissimus* are found. Further south, the series is more complete. Sands with *Ostrea crassissima* at Mont-de-Marsan, possibly Helvetian, overlie Tortonian beds of a deeper water facies with a rich fauna of turrid gastropods. These are the blue **marls of Saubrigues** whose precise age is still uncertain. By this time the Aquitaine gulf was very much reduced in size and the deposits of its margins were already above sea level. This was the prelude to the final drying-up of the Aquitaine basin, which had survived for 10 million years longer than the Paris Basin.

To the east, continental sedimentation continued with the deposition of lacustrine limestones in the region of Condom. Still further east, molasse deposits (e.g. the **Armagnac molasse**) become progressively coarser grained in the direction of the Pyrenees.

Finally, it may be noted that important Helvetian vertebrate remains have been found in a clayey and marly limestone intercalation in this molasse at Sansan (Gers). These include *Amphicyon, Anchitherium* and *Mastodon angustidens*. Similar mammalian deposits of Tortonian age are known at Simorre, and of late Miocene age at Saint-Gaudens and Orignac, where lignites of "Pontian" age contain the first *Hipparion* (see Fig. 2.6, p. 35 for location of these deposits).

III. – THE CIRCUM-ALPINE TROUGH AND ITS MARGINS

The Miocene and Pliocene sediments on the border of the Alpine chain provide two good examples of a sedimentary cycle.

On the Languedoc coast near Montpellier and on the coast of Provence at Carry-le-Rouet west of Marseilles, the **early Miocene** Aquitanian transgression predated that of the stratotype in Aquitaine. It commenced as soon as the link between the Pyrenees and the Maures massif was broken, and flooded northwards towards Arles.

The succession begins with Aquitanian marls and sands with *Ostrea aginensis* which rest on lacustrine limestones with *Helix ramondi* marking the end of the Oligocene (Chattian). They are followed by lower Burdigalian sands with the echinoid, *Scutella paulensis*, and *Pecten paulensis* which take their name from the locality of Saint-Paul-Trois-Châteaux. The Sausset molasse, formerly believed to be lower Burdigalian, has been shown to be Aquitanian on the evidence of its microfauna (DEMARCQ, 1970).

The sea advanced rapidly northwards, entering the Digne basin, ignoring the Rhône valley, and turned to the northeast near Crest (Drome) reaching, in the late Burdigalian, the Helvetic depression (which had up to now been filled by the lower Freshwater Molasse) by way of the Vercors and Chartreuse. At this moment the Mediterranean became linked to the Vienna basin and thence to the Balkans (Figs. 6.1 and 7.10, p. 173) through the molasse trough of Switzerland.

The Burdigalian sediments preserved in the valleys of the northern sub-alpine chains are sands or sandstones, often conglomeratic, with *Pecten praescabriusculus*. Towards the south, in the enlarged gulf, calcareous sands and molasse with algae, bryozoans, pectens and echinoids (*Scutella, Cidaris, Clypeaster*) were deposited. These rocks are massive and easy to work, and form excellent building stones. They are worked mainly at St.-Paul-Trois-Châteaux, Fontvieille (Fig. 6.9), south of Beaucaire and at Les Baux near the Alpilles (Figs. 6.10 and 6.11).

6.9 Quarry in Burdigalian molasse at Fontvieille (photo: Pomerol).

In the last locality, the molasse has been sculptured by atmospheric (though not eolian) weathering into strange shapes which contrast with the massive beds of the Urgonian cliffs.

In the **middle Miocene** (Helvetian plus Tortonian, sometimes united as the **Vindobonian**), the sea remained in the whole of the circum-Alpine depression and extended further across the Rhône basin than it had done in the early Miocene. This basin extended to the Massif Central between Montélimar and Lyons and beyond to form a long narrow inlet as far as Lons-le-Saulnier, in which the sandy substratum of the Bresse region was deposited. This facies of sand and sandstone molasse indicates a period of orogenic calm. Further south, as in the Burdigalian, the deposits are somewhat deeper water, marine argillaceous and micaceous sands containing a giant oyster, *O. crassissima*, 40cm long (Fig. 6.3). These sediments resemble the *Schlier* of the Vienna basin (type region of the Vindobonian). They are succeeded by the blue marls of Cabrières d'Aigues (Vaucluse), which contain a Tortonian fauna.

From the end of the Tortonian, throughout the **late Miocene** (often improperly called "Pontian"), and into the **Pliocene** ("Ponto-Pliocene") the circum-Alpine trench became progressively filled with detritus from the rising mountain chain. More than 2000m of sandy and pebbly sediment was accumulated, over-lying lagoonal sands and lacustrine clays. In the valley of the lower Durance, the detritus collected as the fanglomerates of Riez-

6.10 **The scenery near Les Baux:** Miocene molasse with honeycomb structure produced by atmospheric weathering (photo: Pomerol).

Valensole (Fig. 6.6). Further east, at Mont-Lubéron, lacustrine marls and bone beds contain a large mammal fauna including *Hipparion gracile*.

In Switzerland in this same period the "**upper freshwater molasse**" was being formed. At Oeningen, near Lake Constance, this horizon has yielded an extremely rich flora of tropical plants and many insect and vertebrate remains. The latter include fishes and mammals (rhinoceros, tapir, *Anthracotherium*, insectivores and carnivores, including *Machairodus* with oversize upper canines (Fig. 5.11, p. 148).

The last major phase of Alpine folding occurred during the "Ponto-Pliocene". This resulted in the folding of the external zones, the formation of anticlines and synclines in the nappes of the internal zones, the completion of the uplift of the external crystalline massifs which had started in the Miocene, and the subsidence of certain peripheral zones, such as the plain of the River Po and the Mediterranean coast.

As a result of these movements, the sea remained close to the Alpine chain, taking advantage of the least depression to penetrate between the mountains. The most spectacular of these Pliocene incursions is undoubtedly that which affected the valleys of the Rhône and the lower Durance. These became deep rias of Cantabrian or Breton type carved out of the Miocene molasse.

Blue clays of the **Plaisancian facies**, containing *Ostrea cochlear* and *Venus multilamella* were deposited in the deep water of these rias. This marine series was covered by continental beds with mammalian remains in the Villafranchian, at the Pliocene-Quaternary boundary. Remnants of these deposits occur on the Chambaran plateau, south-east of Lyons, and on the plain of Crau, the ancient delta of the Durance.

The Rhône ria almost reached Lyons, and was continued to the north by the *lake of Bresse*, which was filled with a considerable thickness of marls containing *Viviparus* like those of the lakes of eastern Europe. For a short time, the *Rhine flowed into this lake* as shown by the presence of pebbles of Alpine origin. At the southern end of the lake, in the valley of a tributary, the *Meximieux travertine* yields a Pliocene flora indicating a climate rather colder than that in which the Oeningen Miocene deposits were formed.

By Plaisancian (middle Pliocene) times, the Mediterranean coast was approaching its present-day outline, although the sea still penetrated into the rias of the coast of the Maritime Alps. Here the Villafranchian terrestrial conglomerates cap the uplifted Plaisancian blue marls. Along the Languedoc coast, as far as the foot of the Cretaceous hills, *sands of Astian facies*, rich in molluscs and mammals, were deposited near Montpellier. Between the Corbières and the Pyrenees, another ria opened into the plain of Roussillon where, at Serrat-d'en-Vacquer, the type locality of the **Ruscinian** stage has been chosen. This Early Pliocene continental stage marks the transition between the Turolian = Pikermian (late Miocene) and the early Villafranchian (late Pliocene) (see Table XIII, p. 226).

The *Pliocene is marked*, in the continental areas, *by great outpourings of basalts* (Cantal, Velay, Aubrac, Coirons), whose flat surfaces, since altered, provide favourable environments for the raising of cattle. Plant remains preserved in volcanic ashes include ginkgos and pomegranates, indicating a *climate warmer than that of the present*. The faunas comprises *Mastodon* (*M. arvernensis*), *Rhinoceros*, the last hipparions and the last (European) tapirs.

Finally, we may recall that the erosion products of the Massif Central were distributed, in the early Miocene (Burdigalian) in the Bourbon, Orléans and Sologne areas and in the Paris Basin. Augite continued to reach the latter area right up to the Quaternary.

6.11 **Entrance to underground quarries in the Miocene molasse at Les Baux (photo: Pomerol).**

Table XII
The principal stages of the European Miocene

The Diestian, formerly attributed to the Pliocene, corresponds to the upper part of the Deurnian. The term "Falunian" is now disused. Rigorous correlation between the different stages is not possible. The "Helvetian" and "Tortonian" stages were formerly included in the Vindobonian stage (the Vienna molasse, Austria) (see also Table X, p. 138)

Epoch	Belgium	Armorican Massif	Aquitaine Switzerland Spain	Italy	Central and Eastern Europe	Continental stages
Pliocene	Scaldisian	Redonian		Tabianian or Zanclean	Dacian Pontian s.s.	Ruscinian
— 5.3 MA —						
MIOCENE	(Diestian) Deurnian Anversian Houthalenian (Bolderian)	(Falunian)	Andalusian Helvetian Burdigalian Aquitanian	Messinian Tortonian Serravallian Langhian	Pannonian Sarmatian Badenian Carpathian (Karpatian) Ottnangian Eggenburgian	Turolian Vallesian
— 23 MA — **Oligocene**					Egerian	

The Neogene in Europe

7

I. – NORTHERN EUROPE

In the Miocene, the North Sea was more restricted than in the Oligocene, but was generally transgressive. Glauconitic, shelly sands were deposited in Denmark, north Germany and Belgium, but the sea did not reach the English coast.

In Belgium the type sections of the stages were chosen in the Campine (Fig. 7.1). The Houthalen sands (**Houthalenian**, Ghent Symposium, 1961) have a Burdigalian fauna (Aquitanian for some authors) and correspond in part to the former Bolderian (sands of Bolderberg Fig. 7.2) (DUMONT, 1849), a term which has now fallen into disuse (Table XII). Further west, they are succeeded by the Antwerp sands (*sables d'Anvers*) of middle Miocene age. These are highly glauconitic sands containing *Panopea menardii* and abundant *Glycymeris pilosa* (Fig. 7.3) (the **Anversian** of COGELS and VAN ERTBORN, 1879) of middle Miocene age. They are overlain by the upper Miocene Deurne sands, which are very fossiliferous (Fig. 7.4) with *Terebratula maxima* (**Deurnian**, DE HEINZELIN and GLIBERT, 1957) (the **Diestian** and associated formations of DUMONT, 1849, formerly thought to be of Pliocene age). These "Diestian" sands are transgressive, extending as far as Brabant, but are unfortunately barren. They may represent, in certain cases, a facies formed by the alteration of sands of various ages, as is the case with the sands which cap the hills of Flanders, which are outliers of middle and upper Eocene beds resting on the Ypres clay.

The Pliocene rests conformably on the Miocene in the Antwerp basin and begins with the **Kattendijk sands** with *Isocardia cor*. These are overlain by the **Luchtbal shelly marls** containing *Pecten subgrandis* and *Cardita scaldensis* and the **Kallo sands** with *Neptunea contraria*. These three horizons form the **Scaldisian** (from the River Schelde, DUMONT, 1849), which in its upper part (the Kallo sands) is regressive.

There is a gradual disappearance of a number of southern species and the appearance of other, more boreal ones. This is shown particularly well by *Globigerina pachyderma* which, in the Deurne sands exhibits dextral coiling, but becomes sinistrally coiled in the Kattendijk sands as the sea temperature fell. This is the first definite evidence of the cooling of the waters of the Neogene northern basin.

The Scaldisian, which is correlated with the Coralline Crag of East Anglia and the Redonian of the Armorican massif, is succeeded by the Merksem sands (the **Merksemian** of the Ghent Symposium, 1961). These sands contain *Mya arenaria* and *Cardium parkinsoni* and are already of Pleistocene age, though some authors believe that the Pleistocene begins with the underlying Kallo sands. Many species, such as *Isocardia cor* from the Scaldisian and even *Glycymeris pilosa* from the Anversian, are still extant.

Sedimentation in the Belgian gulf, restricted essentially to the Antwerp basin, has been more or less *continuous from the early Miocene up to the present day*. The maximum transgression in Belgium occurred in the late Miocene with the deposition of the Deurne sands, of which the Diestian s.s. appears to be a lateral facies, in

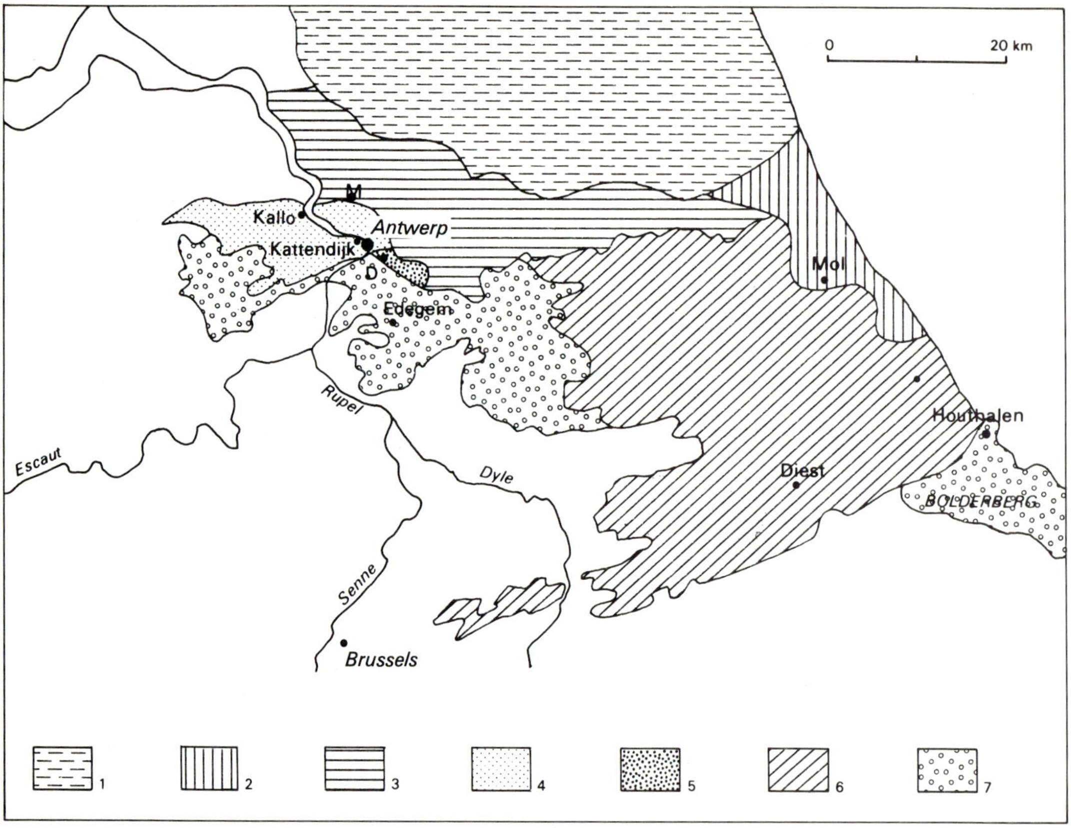

The Neogene in Europe

7.1 **The Neogene and Quaternary of the Campine** (Kempen) area of Belgian. Localities and outcrops (after Tavernier and De Heinzelin, 1962). Quaternary: 1 – Campine clays (Tiglian): 2 – Mol sands (Reuverian); Pliocene: 3 – Merksem sands; 4 – Kattendijk and Kallo sands. Miocene: 5 – Deurne sands; 6 – Diestian; 7 – Anversian and Bolderian (= Houthalenian). (D: Deurne; M: Merksem).

7.2 **White sands with bed of quartzitic sandstone about 1m thick, overlain by lignitic clay (Bolderian = Houthalenian) at Opgrimbie, Campine** (photo: Gulinck).

7.3 **Glauconitic sands with Pecten pilosus at Antwerp, Belgium (Anversian).** Note the position of the valves, convex side upwards, characteristic of a thanatocoenosis (photo: Laga).

7.4 Bioturbation caused by burrowing organisms in the Deurne sands at Borgerhout, near Antwerp (photo: Laga).

contrast to the Mediterranean region where the late Miocene is nearly everywhere a period of regression.

In north Germany, the principal correlative formations are, from bottom to top, the Vierland beds the Hemmoor beds and the Reinbeck beds[1]. East of the Belgian gulf a luxuriant swamp vegetation gave rise to the *lignite basin of north Germany*, which persisted from the Oligocene to the Pliocene (Fig. 7.10, p. 173).

The oldest Miocene deposits of northern Europe occur in Denmark and in the Hamburg region, where sands and shelly sandstones of the Vierland beds (**Vierlander Stufe**) were laid down, whereas in Belgium and Great Britain the sea had almost completely receded at the end of the Oligocene.

The transgression continued in the middle Miocene, reaching Bremen and Osnabrück, and, probably by way of Silesia, extended to the Vienna basin where a characteristic northern fauna has been found. As in Belgium, the sands are glauconitic (Hemmoor beds = **Hemmoorer Stufe**). They contain notable quantities of Aquitanian forms which must, it seems, have arrived via the North Sea, as the Channel was closed at this time. Immense swamps on the edge of this sea supported a dense vegetation of *Taxodium* and *Sequoia* and probably resembled those known today on the edges of the Bay of Florida and parts of the

Gulf of Mexico. The swamp vegetation of the north German Miocene formed thick beds of lignite which are now exploited in very large opencast workings in Pomerania, Brandenburg, Saxony, Silesia and Poland (Fig. 7.10). Lignite also accumulated in the subsiding Rhine Graben.

At the end of the middle Miocene and in the late Miocene, the sea abandoned the plains of north Germany, withdrawing into Denmark and the Hamburg region, depositing the Rein-

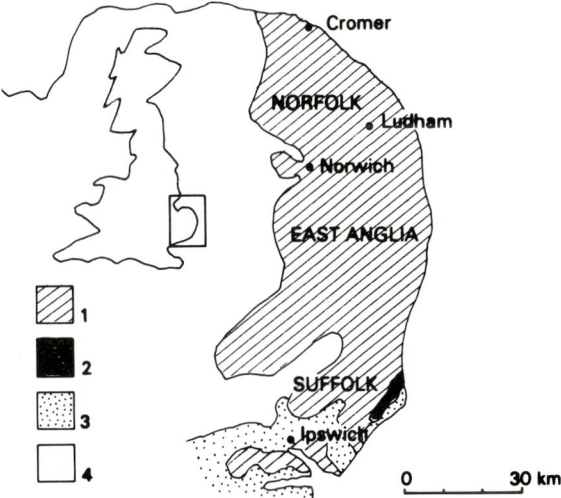

7.5 The Eocene and Plio-Quaternary outcrops of East Anglia. 1 – Pleistocene Crags and Red Crag; 2 – Coralline Crag (Pliocene); 3 – Eocene; 4 – Mesozoic (after Rayner, 1967).

1. The localities chosen as the type areas of these Miocene beds (Stufe = Stage) are in the region of Bremen and Hamburg.

beck beds (**Reinbecker Stufe**), which are contemporaneous with the Anversian. However, in Holland the sea was transgressive and, as we have seen, it entered the Campine and reached Brabant (Deurnian).

In the Pliocene, the sea left Germany and most of Denmark, remaining only in the Island of Sylt on the west coast. By contrast, the Pliocene sea was transgressive in south-east England, where the shelly ''Crags'' were laid down.

The oldest British marine Neogene deposits are the relict deposits of the Lenham Beds, found locally in Kent in pipes in the Chalk, which are of late Miocene age. In Suffolk and Norfolk (East Anglia) the sequence includes the Coralline Crag, which is succeeded by the Red Crag, whose fauna indicates somewhat cooler conditions. These two units are of late Pliocene age, like the Luchtbal sands of Belgium and, probably, the Redonian, as all have a temperate fauna and yield the dextral form of *Globigerina pachyderma*. It seems therefore that the North Sea extended as far as Kent without, however, being in communication with the English Channel gulf (Fig. 6.6, p. 158).

The first cold fauna appeared at the beginning of the Pleistocene, in the Norwich Crag. This horizon is the former **Icenian**[1] of

1. From Iceni, an ancient Celtic tribe (Lankester, 1914).

English geologists. There follows an alternation of temperate and cold faunas with corresponding transgressive and regressive periods. This succession can be traced to the east of Norwich as well as in the *Cromer Forest Bed* series of north Norfolk, on which LEAKEY (1934) erected the **Cromerian** stage, with a fauna corresponding to the Günz-Mindel (or Intra-Mindel) interglacial period.

But we have strayed here into the Quaternary episode, which will be discussed in another chapter. It is difficult to do otherwise in this region where the transgressions of the North Sea, having covered much of Germany in the middle Miocene and then left it, spread southwards into Holland and Belgium, where the maximum extension of the sea occurred in the late Miocene. In the Pliocene and Pleistocene, the basins of Antwerp and of East Anglia were embayments on the margin of the North Sea, like the Baltic sea at the present time. These embayments had no communication with the Atlantic Ocean through the English Channel, which did not exist at that time and is, in fact, a very young feature. It is comparable in age with the eruption of the volcanoes of the Auvergne, but (more significantly) its opening must have been witnessed by Man a mere seven to eight thousand years ago.

II. – SOUTHERN EUROPE

1. General characters

During the Neogene, the evolution of southern Europe exhibited three consecutive and contrasting stages. The first, following the regression at the end of the Oligocene, was the **development of molasse trenches,** of which the best example is the circum-Alpine trough extending from Marseilles to Vienna with its prolongation through Romania to the Carpathian foredeep (Fig. 7.10, p. 173). Similar foredeeps were formed in front of the Betic and Rif cordilleras, the Balkans, the Caucasus, the Dinarides and the Aegean mountains. Deep troughs were also formed between adjacent mountain chains, as for example, between the Albanian and Thessalonian mountains in northern Greece. On the other hand, *neritic*

basins can also occur between two chains, as for example, in the **Pannonian basin** of Hungary, lying between the Carpathians and the Dinarides; the **Dacic basin** between the Carpathians and the Balkan Range; and the **Pontic basin** between the Caucasus and the Taurides. Neritic basins can also occur between a mountain chain and a continent, such as the **Aral-Caspian** basin between the Caucasus and the Russian platform. These various seas communicated through straits corresponding to the cols between the chains. One of the most important of these connecting links was the **Vienna basin** lying between the circum-Alpine trough and the Pannonian basin.

The second stage, following the maximum extent of the Miocene transgression (in the Helvetian), was one of **general uplift** which makes it difficult to select a good stratotype for a late Miocene stage. Recent boreholes throughout the Mediterranean region have shown that the end of the Miocene was an arid period which resulted in the deposition of several hundred metres of evaporites. These have been found both on the present land mass (e.g. in Sicily and in the northern Apennines) as well as in borings in the floor of the Mediterranean itself. Only at the western extremity of the Tethys, where the northern Betic and the southern Rif straits became gulfs of the Atlantic, is there evidence of continuity of marine sedimentation from the Miocene into the Pliocene. For this reason, Spanish geologists have recently proposed the establishment in Andalusia of a terminal Miocene stratotype, the Andalusian.

The central European sea, or Paratethys, was separated from the northern, or Boreal, sea at the end of the Oligocene by the closure of the Ural, Polish and Alsace straits. Towards the end of the Miocene period of aridity, Paratethys became at first lagoonal and then lacustrine, as its waters drained towards Tethys and were evaporated. Finally, by the Pliocene, all that remained of Paratethys were the land-locked units of *Lake Balaton* in Hungary, the *Black Sea* (rejoined to the Mediterranean as a result of Quaternary faulting), the *Caspian Sea* and the *Aral Sea*.

The third stage was the **phase of Pliocene decompression**. The sea returned to the Mediterranean basin, which took on its present outline. It was rather less extensive in the east than previously, with the Aegean sea, for example, retaining some lacustrine features.

In the west, however, areas which had remained continental during the Cenozoic, such as that of the Tyrrhenian Sea, were at last submerged. As a result, the pre-existing topography was greatly modified in the Pliocene: this was the "Pliocene revolution" of BOURCART (1960-2). The continuing uplift of the mountain chains and the formation of depressions cutting across them gave the Mediterranean coast its present day characteristics. This is the time of the culmination of the Alps, but it is also the time of the deposition of 2000m of soft clays on the Miocene evaporites. Subsequently, in the early Quaternary, upward

movement of the evaporites produced diapiric domes more than 2000m high.

Two examples from Provence illustrate the changes that took place between the Miocene and the Pliocene:

1) During the Miocene the sea lay to the west and to the north (in the Aix basin) while to the south lay the Tyrrhenian continent (cf. Fig. 7.10) of which only the Maures-Esterel massif (east of Toulon) remains today.

2) Immense accumulations of pebbles and conglomerates on the flanks of the Maritime Alps came from the south in the Miocene from a source situated between Nice and Corsica. Today sediment is supplied from the north.

2. Spain

Almost the whole of Spain remained above sea-level in the Tertiary period. However, in Palaeogene times the Ebro valley was submerged and a thick series of deposits was laid down from which the stratotype of the Ilerdian was selected (cf. p. 128). Elsewhere, as for example in the region of Teruel (about 150km north west of Valencia), continental deposits were accumulated which contain a mammalian fauna which justifies the definition in Spain of two "continental" stratotypes for the Neogene, the **Vallesian** and the **Turolian** (CRUSAFONT, 1951, 1965). The Vallesian type section is in the Valles-Penedés depression, near Sabadell in Barcelona province. A species of *Hipparion* (*H. gracile*) from this section has been dated as 12Ma in a fauna which is post-Vindobonian, but pre-Pontian. The Turolian type section (at Teruel) was proposed by CRUSAFONT (1965) as a replacement for the **Pikermian** (from Pikermi, near Athens) which he had proposed in 1950. It yields *Hipparion concudense* and *Mastodon longirostris*. These two stages correspond approximately to the sequence extending from the late Tortonian to the end of the Miocene (cf. Table XII, p. 164). Beds of early Tortonian age do not contain *Hipparion*.

Only in the south east was Spain traversed by an arm of the sea, the north Betic strait. This is paralleled in Morocco by the south Rif strait. The sequence in the Guadalquivir basin and the nearby sub-Betic chains extends from the Aquitanian to the Pliocene. At Carmona (30km north east of Seville) PERCONIG (1964) proposed a stratotype for a marine formation filling the gap between the Tortonian and the Pontian, which he named the **Andalusian** (cf. (Table XII, p. 164).

These beds contain a rich fauna of molluscs, especially *Chlamys* and *Pecten*. The presence of *Globigerinoides extremus extremus* and ancestral forms of *Globorotalia margaritae* indicates that the formation is of late Miocene age. For some authors, the presence of *G. margaritae* suggests that the Andalusian extends into the Pliocene, but, on the basis of the world-wide distribution of this species, PERCONIG (1971) resists this assertion, believing that this *Globorotalia* appeared already in the late Miocene. The Andalusian fauna has been found in the provinces of Murcia and Alicante and is correlated with the mammals of the Vallesian and of the Turolian-Pikermian in the province of Teruel (MONTENAT).

The north Betic strait extended to the Balearic Islands and probably joined up with the circum-Alpine trough in the early and middle Miocene. In the late Miocene, it was temporarily obliterated in the region between Cartagena and Valencia during the episodes of drying out of the Tethys. This strait was finally closed in the Pliocene when the sea returned to the Mediterranean by way of the Straits of Gibraltar.

3. Italy (Table X, p. 138)

The Oligocene transgression in the Piedmont province of northern Italy came from the east, the Gulf of Genoa being occupied by a continent. The Miocene transgression came from the same direction and sedimentation began in the Aquitanian with the formation of sandstones and conglomerates followed by marls containing *Chlamys, Miogypsinoides, Miogypsina* and *Lepidocyclina*. These marls were followed by Burdigalian sandstones from which *Miogypsinoides* is absent. They are succeeded by blue sandy marls with pteropods, for which PARETO (1865) created the **Langhian** stage, from the hills of Langhe in the valley of the Bormida (Fig. 4.21, p. 129). In its microfauna, the Langhian resembles the Burdigalian. It contains a cephalopod, *Aturia aturi*, which has been found at the base of the

Helvetian in the Austrian *Schlier*. The Langhian is succeeded by the **Serravallian** (p. 139).

The Helvetian proper is in a sandy, neritic facies marking the beginning of the Miocene regression. Molluscs are abundant, *Cardita jouanneti* (Fig. 3.2, p. 60) being typical, but *Lepidocyclina* is absent.

The type section of the **Tortonian** (the upper part of the middle Miocene) is at Tortona in the province of Alessandria (Piedmont) (Fig. 4.21, p. 129) and was chosen by MAYER-EYMAR (1857). This section exposes 300m of grey marls with a sandy intercalation and contains large turrids, *Ancilla glandiformis, Cardita jouanneti, Cerithium rubiginosum* and the planktonic foraminiferids *Globigerina nepenthes* and *Globorotalia menardii*. The beds which overlie the Tortonian in Piedmont are lagoonal or continental and contain hydrocarbons, gypsum, and sulphur, the latter produced by the reduction of gypsum by organic matter (bacteria). This is the *formazione gessosolfifera* (the gypsum-sulphur formation) which is found throughout the Mediterranean region, even in borings below the sea floor. It is of late Miocene age. The fossils in these beds (*Mactra podolica, Ervilia podolica*) are similar to those of the Sarmatian of the Paratethys.

In Italy, this sequence is known as the **Messinian** (MAYER-EYMAR, 1867). A stratotype has been proposed by SELLI (1960) at Caltanissetta, in the south of Sicily near Agrigento (Fig. 4.22). There, the succession from the base upwards is gypsum, limestone, evaporites, diatomite, and grey-blue marls with foraminiferids. The marls also contain small molluscs of the family *Dreissenidae*, analogous to those of the Danube basin. It is difficult to accept a stage defined in a lagoonal sequence and with an impoverished fauna, as of chronostratigraphic or even a biostratigraphic value. There appears to be a strong case for adopting the Andalusian, a truly marine sequence, as the standard chronostratigraphic stage. However, Italian geologists regard the Andalusian as belonging to the Atlantic domain rather than to that of the Mediterranean.

Sicily is well-known for its important deposits of Miocene calcareous molasse, which have been used as building stone for more than

Southern Europe

two thousand years (Figs. 7.6 and 7.7). Sulphur from the "gypsum-sulphur formation" has also been of considerable economic importance in Sicily.

In central and southern Italy, Miocene beds form the *backbone of the Apennines* which were elevated by the Attic phase at the end of the Tortonian. The rocks are largely arenaceous marls and calcareous molasse which are less varied and less fossiliferous than the Miocene of Piedmont and Sicily.

In the **Pliocene,** a new transgression led to the submergence of a large part of Italy (Fig. 7.8). In particular, a gulf penetrated far into

7.6 The Roman amphitheatre at Syracuse, Sicily. The tiers of seats are cut into a hillside of Miocene molasse. Weathering of the more marly beds shows the dip of the strata (photo: Pomerol).

7.7a **The Temple of Concord at Agrigento, Sicily.** It was constructed about 2500 years ago from Quaternary shelly beds ("panchina") of Calabrian age. Erosion has picked out the rhythmic stratification of the stone (photo: Pomerol).

7.7b **Detail of the steps of the Temple of Concord at Agrigento** (Centimetre scale, photo: Pomerol).

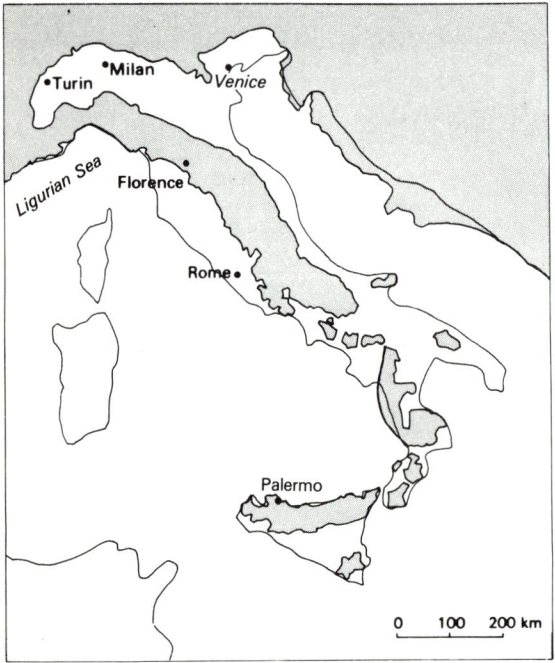

7.8 **Italy at the maximum of the Pliocene transgression** (after Azzaroli and Cita, 1963). Comparison with Fig. 7.9 shows the change in area in a relatively brief time interval.

7.9 **Italy at the maximum of the Pleistocene regression** (after Blanc, 1942).

Piedmont, where the classic Pliocene localities are now to be found (Fig. 4.21, p. 129). Two major stages are recognised in the Italian Miocene (Table XI, p. 140).

The **Tabianian** (MAYER-EYMAR, 1867) or lower stage consists of some 200m of blue pelitic marls in the type section at Tabiano, near Parma, Piedmont. At their base is a sandy conglomerate, resting on the upper Miocene evaporites. Among the characteristic molluscs, *Ficula undata* and *Xenophora testigera* are noteworthy, as are *Sphaeroidinellopsis* and *Globorotalia margaritae* (formerly *G. hirsuta*) among the foraminiferids. The Tabianian is synonymous with the **Zanclean** (SEGUENZA, 1968), from Zancla, the Roman name for Messina. The Zanclean beds are white marls with foraminiferids, known as *trubi* in southern Italy (Calabria, Sicily), which rest on the Messinian. They contain a microfauna of early Pliocene age which has been found by CITA (1972) in the deep-water borings of the "Glomar Challenger". Unfortunately, a stratotype for the Zanclean has not yet been defined.

At the top of the type section of the Tabianian, the contact with the succeeding stage, the Plaisancian, is clearly visible.

The **Plaisancian** (ex-Piacenzian, MAYER-EYMAR, 1857) or upper stage consists of blue marls with a variable sand content. In the Po valley (Plaisance) these beds are about 275m in thickness. The marls are rich in molluscs (*Turritella communis, Natica millepunctata, Murex trunculus, Conus ponderosus*). *Globorotalia crassula*, which is characteristic of the middle Pliocene, is rare, and the zonation is based on *Globogerinoides obliquus extremus* in the lower part and *Globorotalia inflata* in the upper part. The blue marls of the type section are overlain by a yellow sand of the Astian facies. The **Astian** (DE ROUVILLE, 1853) takes its name from the plain of Asti in Piedmont where yellow, fine-grained sands with small lenses of marl or limestone reach a thickness of about 100m. These beds also are rich in molluscs, but contain few planktonic foraminiferids. Once considered to be a separate stage, the Astian is, in reality, only a lateral facies of the Plaisancian. Further east, the sands thin out and finally disappear, thus tracing the eastwards regression of the late Pliocene sea, which had begun much earlier in the west. Thus the Astian sands are diachronous and have no chronostratigraphic value. The Pliocene, a relatively brief epoch (3-4Ma), would thus include two stages, the Tabianian and the Plaisancian. However it might be more logical to accept that such a subdivision is unnecessary in the context of the much longer mean duration of Tertiary stages in general.

The Po valley is the site of yet another stratotype, that of the **Villafranchian.** This is a continental stage, the upper part of which is correlated with the marine Calabrian stage, at the base of the Quaternary. This correlation implies that the lower Villafanchian is Plaisancian. The 'heresy' of a stage sitting astride the boundary of a period is, however, less important for continental formations than for marine formations.

4. Paratethys

Paratethys was the domain, at first marine and later lagoonal, which occupied central and eastern Europe together with its prolongation from Vienna across the Caucasus to the Aral sea. It was initiated when the links with the northern oceans by way of the North Sea and the Ural straits were broken at the end of the Oligocene. The Paratethys was thus isolated from the northern seas and separated from the Mediterranean proper (or Tethys) by the Dinarides and the Balkan Range. This separation was not complete, however, since there was a permanent connection through the Aegean strait to the east and a temporary connection through the Vienna basin and the circum-Alpine trough in the west. There may have been additional connections through the Dinarides and the Hellenides (Fig. 7.10).

Four great inter-connected basins can be recognised in the Paratethys. From west to east, they are as follows:

1) The **Pannonian basin** (from Pannonia, the region including Hungary and the Vienna

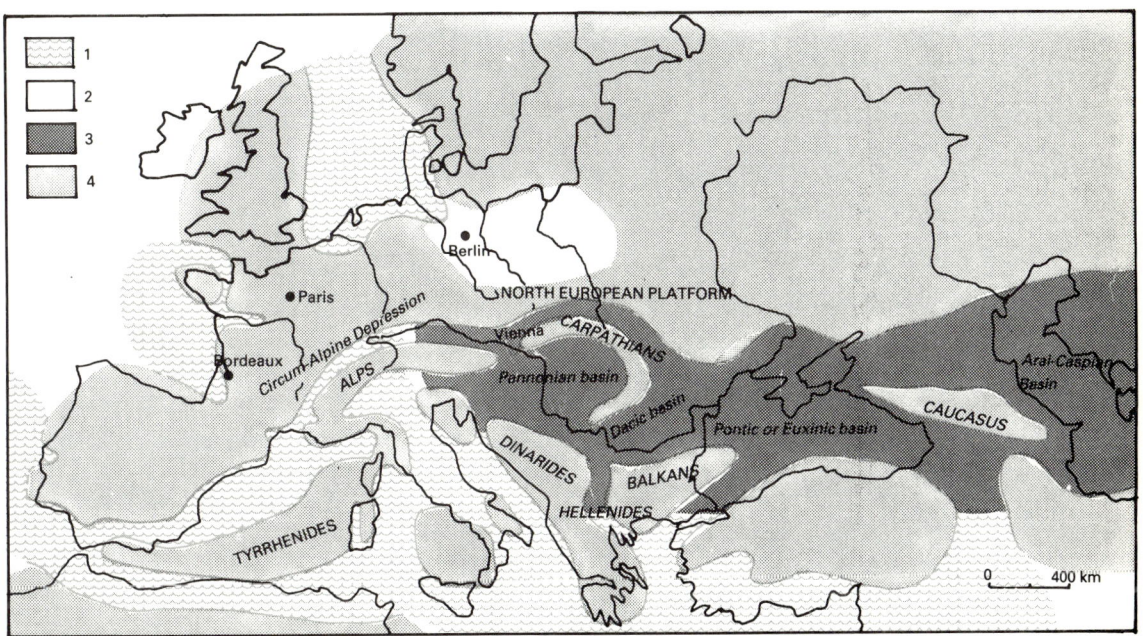

7.10 **Europe in the middle Miocene.** 1 – North Sea, Atlantic, Tethys; 2 – lignite-bearing continental formations of Germany and Poland; 3 – Paratethys; 4 – land areas.

basin) lying between the Carpathians to the north and east, the Dinarides to the south and the Alps to the west. This basin occupied what is essentially the Hungarian plain, but all that remains of it now is *Lake Balaton*. It had an extension to the east, the *Transylvanian basin*, from which it was separated by the Apusen mountains. The basin communicated to the east with the Dacic basin through the straits of the Iron Gates; to the north west with the circum-Alpine trough through the Vienna basin and to the south with the Tethys through the poorly known trans-Hellenic and trans-Dinaride straits.

2) The **Dacic basin** (from Daces, a tribe of the left bank of the lower Danube) lay between the Carpathians and the Balkan Range and occupied the lower valley of the Danube in southern Romania. The present-day massif of Dobrogea formed an island. This basin is now entirely continental.

3) The **Euxine or Pontic basin** (from Pont-Euxin = Black Sea) was bounded on the north by the Russian platform, on the east by the Caucasus, to the west by the Balkan Range and to the south by the Taurus mountains. The southern part of the Crimea formed an island. All that remains of this basin are the Black Sea and the Sea of Azov.

4) The **Aral-Caspian basin** was also bounded on the north by the Russian platform. On the south it was limited by the northern mountain chains of Iran and it extended eastwards as far as the present Aral Sea.

The Paratethys was occupied from early Miocene times by a sea of normal salinity in which were deposited highly fossiliferous sands and molasse. The intra-Alpine Vienna basin was not, however, invaded until the middle Miocene (Helvetian). As mentioned earlier (p. 161), communication with the Rhône valley was then established through Switzerland.

In Austria and Bavaria, very thick, micaceous blue marls, with *Pecten denudatus, Solenomya doderleini, Cardita jouanneti* and the nautiloid *Aturia aturi*, were deposited. This is the **Schlier** facies, which also occurs in the Helvetian in Switzerland. Above these come beds which are traditionally attributed to the Tortonian, showing two different facies. One is an open sea facies, the **Baden marl** containing

turrids and *Ancilla glandiformis*. The other is the **Leitha limestone** with *Lithothamnion*, bryozoans, echinoderms (*Clypeaster, Echinolampas*) and large bivalves (*Spondylus, Pecten*). In the late Miocene, communication with the west was interrupted and the succession of faunas (*Cerithium* beds, overlain by *Congeria* beds) is related to that of the region to the east. In that region, the lower and middle Miocene are of similar facies to those of the Pannonian basin. However, thick beds of salt (Wieliczka) and potash (Kalysz) occur in southern Poland and in Romania (the Slanic salt deposits). Locally, oil is associated with these salt deposits, and with the underlying flysch deposits of Eocene age.

From the late Miocene, the Paratethys was completely isolated and its fauna evolved in a closed environment (Fig. 7.12). Because of this, it has long seemed necessary to erect a purely regional stratigraphy, illustrated by the old stages Sarmatian and Pontian. They were regarded as of late Miocene age overlying the last marine stage, the middle Miocene **Vindobonian**, a term proposed by DEPÉRET (1895) (from Vindobon = Vienna) to group together the Helvetian and the Tortonian which he considered were two facies of the "second Mediterranean stage". However, Austrian geologists have never accepted this proposal since it is not applicable to the Vienna basin where palaeogeographic changes occurred in the middle Miocene. As a result SENES and others have recently proposed the following stages for the Paratethys (Table XII, p. 164):

Early Miocene

Egerian (BALDI and SENES, 1968) (from Eger in Hungary). This stage has a characteristic molluscan fauna (*Chlamys incomparabilis, Flabellipecten burdigalensis, Mytilus aquitanicus*) and also contains *Miogypsina* and *Globigerinoides*. The Egerian corresponds approximately to the "Chattian-Aquitanian" of central Europe and is regarded by some authors as a transitional sequence between the Oligocene and the Miocene.

Eggenburgian (STEININGER and SENES, 1968) (from Eggenburg in Austria). At the base of this stage there is a well-known calcareous molasse with *Chlamys gigas, C. holgeri, Pecten*

7.11 **Fossils of the Paratethys (Mio-Pliocene = "Pontian").** 1 and 2 – *Dreissenomya aperta*; 3 – *Unio davilai*; 4 – *Didacna praetrigonoides*; 5 – *Limnocardium apertum*; 6 and 7 – *Congeria zsigmondy*; 8 – *Paludina eburnea*; 9 – *Lyrcea (Melanopsis) alartimiana*; 10 – *Neritina prevostiana* var. *gizelae* (photo: Fay).

hornensis, with *Miogypsina intermedia* just above it. The basal beds of the Eggenburgian contain *Mastodon angustidens*.

Middle Miocene

Ottnangian (PAPP and RÖGL, 1968) (from Ottnang in Upper Austria, between Salzburg and Linz). Micaceous marls of the *Schlier* facies contain abundant *Globigerina ciperoensis* and, at the top of this sequence, species of *Uvigerina* ancestral to those of the Carpathian stage. The bivalves are thin-shelled (*Pecten denudatus*,

175

Solenomya doderleini) and are accompanied by the nautiloid *Aturia aturi* which is also found in the marls of Tortona. In the upper part, which is equivalent to the middle Helvetian, the first brackish water molluscs appear (*Oncophora, Limnopagetia*).

Carpathian (CICHA and TEJKAL, 1959). The stratotype of this stage was chosen near the village of Slup in the Carpathians of Czechoslovakia. This stage shows an influx of Mediterranean molluscs of Helvetian age (*Vaginella, Ringicula, Pirenella, Chlamys fasciculata*) and, among the foraminiferids, *Uvigerina graciliformis* and *U. bononiensis*.

Badenian (PAPP and CICHA, 1968) (from the Baden marls in the Vienna basin). The base of this stage is characterised by the appearance of the genus *Praeorbulina*, which is followed by *Globigerina nepenthes* of middle Miocene age. Among the molluscs are *Flabellipecten besseri, Chlamys malvinae, Ancilla glandiformis* and abundant turrids.

Late Miocene

Sarmatian (SUESS, 1866) (from the country of the Sarmates in southern Russia). This stage contains a lagoonal fauna rich in individuals but poor in species (*Cerithium pictum, Ervilia podolica, Mactra podolica, Tapes gregaria*). The Sarmatian corresponds to the latest part of the Miocene in which the basins of the Paratethys were still united. The genus *Hipparion* appeared in the Sarmatian (Vallesian, see p. 169). Towards the end of this stage, the Paratethys began to break up into more or less isolated basins containing endemic faunas.

Pannonian (ROTH, 1879) (from Pannonia, Hungary). The Pontic basin now became totally isolated and in it developed large congerids, mactrids and cockles related to the common *Cardium edule*. In the Ukraine, the latter part of the Pannonian is termed the Meotian (from the Meotian marshes, the old name for the sea of Azov). The corresponding continental stage is the Turolian = Pikermian, which continues into the Pontian s.s. (see below), the last stage of the Miocene if 5Ma is taken as the date of the upper limit of this epoch.

Pontian[1] **s.s.** (BARBOT DE MARNY, 1869) (from Pont-Euxin, the Black Sea). This stage was based on the Odessa limestones. The four basins of the Paratethys were now re-united, but the large congerids were more abundant in the west than in the Odessa limestones of the type area. (It should be noted that the Pontian s.l. may cover a longer or shorter time). Some authors suggest that the Pontian should begin with the appearance of *Hipparion* at about 11Ma, in the first half of the late Miocene (late Tortonian).

Pliocene

Dacian (TEISSEYRE, 1907) (from Dacia, a region of Romania, north of the Danube). This was a period pf regression where a freshwater fauna (*Unio, Viviparus*) predominated in the Pannonian basin, while a laguno-marine fauna, similar to that of the modern Caspian sea (*Limnocardium, Congeria*) persisted in the east.

Romanian (from Romania). In this stage two separate domains developed. To the east lay the Euxine (Pontic)-Caspian-Aral basin with a fauna like that of the modern Caspian sea. Westwards, a series of lakes included the Dacic lake in Bessarabia, the Aegean lake (Sea of Marmara and the northern part of the Aegean sea) and the Pannonian lake. The freshwater fauna was characterised by the evolution of *Viviparus, Paludina, Unio* and the *Dreissenidae*. This facies is still found in Lake Baikal and in the rivers of Yunnan.

The isolated region of the Paratethys thus displays a sequence of marine, lagoonal, lacustrine and finally, in the west, continental environments which allow the reconstruction of the palaeogeography. The evolution of certain genera such as *Congeria* and *Viviparus* (Fig. 7.11) in a closed region can also be followed. The maximum regression occurred at the end of the Pliocene and the beginning of the Quaternary, except in the Black Sea area where a series of faults caused a deepening of the waters and the opening of the Bosphorus and the Dardanelles in the late Quaternary.

---

1. The naming of this stage is often, though incorrectly, attributed to La Play (1842) who, in fact, used the term "Pontic" only in a geographical sense.

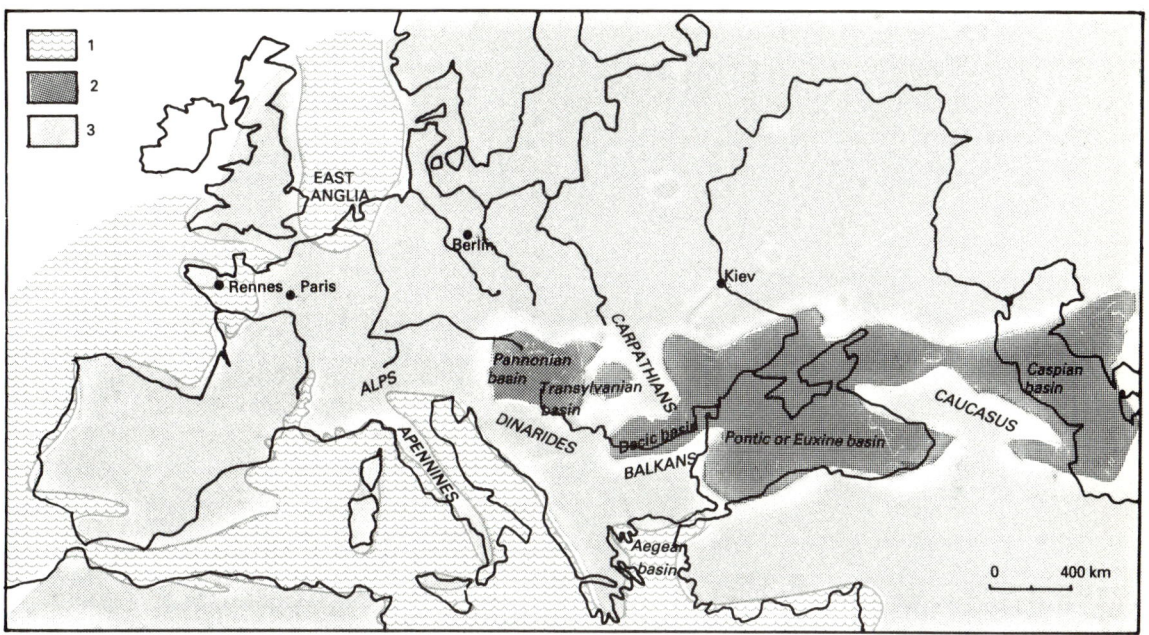

Southern Europe

7.12 **Europe in the Pliocene.** Comparison with Fig. 7.10 shows the reduction in size and the isolation of the Paratethys, while the other marine areas were approaching their present configuration. The English Channel was not yet open and Brittany was an island at the beginning of the Pliocene (Redonian). 1 – open marine areas; 2 – Paratethys; 3 – land.

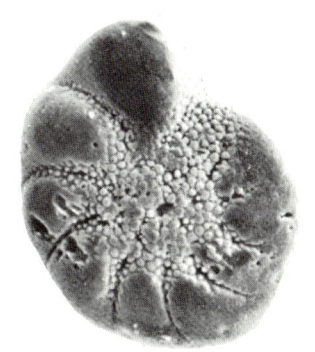
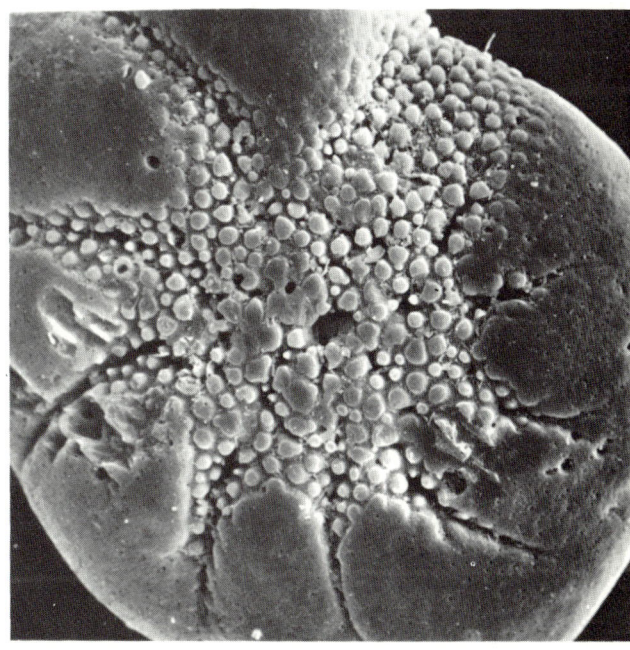

7.13 **Eocene benthic foraminiferid;** *Nonion graniferum*, left: x 180; right x 450 (photo: Le Calvez).

177

7.14 **Sandstone microfacies with Nummulites planulatus** (x 25). This is a thanatocoenosis resulting from the accumulation of nummulites in the Cuise sands (Eocene of the Paris Basin). Note the flattened form of this species. Note also the coexistence of the form A or gamont form (smaller, more inflated, with large initial chamber) and form B or schizont form (larger, flattened, with small initial chamber) (photo: Blondeau).

In North Africa, Tertiary marine formations occur in two very different regions; the northern coastal chain (the Maghrebides) which extend from the Rif in the west to the Kroumirie hills in the east (Fig. 8.1); and the epicontinental basins of the Atlas mountains of Morocco, Algeria and Tunisia.

The Maghrebides form part of the ring of structures resulting from the **circum-Mediterranean Alpine movements,** which also include the Betic cordillera of southern Spain which is linked across the Straits of Gibraltar to the Moroccan Rif, the coastal Atlas mountains of Algeria and northern Tunisia, and Sicily and the Sicilian-Calabrian arc. These structures are of considerable size as is shown by some distances round the perimeter of the ring: Minorca to Cadiz – 800km; Tangier to Tunis – 1600km; Tunis to Trapani – 200km; and Trapani to the Gulf of Taranto – 500km (Fig. 8.1).

DURAND-DELGA (1969) has described the structural geology of the region as follows:

"This orogenic structure takes the form of a much-flattened ring some 2000km in length from the Atlantic to the Ionian sea. It has two principal parts, the Betic cordillera and the coastal Atlas mountains of North Africa, which are separated by distances varying from 200km (from Almeria in Spain to Melilla in Morocco) to 400km (from Majorca to Algiers). This structure is the size of the Appalachians, yet it forms only the western termination of the Alpine chains, which run through Iran, the Himalayas to the East Indies and Timor.

Superficially, it appears that the northern part of the Alpine chain, the Pontides, the Balkans, the Carpathians and the Alps can be followed into the Betic cordillera. Similarly, the southern part, the Taurus, the Dinarides and the Apennines can be extended into the Calabro-Sicilian arc and the coastal Atlas of

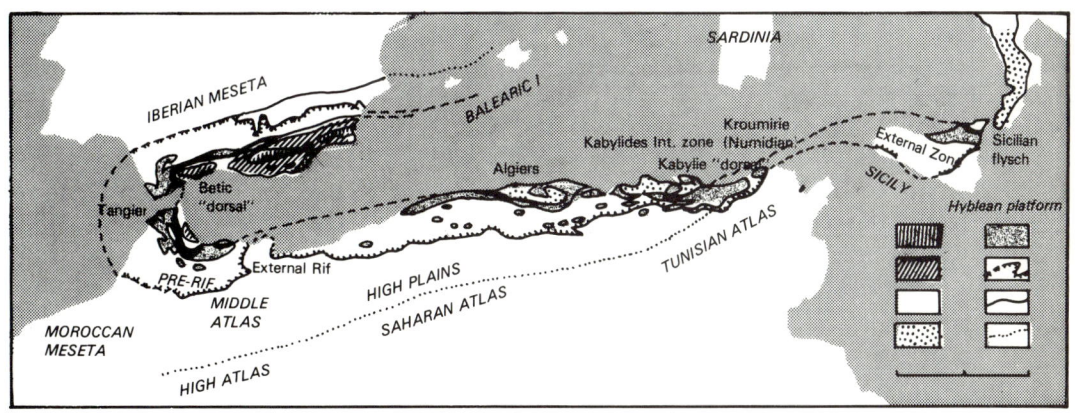

8.1 The circum-Mediterranean Alpine orogeny. *Internal zones:* 1 – In Spain and in the Rif, Nevado-Filabrides; 2 – Alpujarrides and Sebtides; 3 – Malaguides and Ghomarides; 4 – Kabylie massif (Algeria) and Calabro-Sicilian arc, ancient shield; 5 – Cretaceous and Palaeogene allochthonous flysch. *External zones:* 6 – margin of the strongly displaced units (thrust faults); 7 – boundary between Prebetic folding and Iberian meseta; 8 – boundary between the High Plains of Algeria and the folded Atlas Mountains of the Sahara and Tunisia; in black, the *Dorsale* zones – Betic, Rif, Kabylie, Sicilian (after Durand-Delga, 1967).

Barbary. However, this is a gross over-simplification. Although the Apennines and Calabria are near neighbours, their structure and history are very different. Between Corsica, which is the southern termination of the Alps, and Minorca, which is the eastern limit of the Betic system, there is a great hiatus, resulting from the presence of the old craton of Sardinia. There is no way in which connection could be made across this mass unless rotation of the Corsican – Sardinian massif is assumed.

The **'core' of the circum-Mediterranean Alpine orogeny is in fact buried beneath the waters of the western Mediterranean.** Nevertheless, it is convenient to place there a homologue of the 'Zwischengebirge' of

Anatolia (Turkey) or the Pannonian basin of Hungary. Here there are good indications of the **intensity of Recent, Quaternary and Neogene fracturing** in the western Mediterranean which are clearly later than the tangential Alpine tectonics of Eocene to late Miocene times. This explains the apparent absence of morphological unity in the circum-Mediterranean orogeny since large pieces of the structures are below sea-level or are buried by recent formations. However, the elements still observable show the contrast between the internal zones situated in the interior of the ring and the external zones situated around its periphery.''

I. – MOROCCO

The Rif (Fig. 8.2)

In place of an attempt to reconstruct a still rather uncertain palaeogeography, the principal units which can be observed in going from the Mediterranean to the south Rif trench are outlined below[1].

On the edge of the Mediterranean, *the most internal (northern) zones* which are exposed, the Sebtides and the Ghomarides (which correspond to the Algerian Kabylides) are formed of Palaeozoic sandy shales and metamorphic rocks. They are flanked on their southern side by the median zones which are made up of two main parts:

1) The *Dorsale rifaine*[2], a narrow structural unit, well-defined but discontinuous, which is represented in Algeria by the *Dorsale kabyle*. It is made up of Mesozoic rocks capped by a marly limestone of Palaeogene age. It was folded and eroded at the end of the Lutetian, and buried under a thick succession of Priabonian and Oligocene sands and marls. Finally, it was broken up by faults.

2) The circum-Mediterranean allochthonous flysch deposited from the end of the Jurassic to the Aquitanian in two troughs, the more internal of these being the Mauretanian trough,

the other the Massylian. They are better differentiated in Algeria than in Morocco.

The nappes of the flysch going from the more internal to the external (i.e. from north to south) are:

1) **The Beni-Ider nappe,** which begins with Cenomanian sediments and continues with reworked Palaeocene and lower Eocene deposits with *Microcodium*. This is followed by a marly limestone flysch with which is associated a conglomeratic facies. In the upper Eocene, beds of limestone with nummulites and discocyclinids are intercalated in purple-blue marls which become sandy in the upper part. The highest beds of this nappe are Oligocene sandy and micaceous grey marls alternating with grey limestones. This facies varies from a true flysch to a coarse molasse.

2) **The Jebel Tisiren nappe.** Although tectonically above the Beni-Ider nappe, it is made up of older rocks (upper Jurassic to lower Cretaceous).

3) **The Numidian nappe** is of Oligocene and Aquitanian age and is well developed in eastern Algeria and in Tunisia. In the Rif it forms high mountains and covers an area of about 1000km[2]. Around Tangier, the basal beds consist of 100-200m of mottled argillites rich in ironstone concretions with traces of crustacean(?) burrows (*Tubomaculum*). At the top, there is an alternation of sandstones and shales which may reach a total thickness of 1km.

---

1. This chapter (of the French text) has been reviewed and revised by Professor Durand-Delga, University of Toulouse.
2. There appear to be no accepted English forms for some of the North African tectonic units, which are therefore used in the French forms. – *Ed.*

The origin of these median units (the *Dorsale rifaine*, Cretaceous and Palaeogene flysch, Numidian sandstones) is still uncertain. They may be ultra-Rif, that is to say, more internal than the Sebtides (MATTAUER, 1963, ANDRIEUX, 1971) or more external then the Ghomarides, having been carried back as a result of subduction movements (DURAND-DELGA *et al.*) (Fig. 8.3). Though it is not possible to discuss all the available evidence, the second hypothesis

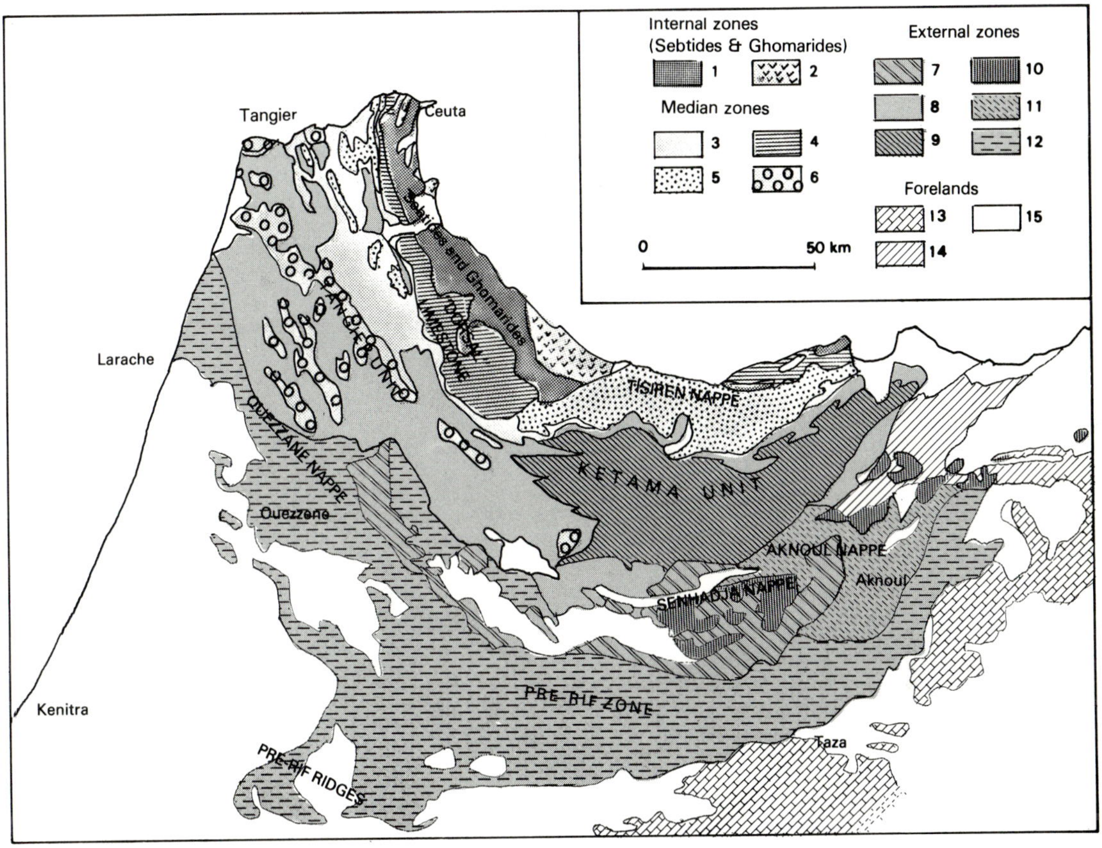

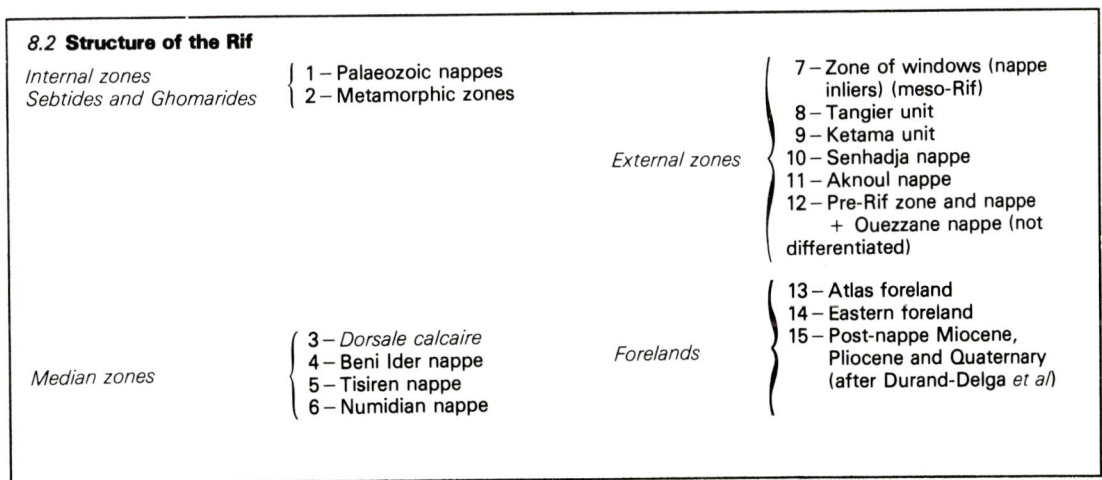

8.2 Structure of the Rif

Internal zones *Sebtides and Ghomarides*	1 – Palaeozoic nappes 2 – Metamorphic zones		7 – Zone of windows (nappe inliers) (meso-Rif) 8 – Tangier unit 9 – Ketama unit 10 – Senhadja nappe 11 – Aknoul nappe 12 – Pre-Rif zone and nappe + Ouezzane nappe (not differentiated)
		External zones	
Median zones	3 – *Dorsale calcaire* 4 – Beni Ider nappe 5 – Tisiren nappe 6 – Numidian nappe	*Forelands*	13 – Atlas foreland 14 – Eastern foreland 15 – Post-nappe Miocene, Pliocene and Quaternary (after Durand-Delga *et al*)

seems to be more acceptable (cf. Fig. 8.7).

While the rocks of the internal and median zones are part of the cover between the European and African shields, **the external Rif zones are**, according to DURAND-DELGA, closely linked to the latter block. The external Rif, separated from the Tell by the Melilla disturbance (Fig. 8.1) is not yet well understood. The autochthon which is related to these structures, appears in the "zone of windows" (pre-nappe Cretaceous to Miocene meso-Rif), in the Ketama unit (Jurassic to upper Cretaceous) and in the Tangier unit (upper Cretaceous to Oligo-Miocene), which together constitute the Intra-Rif.

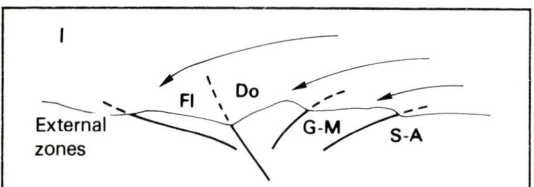

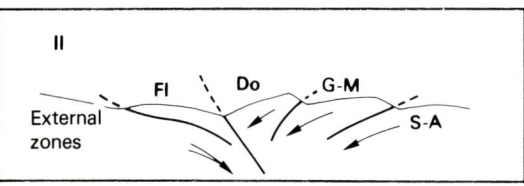

8.3 Structural interpretations of the Maghrebides.
S-A Sebtides-Alpujarrides; G-M Ghomarides-Malaguides; Do *Dorsale*; Fl Allochthonous flysch (after Durand-Delga). Although authors agree on the geometry of the tectonic units to the north and to the south of the Straits of Gibraltar and on their close similarity, the interpretation of the whole structure is disputed.
1 – According to Mattauer and Andrieux, who had worked in Morocco, the Ghomarides (G-M), *Dorsale*(Do) and Flysch (Fl) were more internal than the Sebtides (S-A). The flysch basin was in the region of the present Sea of Alboran[1]. The mechanism invoked for the formation of the nappes is essentially gravitational.
2 – Durand-Delga and Didon, on the other hand, believed that the internal zones were initially underthrust (single arrows) towards the exterior. Later, the external zones (double arrows) were driven under the internal ones, which were sheared at their base. An underthrust, with a contraction of several hundreds of kilometres (under each branch of the orogeny) separated the Internides and the Externides. The mechanisms invoked are essentially crustal (*i.e.* subduction).
1. Alboran: A Spanish island in the western Mediterranean about 220km east of Gibraltar.

In the **Tangier unit,** Senonian marls containing calcareous nodules with a yellow patina, which also occur in the epi-Tellian nappes of Algeria, are overlain by Palaeocene to lower Eocene marls and calcareous marls with black cherts. These are followed by black marls with yellow nodules of the Lutetian. These beds, with a thickness of less than 100m, are overlain by 1000m of sandy marls of late Eocene to early Miocene age.

All these units have undergone considerable displacement in relation to the true autochthon of the middle Atlas. In addition there are a number of true nappes derived from the external regions. These include the Ouezzane nappe of Palaeocene to Tortonian rocks which was moved in two stages; the Rif nappes of Senhadja and of Aknoul (Trias to Eocene), probably coming from the "Nekor scar" (*cicatrice du Nekor*) between the intra-Rif and the Meso-Rif, and the more external pre-Rif nappe (Jurassic to Miocene) which was emplaced in a submarine environment in the Tortonian. The thick Miocene marls of this nappe contain many planktonic foraminiferids and miogypsinids. They are remarkable for the occurrence of reworked fragments of Trias sediments, Eocene nummulites, Priabonian and Oligocene large foraminiferids and planktonic forms of Cretaceous and Eocene age.

According to DURAND-DELGA, the sequence of tectonic phases affecting the Rif and the Tell in the Tertiary was as follows:

1) *A late Lutetian phase* where more recent Palaeogene formations rest with a slight but definite discordance on older ones. The preliminary movements probably began in the Palaeocene.

2) *Oligocene phases* which are recognised by the occurrence of conglomerates in the *Dorsale* regions. The Numidian sandstones are linked to an immediately pre-Aquitanian phase.

3) *Miocene phases* corresponding to the progressive emplacement of the nappes commencing with the most internal (intra-Aquitanian or intra-Burdigalian) and affecting the pre-Rif external zones in the Tortonian.

Following the formation of the nappes, broad folding of the internal and external zones occurred from the upper Miocene onwards.

The Atlas

The palaeogeography of the Atlas region in the Tertiary is the outcome of a slow evolution which commenced in the Palaeozoic.

As in Tunisia, the Eocene was a period of the formation of sandy or marly *limestones, very rich in phosphates* and fish remains. The sea penetrated eastwards into the gulfs of Bahira, Haouz and Sous (Fig. 8.4). The phosphate formations were replaced to the east by a neritic facies containing oysters (*Ostrea multicostata*) and gastropods. At the end of the Ypresian or at the beginning of the Lutetian, the phosphate beds were succeeded by a tabular limestone containing *Thersitea*, a large gastropod restricted to North Africa (Fig. 8.5). Nummulites are absent so that it is customary to distinguish the "Nummulite sea" of the Rif area from the "*Thersitea* sea" of the Atlas region. The sea retreated at the end of the middle Eocene as a result of the late Lutetian

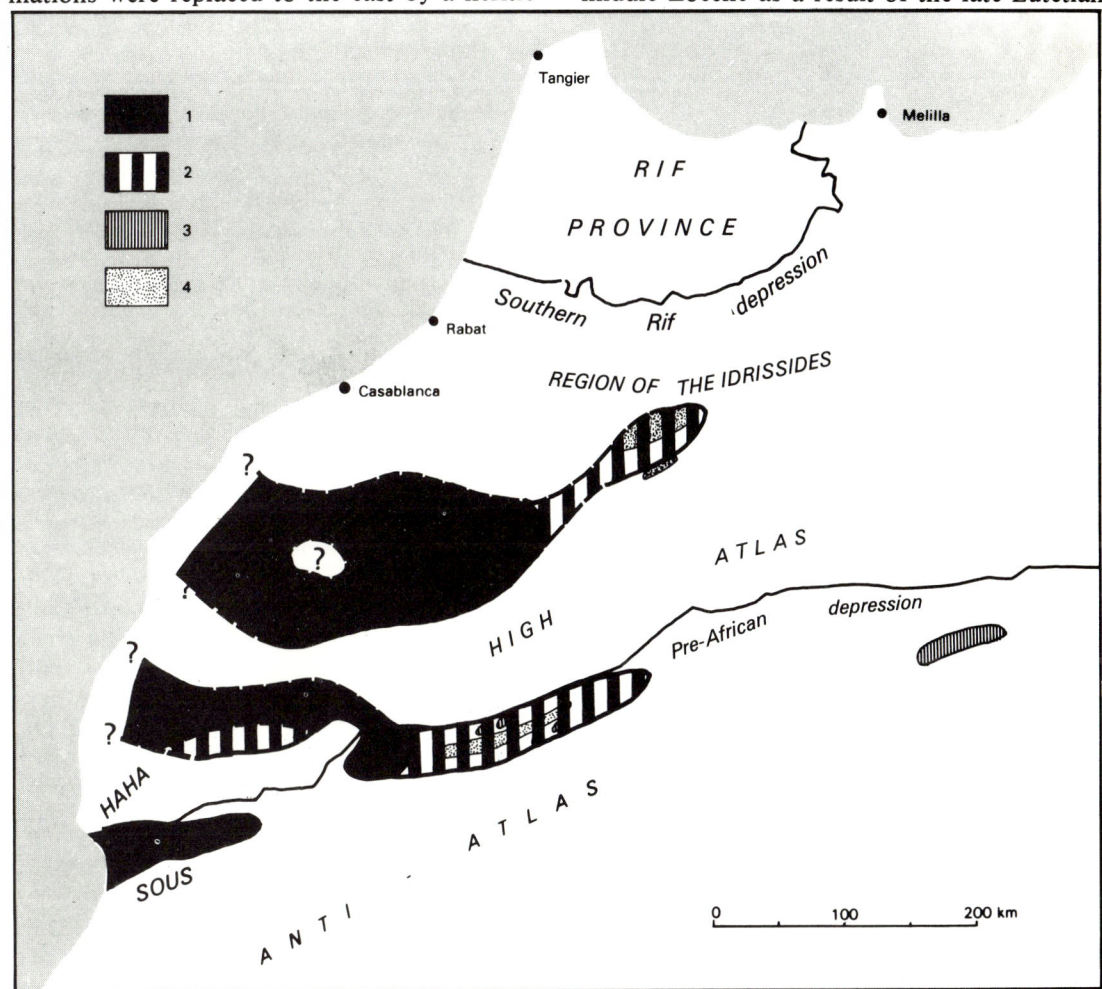

8.4 **Palaeogeography of Morocco during the Eocene and early Oligocene** (after Choubert and Faure-Muret, 1960-2, 1974). *Eocene:* 1 – phosphatic and calcareous facies with *Thersitea*; 2 – facies with echinoids, oysters and gastropods (floor of gulfs); 3 – Lower Eocene lacustrine limestone. *Lower Oligocene:* 4 – continental: red sandstones of Ajt Ibrhirene (Oued Dades), or lagoonal: Feleddi series (Middle Atlas) and middle Moulouya red beds with gypsum.

8.5 Thersitea from the Palaeogene of North Africa (8cm) (photo: Fay).

tectonic phase, and marine sedimentation was replaced in the east by lacustrine limestones and evaporites on the plains with conglomerates and sandstones on the hillsides. This situation continued into the Oligocene, but was brought to an end by peneplanation during the Miocene (infra-Tortonian).

After the phase of **Miocene tectonism** which saw the emplacement of the nappes in the Rif, the sea rapidly invaded the northern margin of the Atlas and spread southwards across southern Morocco and part of Algeria, which was broken up into a series of islands during the Mio-Pliocene (Fig. 8.6). As a result **the southern Rif strait** was opened up, a counterpart of the North Betic strait in the Guadalquivir depression. These straits permitted communication between the Mediterranean and the Atlantic at a time when the Straits of Gibraltar were closed (the latter was not opened until the Pliocene). Three small

gulfs were also formed on the Atlantic coast: Mohammedia (Fedala), El Jadida (Mazagan), and Agadir where CHOUBERT and FAURE-MURET (1962) have identified three horizons:

1) The basal molasse with *Clypeaster* and *Pecten*, together with conglomerates, is, according to HOTTINGER (1966), of Tortonian age.

2) Blue marls of late Miocene age, about 2000m thick, formed in a subsiding basin.

3) Sands and shelly marls at the top of the sequence formerly attributed to the Sahelian, but now shown to be Pliocene.

The sea slowly withdrew from the southern Rif strait in the late Miocene depositing thin beds containing *Ostrea crassissima* as it left the area. As the sea level fell and new slopes were exposed "Ponto-Pliocene" molasse was laid down which contains *Hipparion, Mastodon*, rodents and freshwater molluscs.

Volcanism occurred during the phase of development of a lacustrine limestone in the Plio-Villafranchian and, at the same time, the sea invaded the Atlantic coast of Morocco (Fig. 8.6). This episode, the *Moghrebian*, has yielded a warm water fauna, which nevertheless, contains *Hyalinea balthica*. According to CHOUBERT (1957-65) this allows the Moghrebian to be correlated with the Calabrian. The mammalian fauna, however, indicates that these beds are still lower Villafranchian. At the end of the Villafranchian the last Atlas phase of tectonism destroyed the palaeogeography of the area causing the lakes to be drained and the sea to retreat.

Finally, evidence of the rise and fall of sea level on the Atlantic coast in middle and late Pleistocene times has been preserved in the form of raised beaches. The Sicilian (+50m), Tyrrhenian I and Tyrrhenian II or *Ouljian* (+12m) and finally the climatic optimum of the Flandrian (+2m), stages have been identified.

II. – ALGERIA

Throughout the Tertiary, the sea occupied only the northern border of central and western Algeria, though it extended far across eastern Algeria and Tunisia. However, the High Sahara, the Saharan Atlas (with the exception of the Aures massif) and the central and western High Plateaux remained emergent[1].

---

1. Based on unpublished information kindly supplied to the author by M. Kieken (B.R.G.M., Orléans).

Since the recognition (in 1950-1955) of the great nappe structures of North Africa, it has been accepted that the northern limit of the southern Algerian continent, west of Bejaia (Bougie) must have lain some 100km further north. In the littoral zones of Algeria, likewise, all the regions associated with flysch and earlier than Miocene II (i.e. middle-upper Miocene and post-orogenic) are allochthonous. The Algerian allochthons occur as three great nappes (fig. 8.7) as follows:

1) The flysch nappes of north and south Kabylie (formed by gravitational sliding –

"nappes d'écoulement"). This group includes the Numidian nappe, the Massylian flysch nappe and the Mauretanian flysch nappe.

2) The Tellian or sub-Tellian nappes (also formed by gravitational sliding) which include the ultra-Tellian nappe, the epi-Tellian nappe, the meso-Tellian nappe and others.

3) The Kabylides (over-riding nappes still attached to their roots) and including the *Dorsale calcaire* and the old massifs.

It may be noted that the term "trans-Tellian" now replaces the older term sub-Numidian which was a source of

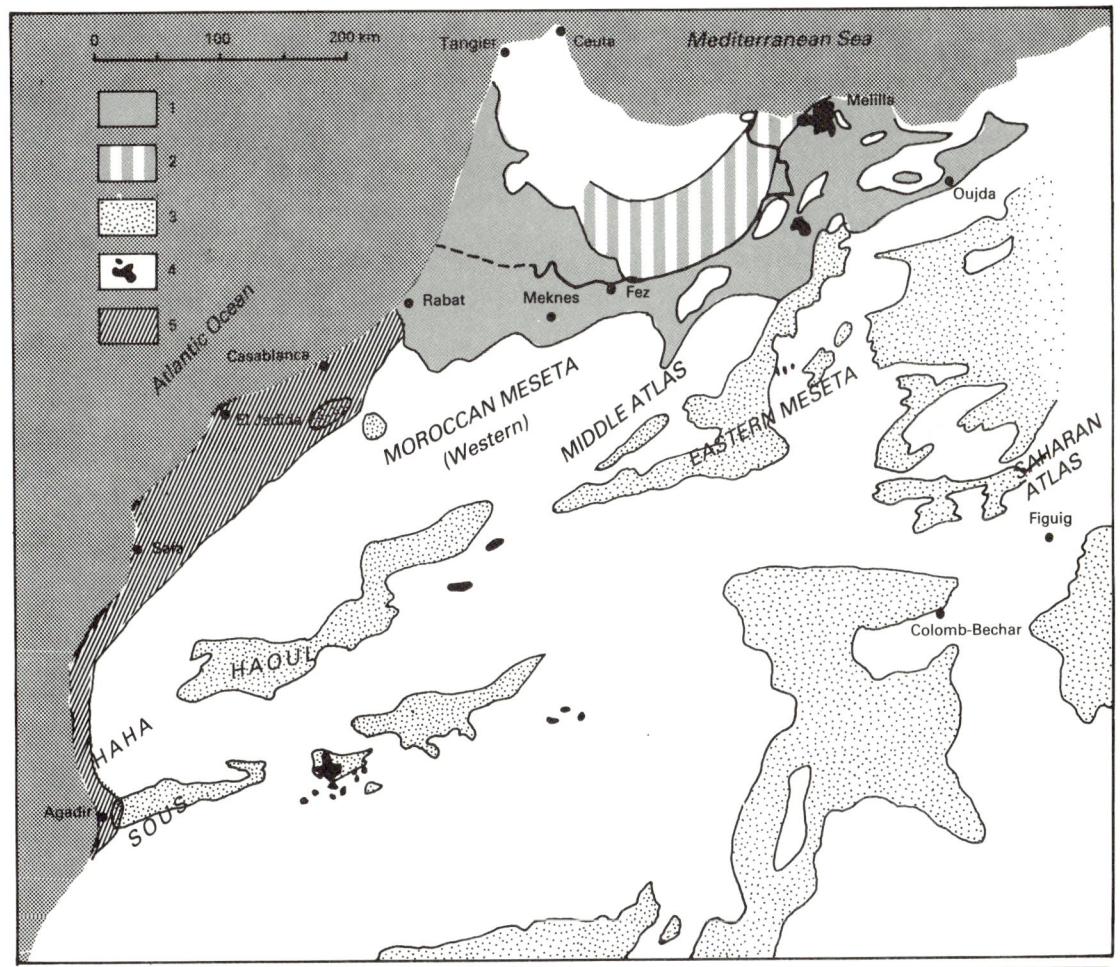

8.6 **Palaeogeography of Morocco in the Neogene** (after Choubert and Faure-Muret, 1960-2): 1 – Mio-Pliocene sea of the Southern Rif Straits, the gulfs of Casablanca, el Jadida (Mazagan) and Agadir (post-nappe Miocene); 2 – Eastern Rif archipelago; 3 – continental Mio-Pliocene; 4 – Mio-Pliocene volcanoes; 5 – Moghrebian transgression and post-Moghrebian consolidated dunes.

confusion. The ultra-Tellian includes upper Senonian microbreccias and breccias (with Tellian fragments of Triassic serpentines) and Miocene conglomerates with both Tellian and "coastal" fragments.

It is necessary to define the two adjectives used above. For simplicity, a "Tellian" zone can be distinguished which corresponds not only to the Tell mountains (geographically) but also to areas of Tellian facies which exist near the present coast (e.g. at Babors, Bou Maad and Fillaoussène). The "coastal" zone, on the other hand, refers to the *Dorsale calcaire*, the old massifs and all the allochthonous flysch whether now situated to the north of the Kabylie mountains (and effectively near the present day coast) or to their south (and thus far inland).

1. Palaeocene

In the Low Sahara (south of the Aures mountains), the Danian appears to be missing. The Palaeocene is represented by 40-80m of white limestone with oysters and chert nodules. The proximity of a shoreline is indicated by red clays with pebbles of Senonian limestone.

Further south (Busson *et al.*, 1955) the base of the Tertiary is formed by 80-150m of marls and dolomitic limestone with algae (Dasycladaceae). These rest on an anhydrite bed, which can be traced as far as Hassi Messaoud and in turn, rests on Maastrichtian limestones with *Laffiteina* (Fig. 8.8).

In the Aures mountains, the series begins with black marls with *Ostrea overwegi, Cardita*

beaumonti and *Roudaireia adrui* (Danian) and then passes up into marls and limestones with *Turritella*, oysters, chert and phosphate nodules. At Kasserine (Tunisia) the top of the Palaeocene consists of 20-30m of phosphate beds (with 40-60% of calcium phosphate), rich in fish remains.

On the northern and eastern borders of the southern Algerian continent (see Fig. 8.8), the Palaeocene is represented by 30-40m of black marls with *Nodosaria* and ostracods. Oysters are common only in the Hodna mountains where limestones with chert and phosphate again occur. Phosphates are also well-developed on the northern edge of the massif. All these deposits indicate a very shallow sea, which was often isolated as shown by the frequent occurrence of gypsum.

Towards the north, the sea was deeper and the Dano-Palaeocene consists of black marls with lenses of glauconite and globigerinids and *Globorotalia* as well as the benthonic fauna. This formation was part of the Tellian allochthon and became incorporated in the meso-Tellian nappe.

In the epi-Tellian nappe, the Dano-Palaeocene marls are impoverished in glauconite and only the microfauna, which is essentially pelagic (the zone of *Globorotalia velascoensis* is well-developed), enables them to be separated from the Maastrichtian marls, In the ultra-Tellian nappe, the Palaeocene is represented only by rare remnants of this facies.

In the Mauretanian nappe, the Palaeocene often begins with coarse conglomerates with

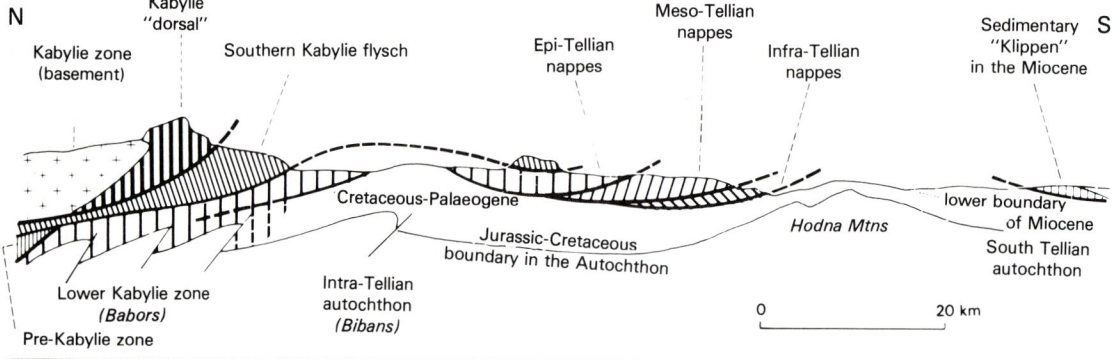

8.7 **Hypothetical structural relationships within the Tellian units and between the Tellian units and the more internal units** (see also Fig. 8.3).

blocks of Lias which may be metres across. These are overlain by sandstones or sandy limestones with rare pebbles of Tithonian age and bands of pale-coloured chert. They also contain *Microcodium* and *Melobesia*. The pebbles appear to have come from the *Dorsale* which must have been emergent. Sandy limestones with *Microcodium* and sandy marls are found locally in the Massylian flysch. The thickness of these deposits is small and they only average about 20m.

In the Kabylie ridge, marls with a Danian microfauna are reported at Chenoua (west of Algiers).

2. Early Eocene (Fig. 8.8)

The lower Eocene is represented throughout Algeria by cherty limestones which, when neritic contain *Nummulites globulus, N. guettardi, N. irregularis* and *N. gizehensis,* which mark the passage up into the Lutetian. The presence of *Truncorotalia aragonensis,*

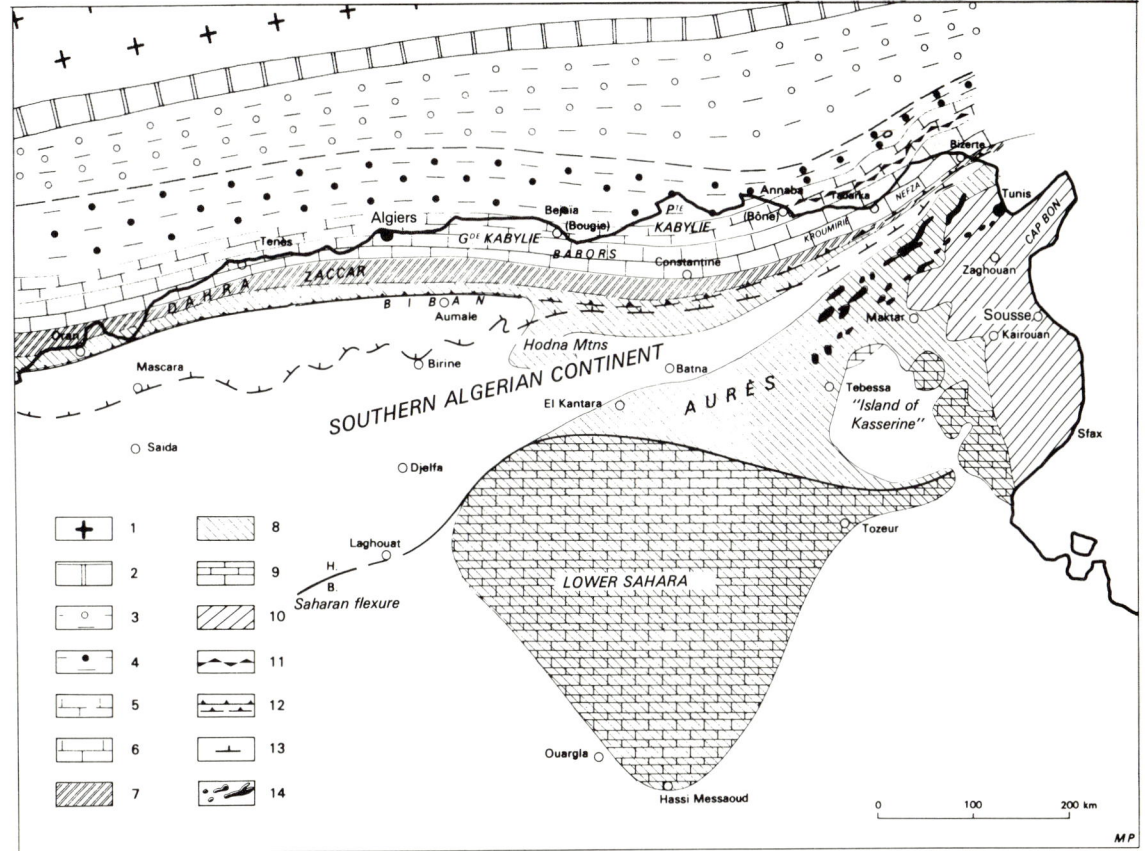

8.8 Palaeogeography of Algeria and Tunisia in the early Eocene (after Kieken and Rouvier, original). 1 – Kabylie and Palaeozoics; 2 – *Dorsale calcaire;* 3 – Mauritanian flysch: red marls and microbreccias with *Nummulites;* 4 – Massylian flysch: microbreccias, marls with *Cuvillierina* in Algeria; calcareous micro-breccias, conglomerates (cf. Adissa formation, Ypresian) in Tunisia; 5 – Ultra-Tellian: marly limestones, calcareous microbreccias, conglornerates + facies with Tellian affinities in the southern part; 6 – Epi-Tellian: limestones with *Globigerina* and scattered flints (the Suessonian facies of Kieken and Rouvier); 7 – Meso-Tellian: limestones with mixed microfaunas, flints at the base and rare *Nummulites* in Algeria; limestones with mixed microfaunas (*Globigerina* at base and *Nummulites* at top) in Tunisia; 8 – Algeria: limestones with *Nummulites* and benthos (dominant), with thick beds of flint and some phosphate layers; Tunisia: limestones with *Nummulites* or mixed microfaunas (Ypresian-lower Lutetian); 9 – dolomitic limestone with algae, nummulites and ostracods, with flint at the top; 10 – zone of *Globigerina* facies in Eastern Tunisia; 11 – hypothetical ridge: boundary between Massylian flysch trench and Tellian trench (Tunisia); 12 – thrust plane of the Tellian nappes; 13 – Tellian nappe-front; 14 – diapirs.

Globorotalia acuta and *Globigerina soldadoensis*, found in slightly deeper water (infra-neritic) sediments, indicates the Ypresian age of these beds, which cover wide areas in the Lower Sahara and the Hodna mountains. In the Constantine autochthon and in the meso-Tellian nappe, however, there is a rapid passage to deposits of an intermediate facies in which plankton and benthos are present in nearly equal proportions. East of the longitude of Constantine, the Palaeogene is not known *in situ*.

In the epi-Tellian nappe, cherts are dispersed throughout the Lower Eocene, while the limestones, which are more than 200m thick, contain large numbers of globigerinids. This is in the infra-neritic zone which covered the northern part of the Tell.

Further north, in the region of the ultra-Tellian nappe, clayey limestones with bands of microbreccia and conglomerate lenses were deposited in the lower part of the series. This is the "Adissa" facies, well-known in Tunisia (JAUZIEN and ROUVIER, 1965). However, the trans-Tellian nappe includes only Senonian microbreccias and conglomerates and very coarse grained Miocene detritus. This tends to suggest that the Palaeogene terrain was eroded to provide the detritus for the Miocene sediments.

In the Massylian nappe, the Ypresian consists of calcareous microbreccias (similar to those of the preceding nappe) and marls with *Cuvillierina*. The same facies occurs in the Mauretanian nappe, but it has an abundance of small nummulites, discocyclinids, *Alveolina* and *Assilina*. Microbreccias also occur, intercalated in red marls.

3. Middle Eocene

The middle Eocene was a period of regression. The sea retreated from the Low Sahara, although this area subsided considerably. Between 500 and 600m of gypsum and flaggy limestone were deposited in a trough formed south of the "Saharan flexure", while conglomerates accumulated on the edge of the southern Algerian continent. The evaporites extended as far as the region of Hassi Messaoud in the west and as far as Tozeur in the east. Their thickness decreased southwards: 150m of gypsum occurs between Ouled Djellal and

Duargla but there is only about 40m north of Messaoud. Further south these deposits pass into continental sands (BUSSON et al., 1955).

In the Aures mountains, the Lutetian has been largely removed by erosion, but further north in the Hodna basin some 500m of Lutetian gypsum has been preserved. North of the Hodna mountains, in the region of Ain Tagrout, and south of Constantine, the middle Eocene consists essentially of brown clayey marls with numerous calcareous lumachelles with many oysters (*Ostrea multicostata* var. *strictiplicata, O. roncana*), ostracods and arenaceous foraminiferids. This formation, often with a thickness of more than 500m occurs in the meso-Tellian nappe, in which it covers immense areas.

In the region of the epi-Tellian nappe, these platform deposits are replaced by blackish marls with nodules of yellow limestone. Only the micro-fauna of these marls (*Globorotalia spinulosa, G. centralis, G. lehneri, Truncorotaloides topilensis*) enables them to be separated from the upper Eocene.

In the trans-Tellian nappe, the middle Eocene is almost unknown except in the Adissa unit (Tunisia), where it consists of black marls with microfossils. In the Massylian and the Mauretanian flysch, pink marls with conglomerates and microbreccias are attributed to the Lutetian. The microbreccias have a fauna of large granulate nummulites together with *Fabiana, Linderina* and *Assilina exponens*.

South of the *Dorsale*, variegated marls occur, while to the north, massive limestones with nummulites and *Alveolina* are overlain by conglomerates. The microfauna is similar to that occurring in the flysch.

If it is accepted that the Kabylides were originally to the north of the flysch trench, then the continent of the Kabylie was symmetrical with the southern Algerian continent and the margins of the Eocene trench can be clearly recognised.

4. Late Eocene (Fig. 8.9)

The marine regression continued in the late Eocene and beds of this age are unknown in place in eastern Algeria. In the meso-Tellian nappe, the Bartonian consists of marls which in the lower beds contain only benthonic forms.

In the upper beds, however, they contain thick glauconitic sandstones with *Globorotalia cocoaensis* and *Hantkenina dumblei*.

In the epi-Tellian nappe, Bartonian marls occur only very locally, having been largely removed by erosion in the Oligocene and Miocene. In the rare exposures of these marls they can be seen resting on upper Cretaceous beds of the trans-Tellian nappe. In the Massylian and Mauretanian flysch, the upper Eocene is represented by red-violet argillites and arenites. The Mauretanian flysch contains *Nummulites striatus* and *N. incrassatus*. Microbreccias, conglomerates and green sandstones derived from Palaeozoic sources indicate that the Kabylides were emergent in the late Eocene.

5. Early and middle Oligocene (Stampian)

During the Oligocene, the whole of southern Algeria was above sea level. Conglomerates, sandstones and red clays which formed on this

landmass have been preserved in synclinal basins as, for example, at Hodna.

In the meso-Tellian region, continental sediments pass upwards into marine sandstones with *Pecten arcuatus* (the "Boghari" facies). In Oran, these sandstones contain *Nummulites fichteli, N. vascus, Nephrolepidina,* and *Eulepidina*. At the top of the sequence there are very thick (about 1000m) marls and sandstones (1000m) with *Globorotalia opima opima*.

This formation appears in the lower part of the epi-Tellian nappe, but, elsewhere, most of the Oligocene has been removed by erosion. Small patches do, however, remain on Djebel Megriss and in the region of Lamy. The Oligocene has not been found in the trans-Tellian nappe.

North of the area of origin of the epi- and trans-Tellian nappes, the following occur:

1) In the Massylian area there is a complex sequence of micaceous sandstones containing many metamorphic fragments, which is transgressive over various levels of the Cretaceous.

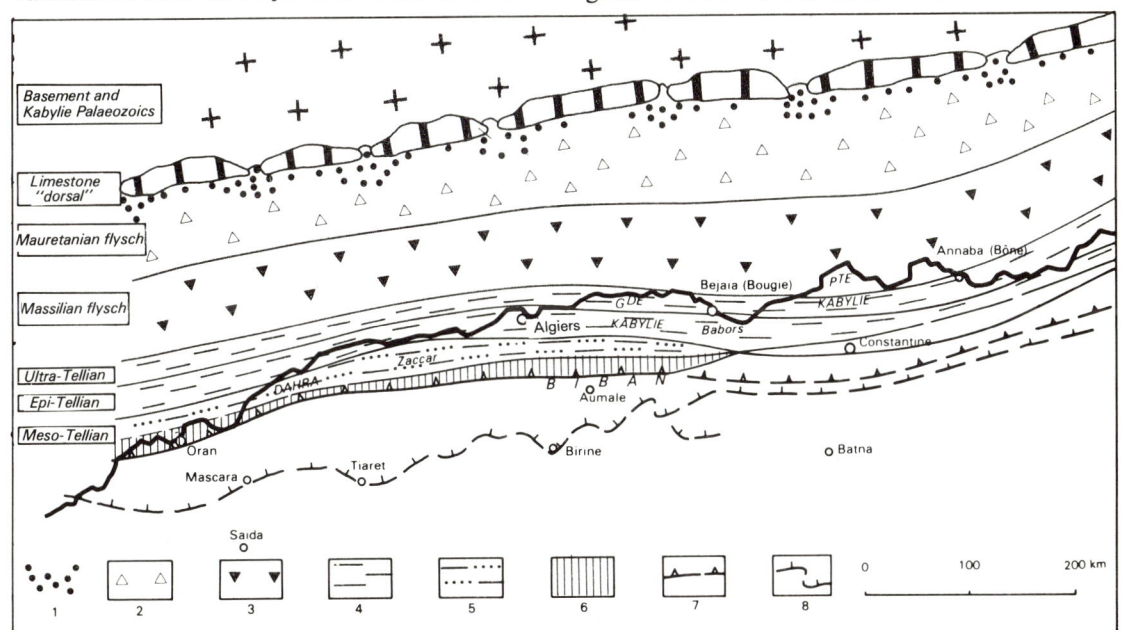

8.9 Palaeogeography of Algeria in the late Eocene (after Kieken, 1970).
1 – Conglomerates; 2 – Mauritanian flysch: red clays with calcareous microbreccias and conglomerates, *Nummulites*; 3 – Massylian flysch: red clays with calcareous microbreccias, *Chapmanina*; 4 – Ultra-Tellian and Epi-Tellian: marls with nodules of yellow limestone, plankton dominant; 5 – Meso-Tellian: bedded glauconitic marls, benthos dominant; 6 – Clayey marls with agglutinating foraminiferids; 7 – Thrust plane of Tellian nappes; 8 – Tellian nappe-front.

RAOULT (1969) has drawn attention to the presence of *Eulepidina, Nephrolepidina*, large rotalids, *Nummulites intermedius* and *N.* cf. *vascus*, an assemblage which, in the absence of *Miogypsinoides*, can hardly be younger than Stampian.

2) In the Mauretanian area, more than 500m of micaceous sandstone with a fauna identical to that of the Massylian area occur. Also present are fragments of mica-schist, Jurassic rocks and many plant fragments. There is a notable absence of pebbles of Tellian origin and it is suggested that the detritus comes from the shield of the Kabylie.

3) Further north, the *Dorsale* is covered on its southern border by conglomerates, micaceous sandstones and microbreccias. This is RAOULT's "Nummulitic II".

The Massylian and Mauretanian flysch sequences and the deposits south of the *Dorsale* are all of early Oligocene age. By contrast, the upper part of the Boghari sandstones contains rare *Miogypsinoides complanata*, indicating a late Oligocene age (MATTAUER, 1963). The sequence appears therefore, to cover the whole of the Oligocene.

6. Oligo-Miocene (Fig. 8.10)

KIEKEN used the term "Oligo-Miocene" for two formations of which the precise age is difficult to establish. These are the Oligo-Miocene Kabylie series and the Oligo-Miocene Numidian facies.

a) **The Oligo-Miocene Kabylie series** is discordant on the Kabylie shield. It was formerly believed to be of the same age as beds which rest discordantly on the south side of the *Dorsale calcaire*. However, recent studies of the area and the microfaunas suggest that they were

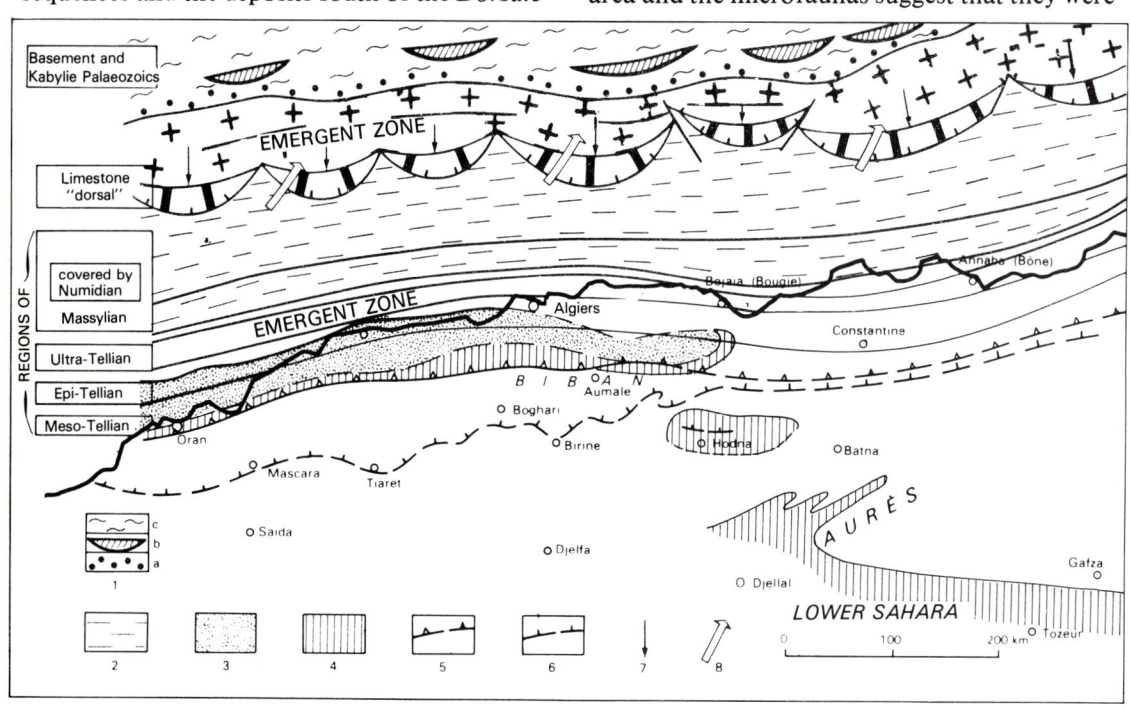

8.10 **Palaeogeography of Algeria at the end of the Aquitanian** (after Kieken, 1970).

1 – Oligo-Miocene of Kabylie Massif:
a) conglomerates
b) olistoliths (southern Kabylie flysch)
c) clay with boulders ("argiles à blocs")
and fissile clays ("arqiles écailleuses");

2 – Numidian and "Boghari sandstone";

3 – Sandstone and marls with mixed microfauna;
4 – Continental clays and sandstones;
5 – Thrust plane of Tellian nappes;
6 – Tellian nappe-front;
7 – Kabylie thrust;
8 – Direction of reversed thrusting.

probably not formed simultaneously. In addition, according to recent surveys, the Oligo-Miocene Kabylie series contains olisto-stromes and olistoliths. These discoveries are very important and form the basis of a new theory to explain the tectonics of this region, as will be shown later.

North of the Grande Kabylie, the series begins with conglomerates and limestones with *Miogypsinoides complanata, Nephrolepidina* and large rotalids of latest Oligocene-early Aquitanian age. This basal formation probably plunges under the Massylian flysch which also contains micaceous sandy flysch with *Lepido-cyclina* but without *Miogypsina* (Stampian). This flysch is overridden by Tellian rocks (RAYMOND, 1970) and the whole is covered by shales containing blocks of Numidian sand-stone and also Massylian and Tellian frag-ments. These shales are of early Burdigalian age as is shown by the presence of *Globigerinoides trilobus* and *G. immaturus*.

North of the Petite Kabylie, the shield is also covered by conglomerates and micaceous sand-stones with fragments of shield rocks and of the *Dorsale*. These beds contain *Lepidocyclina*, large rotalids and *Miogypsinoides* of Chattian age. Above them come cherty beds ("silexites") and reddish, micaceous shales with *Miogypsina*. These latter beds contain olistoliths consisting essentially of fragments of Massylian and Mauretanian flysch. In general, the beds which are transgressive over the Kabylie shield are either latest Oligocene or belong to the Aquitanian-early Burdigalian zone of *Globigerinoides trilobus*.

b) **The Numidian facies,** although wide-spread across all of Barbary, is almost unfossiliferous and, therefore, is difficult to date. The basal beds are clays with coprolites (*Tubotomaculum*) overlain by more than 1000m of alternating shales and poorly sorted sandstones. The sandstones are unfossiliferous but the shales contain, as well as abundant arenaceous forms, rare planktonic foramini-ferids, *Globigerinita dissimilis*, and *Globigerina venezuelana*, species which have a great vertical range. Very occasionally, the shales contain thin lenses of micaceous sandstone with *Lepidocyclina* and *Miogypsinoides complanata*. South of Biban (near Madjana) the Numidian sandstones rest on a micaceous

sandstone with Stampian nummulites and lepidocyclinids. The deposition of the Numidian series was therefore later than that of the micaceous sandstones which cover the Massylian and Mauretanian regions.

MAGNÉ and RAYMOND (1972) have also described a "supra-Numidian" series of argillaceous marls and marly limestones which rest on the Numidian sandstones and contain *Globigerinoides trilobus, G. immaturus* and *G. cf. primordius* of late Aquitanian and early Burdigalian age. This series is similar to the Babouch formation of Tunisia, which has yielded a microfauna of the same age (GLACON and ROUVIER, 1967). CAIRE (1957) has described a similar series south of Biban.

7. The palaeogeography of the Oligocene and the Aquitanian (Fig. 8.11)

It is difficult to reconstruct the palaeo-geography of this period for it was a period of intense tectonic activity. This difficulty is made greater by the uncertainty of the age of many horizons.

The most recent investigations (DURAND-DELGA, KIEKEN) indicate that the sequence of events can be outlined as follows:

a) **Early Oligocene:** the principal feature appears to have been an east to west trough or furrow in which the Boghari sandstone was formed. This trough apparently occupied the area now covered by the meso-Tellian nappe. It was bordered on the north side by a ridge or fold (in the region of the epi-Tellian and ultra-Tellian nappes) beyond which a series of micaceous sandstones were deposited as far as the *Dorsale* region. It seems probable that the Kabylie shield was for the most part emergent at this time.

b) **Late Oligocene and Aquitanian:** tectonic activity increased at this time and it becomes more difficult to reconstruct the palaeo-geography of the region, especially as the rocks are largely unfossiliferous. In the present state of knowledge, the following outline is proposed:

In the late Oligocene, the sea transgressed over the Kabylie shield, but the *Dorsale* and its Palaeozoic margins remained, at least locally, emergent as is suggested by the presence of Palaeozoic and Jurassic debris in the basal beds. To the south, clays, followed by Numidian sandstones, accumulated on top of the Mauretanian and Massylian micaceous sandstone series.

Sedimentation continued to the north and the south of the Kabylie ridge in the Aquitanian, as indicated by the presence of fragments from this ridge in the Oligo-Miocene Kabylie series. Towards the end of the Aquitanian, the southward movement of the Kabylides, which had

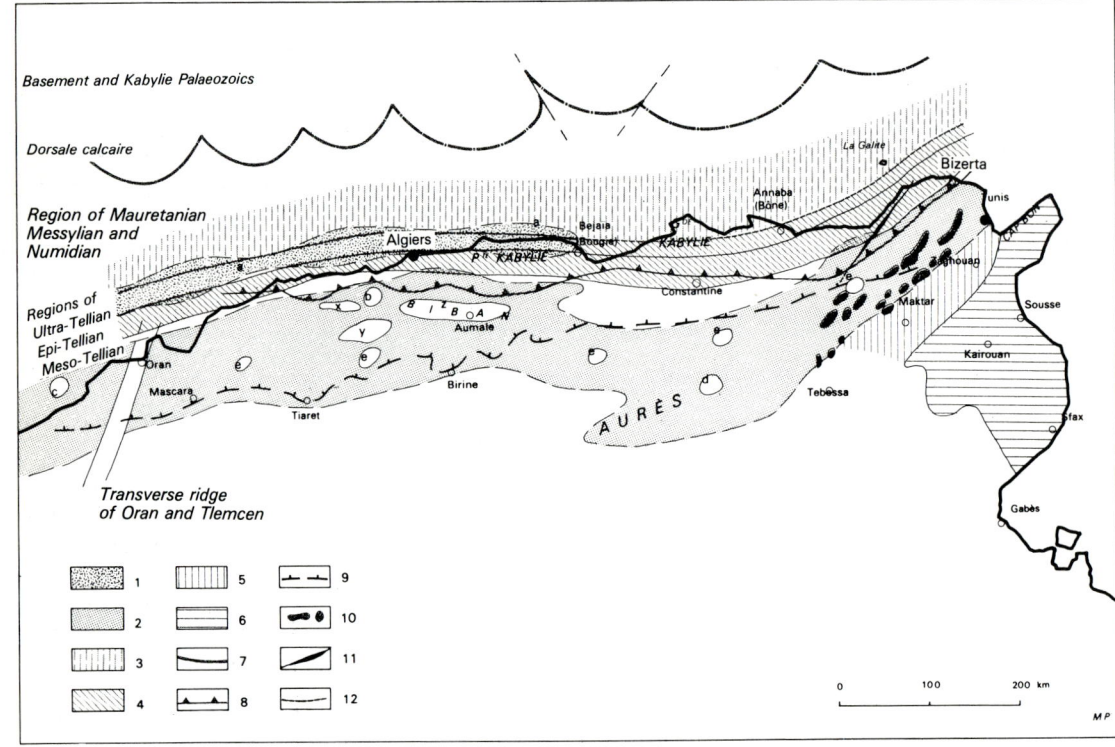

8.11 Palaeogeography of Algeria and Tunisia in the Burdigalian (Miocene I or Pre-nappe) (after Kieken and Rouvier, original).

1 – Conglomerates, coarse sandstones, sandy marls – plankton, oysters, benthos dominant;
   a) Ultra-Tellian area;
2 – Marls and sandstones, mixed microfauna – plankton often abundant or dominant;
   b) Boghari trench (Meso-Tellian area);
   c) Tafna trench (sedimentation continued into the late Miocene);
   d) Aures Basin;
   e) "South Tellian" Basin;
unornamented area = zones which were emerging or had emerged (x, y, z) during the Burdigalian;
3 – Area of early Miocene supra-Numidian of the Babouch type: clays, marls and siliceous horizons of chemical origin ("silexites") overriding the Massylian and trans-Tellian areas;
4 – Lower Miocene transgressive glauconitic sandstone;
5 – Emergent zone;
6 – Zone of deposition concordant on the Oligocene;
7 – Position of the Kabylie front in the Burdigalian;
8 – Limit of flysch outliers (principally Numidian) in the south Tellian zone. In Tunisia: limit of Numidian nappe;
9 – Limit of Tellian nappes;
10 – Principal exposed Triassic diapirs;
11 – Ghardimaou-Cape Serrat structure in which diapir of Trias penetrates into lower Miocene beds;
12 – Root of Numidian nappe.

begun in the Bartonian and continued into the Stampian (as shown by conglomerates in the south of the *Dorsale*), was accentuated and brought about the uplift of the Mauretanian and Massylian areas with their covering of Numidian sandstones. The superstructures of these massifs were carried northwards, essentially by gravity, but also by thrusting. Locally the morphology was such that parts of the Tellian (Dellys unit) were thrust into the north Kabylie trough. Masses, of very variable size, continued to reach this basin until early Burdigalian times.

c) **Pre-Nappe middle Miocene (or Miocene I):** during the late Burdigalian and the Langhian, the sea was restricted to the area south of the mountainous zone which had risen in the Kabylides. The sediments, deposited in east to west troughs, are dated at their base by *Globigerinoides trilobus* and *G. trilobus sacculifer* and at their summit by *G sicanus* and *Praeorbulina. Orbulina* is not present however.

The ultra-Tellian trough, situated between the Kabylie and Tell mountains, subsided rapidly as shown by the accumulation of more than 1000m of Miocene "ultra-Tellian" sediments at Djebel bou Rhida. These beds include conglomerates which were derived from both north and south. In particular, blocks of Numidian sandstone are incorporated in these conglomerates and constitute an important part of the argument for the reconstruction (KIEKEN, 1970) which is put forward here. However, it must be noted in this hypothesis of an "ultra-Tellian" Numidian root zone that towards the end of the Aquitanian, the Numidian was carried towards the south, while at the same time, sheets of Massylian and Mauretanian flysch and of Tellian rocks were being carried towards the north.

South of the ridge formed in the epi-Tellian region, there was the meso-Tellian basin in which the Boghari sandstone was deposited. This sandstone sequence contains, initially, an Oligocene fauna and later one of Miocene age. Still further south, the Biban ridge was the source of the olistoliths interstratified in the Miocene meso-Tellian and infra-Tellian nappes.

The southern Tellian autochthonous basin extends from the Saharan Atlas to the hills stretching from El-Asnam to Constantine. This basin crosses Algeria from east to west, but it is

scarcely recognisable on the geological maps of Algeria since it is largely covered by overthrust nappes. It is the great basin in which the rocks forming the "superstructure" of the Tell and southern Kabylie mountains have accumulated. These rocks reached the basin mainly by sliding, rather than by water-transport.

The Miocene of the southern Tell (Miocene I) began with conglomerates or "pavements" of *Lithothamnion* overlain by marls with sandy intercalations to a total thickness of over 1500m.

d) **At the end of the Langhian:** the Kabylides moved southwards at least 40km, and possibly more than 100km. This tectonic phase caused not only the compression of the flysch deposits and their movement to the south, but also the crumpling and uplift of the zones of the northern Tell. The upper parts of these folds became detached from the substratum and, undoubtedly helped by the distant movement of the Kabylie mass, slipped southwards down the slope formed at the end of the Miocene I phase.

These tectonic movements totally changed the palaeogeography of the whole of northern Algeria and resulted in the formation of an extremely complex mountain chain, the Atlas mountains. It has taken 25 years of patient work to unravel their structure and it is probable that future studies will modify the tectonic scheme accepted today. However, the present reconstruction is not accepted by all geologists working in the Mediterranean region. Some, like MATTAUER (1963) and POLVÊCHE (1956), believe that the flysch nappes have an origin which is not connected with the Kabylie mountains, while others, such as CAIRE (1957), suggest that there were "flysch" troughs both north and south of the Kabylie massifs.

8. The post-nappe Miocene (or Miocene II) (Fig. 8.12).

At the end of the Miocene I (i.e. at the end of the Langhian) the phase of compression of Africa towards Europe was followed by a phase of extension accompanied by the *subsidence of large areas of the Kabylie orogen*. The sea retreated from the southern Tell trough[1] (it

---

1. The sea however, persisted to the west (western Oran and Morocco) where sedimentation continued until the late Miocene, despite the development and emplacement of thick gravity nappes in the middle Miocene. The palaeogeographic position of this trough is quite different from that of the southern Tell trough.

only returned in the late Miocene) and disappeared into the northern grabens.

In Algeria, the post-nappe basins are restricted to the Chelif, the Mitidja and the small coastal basins of Dellys, Babors and Djidjelli.

In the Chelif basin, two cycles of sedimentation can be recognised:

1) The deposits of the first cycle (the Serravallian) begin with conglomerates, sands and limestones with pectinids. These are overlain by blue marls which reach a thickness of 3000m in the centre of the basin and contain *Orbulina universa* and *Globorotalia praemenardii*.

2) The deposits of the second cycle (Tortonian to Messinian) begin with sands or glauconitic sandstones with *Ancilla glandiformis, Cardita jouanneti* and *Pecten fuchsi*. These are overlain by blue marls, with a maximum thickness of 1200m in the centre of the basin, which contain *Globorotalia menardii, Globigerinoides conglobatus* and *Orbulina universa*. Several layers of volcanic tuff, rich in biotite, occur which can be traced over distances up to 300km. On the margins of the basin, the marls pass either into *Lithothamnion* limestones or red continental beds with *Hipparion* or sandstones of fluviatile origin. These beds are probably of Tortonian age.

The subsidence of the basin then slowed down and the Chelif trough was filled with 200 to 300m of "tripoli" marls (diatomite with 90% SiO_2 content). The marls are overlain by brackish water marls with beds of gypsum (10 to 300m) of Messinian age. South of this trench conglomerates and red clays with pebbles fill the valleys of Biban and the Soummam. Similar continental or shallow water sediments occur in basins near Constantine and Guelma.

During the late Miocene, a marine transgression from the Chelif basin worked its way between the Saharan Atlas and the hills of the allochthonous southern Tell region. Its sediments were detrital and amongst them were the Tiaret sandstone and the sandy marls of Hodna. Along with an abundant reworked micro-fauna, GLACON (1970) has identified *Globorotalia crassaformis oceanica, Globigerinoides ruber* and *G. obliquus extremus* of Messinian age.

In the Hodna region, the post-nappe formations attain a thickness of 1100m, showing that the basin was subsiding not only during the formation of the nappes, but also during the late Miocene.

The Messinian beds are followed by 300m of Pliocene conglomerates and it may be noted that the Hodna cuvette and certain shotts (salt lakes) of the high plains of Algeria are still subsiding.

9. Pliocene

In the Pliocene, the sea occupied parts of the basins of Chelif and Mitidja. The initial deposits were thin layers of sand followed by blue marls with *Globorotalia hirsuta* and *Dorothia gibbosa*. These marls reach a thickness of 700m in the centre of the residual trough at Chelif and about 1000m in the Mitidja basin (Plaisancian facies).

In the Mitidja basin, the blue marls pass both laterally and upwards into calcarenites full of *Lithothamnion* and banks of *Ostrea cochlear*. The basin was filled with sandstones with oysters (30m) and, on the edge of the Atlas, by beach deposits (the Astian facies). In the Chelif basin, this facies is represented by about 100m of fine sands and sandstones overlain by dune sands with *Helix*. Above these come continental soils attributed to the Quaternary. The sea reoccupied the region of Mostaganem in the Calabrian (LAFITTE, 1942) with the deposition of 2 to 15m of sandstones with *Glycymeris* and *Lithothamnion*. In a continental environment at the beginning of the Villafranchian, soils and grey clayey sands were deposited in a humid climate. At the end of this stage, conglomerates and red, ferruginous sandstones with *Elephas meridionalis* accumulated. Finally the coastal area was invaded by sand dunes.

The marine Pliocene is also known in the lower part of valleys, east of Algiers, which open towards the Mediterranean. Such valleys include those of the Soummam and the Oued el Kébir (Djidjelli). In these rias blue marls, rich in microfossils, were deposited. In the Constantine basin and in the region of Setif, the Pliocene is continental. It is represented by sandstones, brown clays and lenses of lacustrine limestone containing *Hipparion sitifense, Lymnaea, Helix* and *Hippopotamus*.

In conclusion, it appears that Algeria was, during the Tertiary, a vast sedimentary basin which formed part of the Tethys. From south to north, there is a gradual change of facies from the emergent zones to the basins of flysch deposition with their associated parallel ridges, and a passage from the sediments of the platforms through the deposits of the foredeeps into those of the basins. The original position of the various domains can be restored because the distance which the southern Tellian nappe

has moved is about 100km, while the flysch nappes have moved about 200km. It is probable, therefore, that the Kabylie massif formed the northern border of a great sedimentary basin.

The deposits of this basin were transported as nappes mainly after Miocene I, but they had also been moved at the end of the Jurassic, in the Aptian and in the Eocene.

During the Aquitanian-lower Burdigalian,

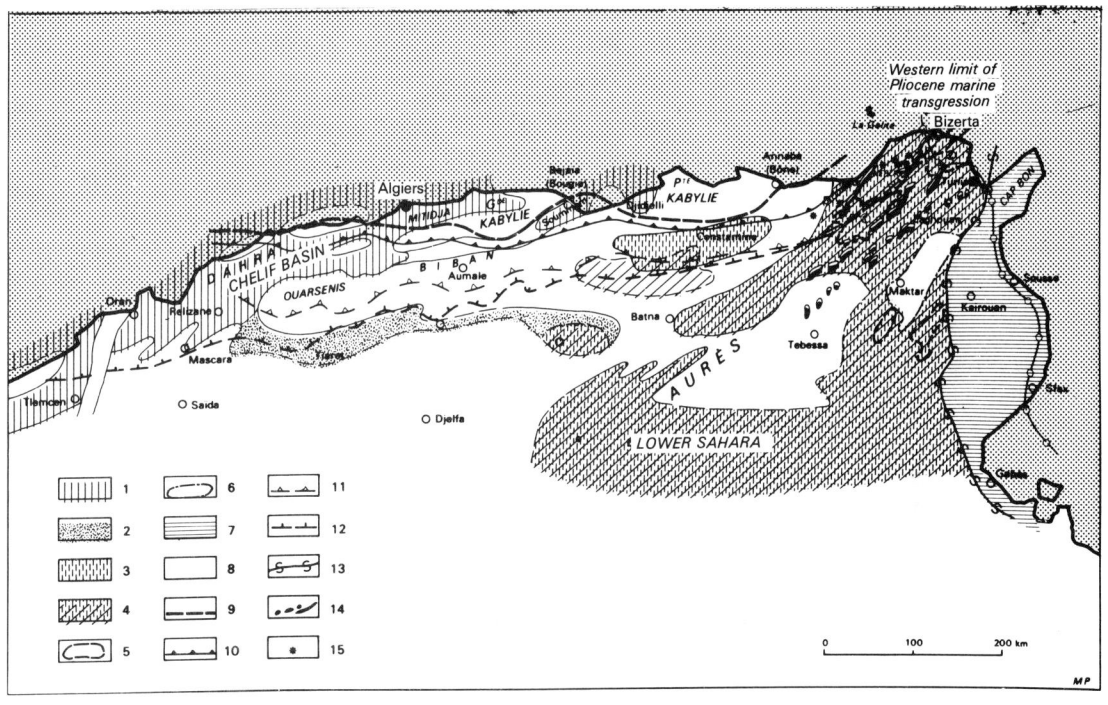

8.12 Palaeogeography of Algeria and Tunisia in the middle and late Miocene (Miocene II) and in the Pliocene.
Tectonic structure as at the end of the Burdigalian (end of Miocene I) (after Kieken and Rouvier, original).
1 – Marine Miocene-Pliocene (middle and upper);
2 – Marine Upper Miocene (2nd cycle of Chelif) transgressive over earlier Miocene;
3 – Continental middle and upper Miocene;
4 – Continental Miocene and Pliocene (middle and upper) (with locally subsiding basins);
5 – In Tunisia: large subsiding basins in the continental Miocene and Pliocene(4);
6 – Stable area with little or no sedimentation (Maktar region);
7 – Zone of subsidence;
8 – unornamented: Emergent zones or zones without important continental deposits;
9 – Kabylie front;
10 – Contact between flysch and Tellian zone;
11 – Limit of flysch outliers (principally Numidian) in the south Tellian zone. In Tunisia: limit of Numidian nappe;
12 – Tellian nappe-front;
13 – Western fold;
14 – Triassic diapirs;
15 – Volcanism in the Mogod-Nefsa-Kroumirie-Galite region.

northward movements carried nappes of flysch and sediments of the southern Tell over to the northern side of the Kabylides. These sheets became broken up into smaller fragments and sometimes may have become mudflows which are interstratified in the *argiles à blocs* (clay with boulders). This is *the first tectonic phase* with an important tangential component. During this phase, the Trias of the Babours zone was locally squeezed out and pushed towards the south. The ultra-Tellian Miocene is discordant on these fragments.

A *second important orogenic phase followed the Langhian*. It resulted in the movement southwards of the Kabylides and the flysch nappes. It also caused the shearing of the cover of the Tell region (the Tellian nappes) and the sliding of this debris towards the southern Tell depressions.

Following this stage of compression and lateral movement, *vertical movements took over* and produced broad swells (Dahra, Grande and Petite Kabylie) and basins, which include those of Tafna, Chelif, Mitidja, Soummam and Djidjelli (Fig. 8.12). These basins were invaded by the sea in the late Miocene, and the desiccation at the end of the Miocene resulted in the deposition of the Messinian evaporites.

As was the case throughout the Mediterranean basin the sea returned in the Pliocene and subsidence accentuated, reaching 5000m in the Chélif basin in the period from the Serravallian (middle Miocene) to the end of the Pliocene. This general subsidence was accompanied by simple folding producing symmetrical anticlines and synclines. The final marine invasion was that of the *Calabrian transgression* which ushered in the Quaternary era. Differential movements continued throughout the Quaternary to the present day and as a result, early Quaternary sediments can now be found at elevations of several hundred metres in the Dahra massif (Fig. 8.12).

However an important phenomenon has escaped our attention: the orogeny of the Maghrebides was born in the Mediterranean from which high mountain massifs arose in post-Langhian times. This roof-zone then subsided, although with some pauses, as is indicated by the Messinian gypsum beds known from the Chélif, and demonstrated both by seismic records and boreholes to occur throughout the Mediterranean (see p. 152).

III. – TUNISIA

Most stages of the Tertiary are represented in Tunisia, but, from the Palaeocene onwards, their composition and their distribution are more and more influenced by tectonic activity. For clarity, it is convenient to group the Tertiary into Palaeocene, Eocene, Oligocene and the Neogene. Furthermore, since shear has strongly affected the northern zones, two regions are distinguished. The first of these is devoid of horizontal movements and includes nearly all of Tunisia. The other region is that of the northern coastal strip which has been strongly affected by the continuation of the Tellian zones of Algeria. In this region, the terminology of the Algerian allochthon will be used, even if the equivalence of structural units cannot be rigorously established[1].

1. Palaeocene

In most regions, except round Kasserine and Sfax (Fig. 8.13) sedimentation was continuous from the Maastrichtian to the Thanetian. This sequence is the El Haria formation (BUROLLET, 1956) to the south of Le Kef. It consists mainly of black, brown or grey shales with thin intercalations of limestone or marl in the lower and middle part and rests on the Abiod formation of late Cretaceous age. The divisions within the El Haria formation can be recognised by the presence of foraminiferids beginning with a lower horizon with *Globotruncana* (late Maastrichtian), a middle horizon with *Globorotalia trinidadensis* (Danian), followed by *G. uncinata* (Montian) and an upper part with the successive zones of *Globorotalia angulata* and *G. pseudomenardii* (Thanetian), even extending into the zone of *G. velascoensis* (Ilerdian). The thickness of the El

---

Haria formation increases northwards with, however, marked variation in the "zone of diapirs" (Fig. 8.8) due to Triassic diapiric structures. In the Hairech Ichkeul zone, along the north coast, the series has the same facies as elsewhere, but is much reduced in thickness, though it is difficult to determine whether this is due to erosion or to changes in sedimentation.

From the northern edge of the Hairech Ichkeul zone to the corresponding autochthon at Nefza the Palaeocene increases from about 100m to 1000m with the appearance of a typically Tellian facies – marls and mudstones with calcareous-dolomitic concretions.

2. Early Eocene

The limestones of the Metlaoui formation (BUROLLET, 1956) (Fig. 8.13) outcrop west of Metlaoui in the Oued Tseldja and are of early and middle Eocene age. The formation consists essentially of carbonates with thick intercalations of gypsum and phosphate in the south. The phosphates are localised around Gafsa, Metlaoui and Tozeur.

In the "zone of diapirs", the lower Eocene with nummulites has been affected by the rise of diapirs. In the areas of Djebel Hairech and Ichkeul, the absence of the lower Eocene indicates the existence of a ridge which separated the Tunisian region in the south from the Tellian region in the north. This ridge is a prolongation of a similar structure in Algeria. North of this ridge in the Tunisian region, limestones with a mixed pelagic and benthonic microfauna appear.

The lower Eocene of the epi-Tellian nappe consists of limestones with *Globigerina* and flints, while that of the trans-Tellian nappe includes marly limestones, microbreccias and conglomerates with Neocomian pebbles. The boundary at this time, between the Tellian trough to the south and the flysch region further north, can probably be placed near the axis of fracturing of this last nappe.

3. Middle and late Eocene

The Souar formation (BUROLLET, 1956) occurring between Sidi Abib and Henchir Souar (about 70km north west of Sousse) spans the middle and late Eocene. It covers much of eastern and north eastern Tunisia and reaches a thickness of more than 1000m at Cape Bon (the north eastern tip of Tunisia). This is a grey, argillaceous formation with frequent intercalations of shelly limestone with *Nummulites gizehensis* and banks of oysters. Further south, the Souar limestone become more marly and thins in the direction of the "zone of diapirs". The upper Lutetian (with *N. gizehensis*) is, however, transgressive across the region of diapirs, and also across the Hairech-Ichkeul ridge in the north.

In the epi-Tellian nappe, marls with calcareous concretions increase to over 1000m in thickness at Ed Diss. The marly sedimentation continues in the Trans-Tellian nappe though calcareous concretions are absent.

8.13 **Sketch map showing some of the localities in Tunisia and eastern Algeria referred to in the text.**

4. Oligocene

The Oligocene was marked by a revival of tectonic activity which resulted in the formation of coarse detrital sediments, often of great thickness (400 to 1000m) in a different area from that of underlying formations. This is the Fortuna sandstone (from a locality near Cape Bon) with *Nummulites vascus, N. incrassatus* and *Lepidocyclina*. These sediments extend from Cape Bon to the Gulf of Gabès and were derived from a craton situated to the east of Tunisia. In the central part of the country (Sahel) these sediments, in neritic and continental facies, have been given the name Cherichera sandstone by CASTANY (1951)

5. Neogene

In the Burdigalian, the sea invaded a deformed Oligocene platform, especially in the north-west where vast islands remained emergent. The most spectacular advance was made in the Haute-Medjerda channel through which the Miocene sea reached Algeria. The Neogene beds of northern Tunisia are distinct from those of the southern part (Fig. 8.11, p. 192) and are separated by a median zone, lying to the south of the region of diapirs, and enclosing the *Dorsale* of Tunisia[1]. The Neogene deposits are absent or discontinuous in this area and JAUZEIN believes that a land mass existed at this time, some 60km wide, stretching from Tunis to Maktar. Along the line of Hairech Ichkeul ridge, Oligocene deposits are very rare, as marine Miocene beds in most areas rest directly on the Eocene or on the Cretaceous. This is evidence that this ridge was a positive area, perhaps in part emergent, at this time.

In the epi-Tellian region the basal Oligocene is represented by thin layers of glauconitic sandstone which mark a clear tendency to emergence of this Zone. Further north, very thick (3000m), late Oligocene, argillaceous sandy flysch was deposited, as in Algeria. The source should not be very far from the Tellian zones. In fact, in the basal part of this formation, there is a conglomerate containing fragments of Tellian rocks, especially Eocene ones. For this reason, it is probable that the southern part of the Numidian facies has spread slightly on to the trans-Tellian area.

The southern (and also the eastern) zone of Tunisia extends from the *Dorsale* to the shotts. In general, the Oligocene is overlain concordantly by the Miocene which can be divided into two parts. One part is the marine group of Cape Bon consisting of the Ain Grab formation (BUROLLET, 1956) which is conglomeratic and sandy at the base, with shell-beds of Burdigalian age at the top. This is overlain by the Mahmoud formation (BIELY, RAKUS, ROBINSON, SALAJ, 1971) which consists of greenish grey clays of Vindobonian age.

The other part is the laguno-continental group of Oum-Douil comprising the Beglia formation with cross-stratified sandstones rich in mammals of Vindobonian age, overlain by the Souaf formation. This includes argillaceous sandstones with lignite and banks of oysters of late Miocene age. Finally, the top of this group is formed by the Segui formation (red sandy clays) of Mio-Pliocene age.

The northern zone includes the region of diapirs and the area to the north which has been subjected to Neogene tectonism. In the zones affected by shear the Miocene can be divided, as in Algeria, into the pre-nappe and post-nappe components (Fig. 8.11, p. 192 and Fig. 8.12, p. 195).

The pre-nappe Miocene includes the Béjaoua group of marly sands with glauconitic patches and, sometimes, organo-detrital limestones. In some places these rocks are conformable on late Oligocene sediments, but more commonly they rest on a varied sub-stratum. They are overlain either by nappes or by younger Neogene deposits. The top of these beds has been dated as Serravallian (*Globigerina nepenthes*) or even basal Tortonian (*Globigerina acostaensis*). This is distinct from the situation in Algeria where the pre-nappe deposits are not younger than Burdigalian or Langhian (see p. 193).

These formations are transgressive over the Hairech-Ichkeul zone and its northern slopes. Beyond, the epi-Tellian domain is emergent, as in Algeria. At the latitude of the "Numidian" roots, argillaceous sediments with chemically-formed silica horizons (or *gaize* – a kind of friable sandstone with a siliceous cement) were

---

1. The term *Dorsale* is not used in the same sense as in Morocco or Algeria where it refers to the limestone chains of the Rif or Kabylie mountains. In Tunisia, it is used in a purely geographical sense.

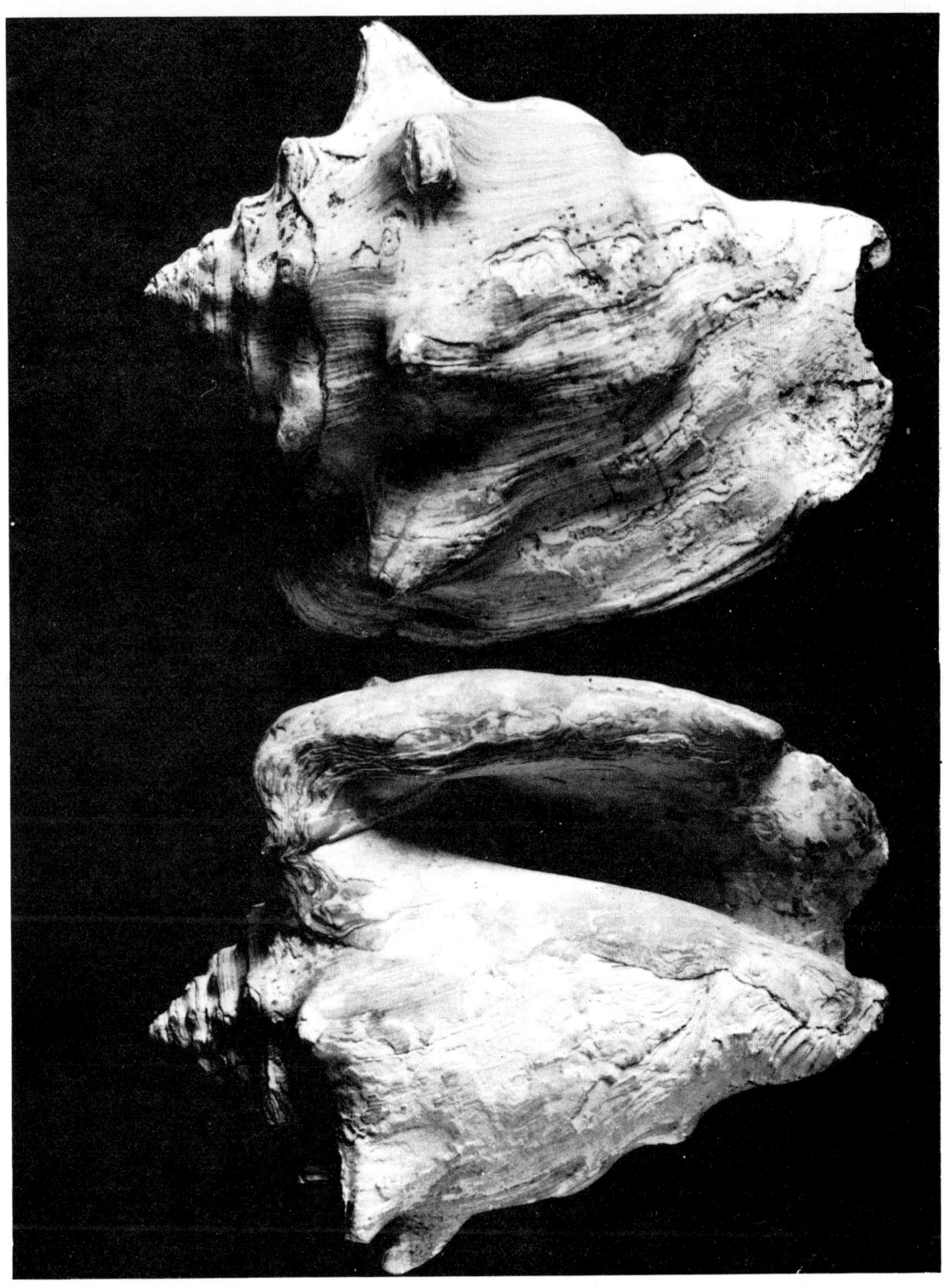

8.15 **Strombus bubonius from the Quaternary of the Mediterranean,** Monastir, Tunisia; Tyrrhenian II, ex-Monastirian. This species indicates a sea-temperature higher than that of the present day (see p. 238).

deposited in continuity with the sandstones of the Oligocene flysch (Babouch formation),

The post-nappe Miocene consists of the Medjerda group (BIELY, RAKUS, ROBINSON). This is a very thick laguno-continental sequence, argillaceous, sandy, and conglomeratic, of Tortonian-Messinian age. It rests on the nappes and their substratum,

including the pre-nappe Miocene, which had been folded by an important tectonic phase in the middle and late Miocene. The post-nappe group is overlain by marine Pliocene sediments (Raf-Raf clays) or by continental Pliocene (the Oued Mellouge group), the latter with mammals which prove their Pliocene age. The fauna is similar to that of the Lake Ichkeul beds of late Villafranchian age. In the Bizerta-Mateur region, the marine Pliocene is represented by the Raf-Raf clays and the Porto-Farina sandstone (Fig. 8.14).

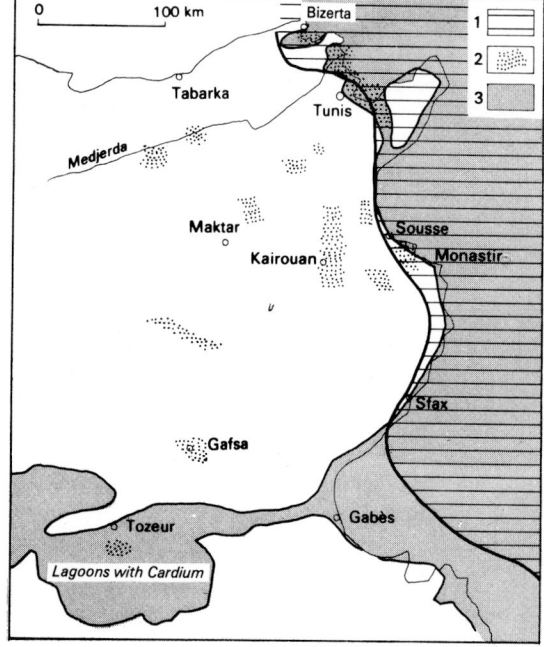

8.14 **Palaeogeography of Tunisia in the Pliocene and Quaternary** (after Castany, 1951, 1954).
1 – Pliocene; 2 – Quaternary zones of subsidence; 3 – Quaternary marine deposits.
The lagoons with *Cardium*, west of Gabes, are now much reduced and are isolated from the sea, forming the Shott-el-Djerid.

6. Quaternary

In the Quaternary, the sea made brief incursions across the coast from Tabarka on the Algerian-Tunisian border to Tripolitania, while the depression of the shotts was occupied by a vast lagoon with *Cardium*. On the coast, there are beaches with *Strombus bubonius* (Fig. 8.15) at various levels between present sea level and +32m with an average level between +6 and +8m (Fig. 8.14). One of the classic sites is that at Monastir, which was the type section of the stage formerly known as the **Monastirian**. Now known as the Tyrrhenian II stage, it is comparable to the Ouljian stage in Morocco (JAUZEIN).

Important tectonic activity occurred in the Quaternary across the whole of Tunisia. These movements had spectacular effects in the southern part of the country, north of the shotts, in the western\plain, in the zone of diapirs and in the region of Bizerta and Porto-Farina. In the latter area dips of 40° to 50° can be seen in beds of Villafranchian age.

Tectonic movements occurred rather later in Tunisia than in other regions affected by the Alpine orogeny, but they were prolonged with considerable vigour until comparatively recent times.

The North Sea in the Cenozoic

9

Seismic studies and borehole information gathered in the search for oil have led to a rapid increase in knowledge of the geological history of the North Sea. This is an epicontinental sea which deepens slowly from south to north. The 50m isobath runs from Hull to the tip of Denmark, skirting the southern end of the Sole Pit trough, which reaches a depth of about 100m (Fig. 9.1). The 100m isobath runs from Aberdeen towards Stavanger but turns southwards before reaching the Scandinavian coast owing to the presence of a trench which reaches a depth of 500m in the Skagerrak. The North Sea basin was entirely emergent during the last glaciation and was partially covered by ice from northern Europe (Fig. 12.3, p. 227). It is a very good example of an intracratonic basin which has subsided and filled with 6000 to 7000m of sediment since the upper Permian or Zechstein. The sequence includes 3500m of Tertiary sediments.

The stratigraphy and structure of the North Sea is largely continuous with that of the countries bordering it. In the Devonian, the North Sea and, indeed, the whole of northern Europe was a region of lagoons in which was deposited the Old Red Sandstone, produced by the dismantling of the Caledonian mountain chains. This landscape was replaced in Carboniferous times by seas, lagoons and landmasses in which were deposited shales and limestones, shales and coals, and red shales and sandstones respectively. At the end of the Carboniferous, the Saalian phase of the Hercynian orogeny affected all these deposits and they were folded, faulted and eroded. They were then overlain by the Rotliegendes, a red sandstone at the base of the Permian, which is an important reservoir for natural gas produced from the coal during the burial of the Carboniferous sediments (Fig. 9.4).

Then follow dolomitic limestones overlain by *many hundreds of metres of an evaporitic series* (the upper Permian = Zechstein) which played an important role in the structural evolution of the North Sea. Under the great load of superincumbent sediment which caused subsidence over much of the basin, the evaporites formed a large number of **diapirs**. These have risen progressively from the Jurassic to the present day, piercing Mesozoic formations and in some cases deforming Tertiary rocks. These salt domes constitute favourable structures for the trapping of oil.

At the beginning of the Trias, shales and red sandstones of the Buntsandstein (= Bunter sandstone) were formed which resemble the Permian Rotliegendes beds. Like them, they are reservoirs for natural gas which is trapped by the overlying Muschelkalk limestone and the Keuper evaporites, thinner than those of the Zechstein. After gentle folding by the Cimmerian movements, the Jurassic transgression brought an epicontinental sea into the North Sea basin. Initially, sedimentation gave rise to clays followed by marls and sometimes sandstones, which are locally productive of hydrocarbons. Further slight movement in the late Cimmerian phase provoked a discordance between the Jurassic and the Cretaceous, following which marine sedimentation laid down shales and sandstones, succeeded by a continuous and monotonous sequence of Chalk, 300 to 500m thick, which brought the Mesozoic era to an end.

The Jurassic and Cretaceous were periods of relative orogenic calm, but a slight discordance appears at the base of the Tertiary as a consequence of the Laramide movements. This was accentuated in the region of the Sole Pit trough and off the west coast of the Netherlands. There, in contrast to the Pays de Bray

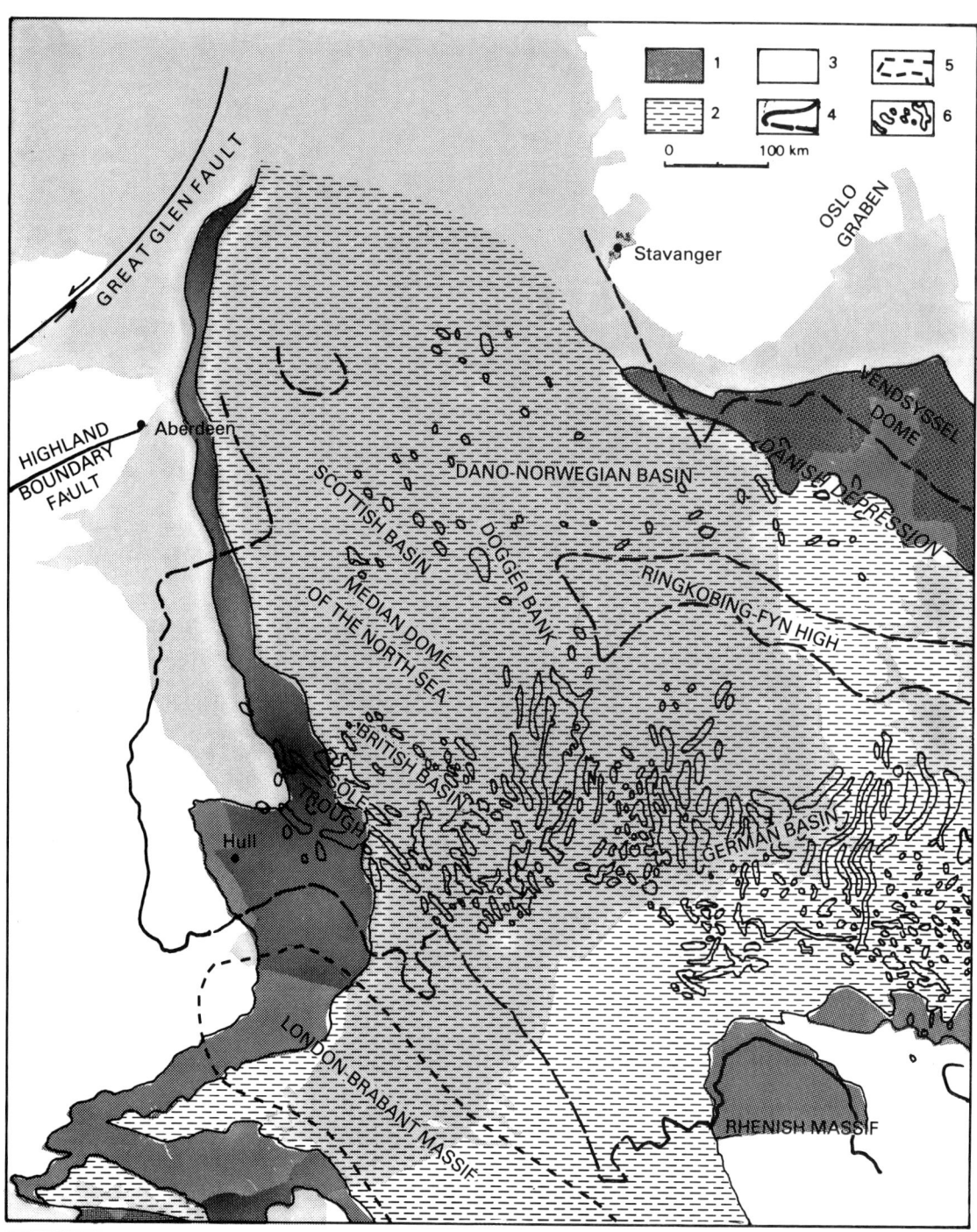

9.1 **Pre-Quaternary Geological Map of the North Sea.** 1 – Cretaceous; 2 – Tertiary; 3 – pre-Cretaceous; 4 – Limit of the present occurrence of the Zechstein; 5 – Approximate boundary of London-Brabant massif; 6 – Salt domes (after Sorgenfrei and Buch, 1964).

area, uplift in the Jurassic and early Cretaceous was followed by *subsidence in the late Cretaceous and Tertiary along axes orientated NW-SE* (Fig. 9.1).

One of the most surprising results of the geological exploration of the North Sea has been the evidence for **considerable subsidence during the Cenozoic** era. Although deposits of this age rarely exceed 1000m on the land (250m in the Paris basin, 600m in the Hampshire basin), they exceed 3000m in the northern North Sea between Scotland and Norway (Fig. 9.2). The principal axis of subsidence changes direction as it is followed from north to south. In the most northerly portion, it is oriented NNE-SSW paralleling the Great Glen Fault in its northern extension through the Shetland Islands and has a Caledonian trend. It then follows a NW-SE Hercynian direction and finally swings SSE before leaving the North Sea in the region of North Holland. The German basin diverges from the main axis of subsidence at the latitude of southern Denmark (Fig. 9.2).

Other troughs, inherited from the late Cretaceous, occurred south of the Sole Pit trough and west of the Netherlands. The Tertiary sea penetrated to the east and to the south into regions such as the Netherlands which still show evidence of subsidence. It is, moreover, remarkable that the Tertiary basin was superimposed, not only on that of the Cretaceous, but also on that of the Zechstein (Fig. 9.1). It is limited to the south-west by the London-Brabant massif which acted as a geographical barrier in Eocene times between the nummulites to the south and a northern sea with a planktonic microfauna (Fig. 9.2 and 9.3).

In the deepest part of the Tertiary basin (near the boundary between the British and Norwegian sectors), the Palaeocene and the Eocene consists of nearly 1000m of argillaceous sands. They are succeeded, after a period of gentle folding (the first Alpine phase), by Oligocene brownish shales between 500 and 1000m thick. In the Neogene 500 to 1000m of grey shales and in the Pleistocene, 500m of argillaceous sands were deposited (Fig. 9.3).

In this region research by British, Norwegian and Danish geologists has led to the discovery of *important reservoirs of oil in the coccolith sands of the Danian*, which had been deformed

by the rise of salt domes (Fig. 9.4). Research has continued vigorously both in the Palaeocene and in younger formations where oil shows have been found, but much of the geological information is still confidential.

It has already been noted that, north of the London-Brabant massif, the Tertiary beds are deposited in a subsiding northern province from which nummulites are absent and where the planktonic foriminiferids which are present show pronounced morphological differences from more southerly species. Consequently, *difficulties arise in correlating between the two basins* which have markedly different palaeogeographies and climates.

As a result new classifications based on nannoplankton and on planktonic foramini-

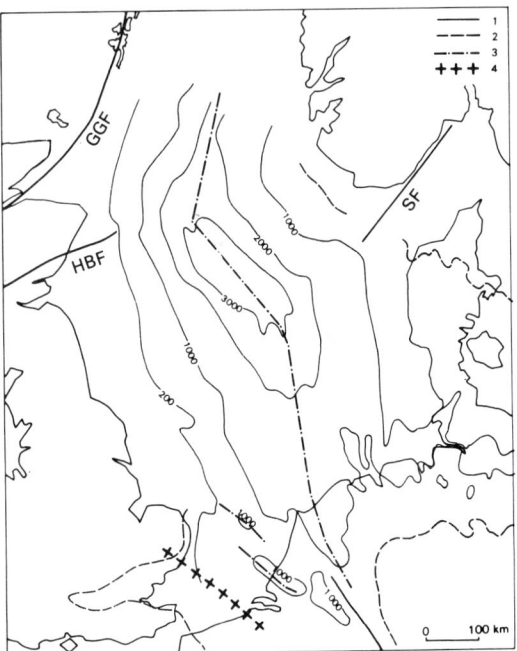

9.2 **Depth of the base of the Tertiary below sea level in the North Sea and neighbouring countries.** 1 – Depth in metres; 2 – Present boundary of Tertiary outcrop; 3 – Principal axes of subsidence in the Tertiary; 4 – Axis of the London-Brabant dome, separating the area to the south containing *Nummulites* from the nothern area with planktonic microfauna. GGF – Great Glen Fault; HBF – Highland Boundary Fault; SF – Skagerrak Fault (after Heybroek *et al*, 1967) (see Fig 1.4, p. 17).

ferids are in preparation.

The identification of foraminiferids has allowed a broad classification of the Tertiary beds of the British section between latitudes 52 and 55°N:

Palaeocene:	*Bulimina trigonalis* and *Discorbis discoides*
Lower Eocene:	*Globigerina soldadoensis*
Middle Eocene:	*Globorotalia pseudomayeri*
Upper Eocene:	*Globigerina ampliapertura*

The Oligocene transgresses from north to south, as does the Miocene. The Pliocene covers the whole area with a generally argillaceous facies which sometimes becomes a

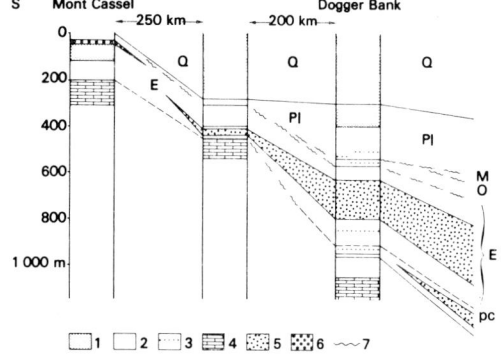

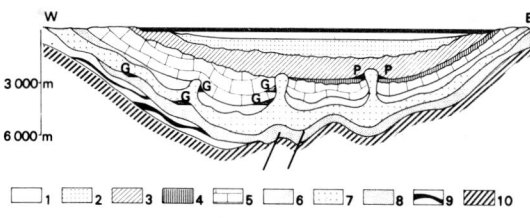

9.3 Sections in the North Sea showing the subsidence during the Tertiary period from south to north and the passage of a nummulite fauna into a planktonic microfauna.

1 – sands; 2 – clays; 3 – sandy clay; 4 – Chalk; 5 – planktonic fauna; 6 – nummulites; E – Eocene; Mi – Miocene; Pc – Palaeocene; Pl – Pliocene; Ol – Oligocene; Q – Quaternary (after Dollé).

9.4 Section across the North Sea. 1 – Quaternary; 2 – Neogene; 3 – Eocene-Oligocene; 4 – Danian; 5 – Upper Cretaceous to Muschelkalk (middle Trias); 6 – Bunter (lower Trias); 7 – Zechstein (upper Permian); 8 – Rothliegende (lower Permian); 9 – Carboniferous; 10 – Devonian to Cambrian; G – gas; P – oil.

shelly sand rich in bryozoans, but poor in foraminiferids.

Morainic deposits and periglacial sandy clays of Quaternary age now cover much of the floor of the North Sea. The outer moraines of the Vistula stadial of the Würm glaciation, are particularly well-known. According to DUJON (1971), the moraines of the Warthe stadial of the Saale = Riss glaciation can also be recognised (cf. Fig. 12.3, p. 227). The Eemian transgression corresponding to the Riss-Würm interglacial stage penetrated the North Sea as far as the Pas-de-Calais and deposited a sand with *Corbicula fluminalis* at Ostend. This may mark the first opening of the Straits of Dover in post-Eocene times.

Until such time as more precise information concerning the stratigraphy of the North Sea is available, a brief account of the Tertiary stratigraphy of Denmark will be given, as it is an integral part of this palaeogeographic province (Fig. 9.5)

Almost the whole of Denmark consists of Tertiary rocks, while Jylland (Jutland) is almost entirely Miocene in age. The northern tip of this province (Vendsyssel) has outcrops of Palaeogene and Cretaceous rocks however, as has the island of Sjaelland (Zealand). The type locality of the Danian occurs on this island at Stevns Klint, where it rests unconformably on the Maastrichtian (see p. 28). The important Danian limestone quarries of Fakse are about 20km west of Stevns Klint. The Danian is transgressive over the Upper Cretaceous of the North Sea, where it becomes sandy, but still rich in coccoliths, in the vicinity of the subsiding troughs. It forms one of the principal oil reservoirs in the North Sea (Fig. 9.6).

From the Late Palaeocene (Selandian = Thanetian) to the Middle Eocene, argillaceous rocks predominate. A provisional biozonation of the Eocene sequence has been established by DINESEN *pers. comm.*:

Upper Eocene (ex-Lattorfian)	Sövind marl	*Globigerina* ex. gr. *bulloides* *Globigerapsis* cf. *index*
Upper/Middle Eocene	Lillebelt clay	*Bulimina aksuatica* *Pseudohastigerina wilcoxensis*
Middle/Lower Eocene	Rösnaes clay	*Acarinina* ex. gr. *pentacamerata*

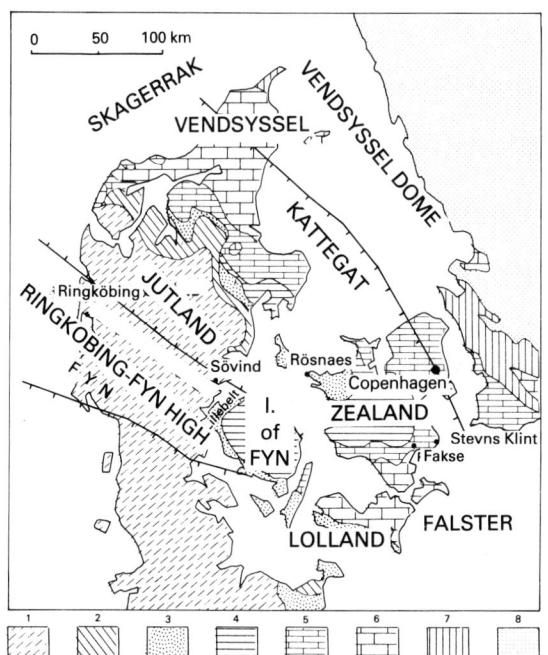

1 2 3 4 5 6 7 8

9.5 Sub-Quaternary geological map of Denmark.
1 – Neogene; 2 – Oligocene; 3 – Eocene; 4 – Upper Palaeocene; 5 – Danian; 6 – Upper Cretaceous; 7 – Lower Cretaceous to Cambrian; 8 – basement (granite and gneiss).

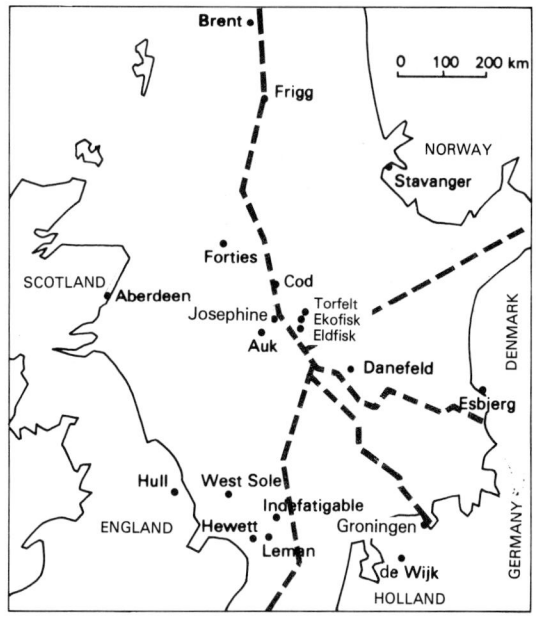

9.6 The principal gas and oil fields in and around the North Sea. The production of gas in the Leman and Indefatigable fields is 10^{10} cubic metres per year. Most of the southern fields produce gas from the base of the Permian (Rothliegende), as does the Groningen field in Holland, or from the Trias (Bunter). The central and northern fields produce oil from the Danian (see Fig. 9.4).

Lower Eocene   Rösnaes   *Globigerina*
                clay   *triloculinoides*

The Eocene begins with the *lowest of 170 horizons of volcanic ash* which occur throughout the Eocene and Lower and Middle Oligocene (SORGENFREI and BUCH, 1964). The radiometric age of the first ash horizon is 51-52Ma. From the Upper Oligocene, the sequence consists mainly of sands and micaceous silts with marine molluscs. Deltaic and continental intercalations with lignite beds are frequent, as they are in the North German basin.

In the west of Jutland, the thickness of the Danian is about 120m, while that of the Upper Palaeocene is 75m. The Eocene is 125m and the Oligocene is 60m thick. The Neogene reaches 400m in borings in western Jutland.

Subsidence of the North Sea continued in the

Quaternary: the moraines of the Saale and Vistula stadials were submerged (see Fig. 12.3, p. 227) and, on the Dogger Bank, the marine Eemian (= Normannian) fills a valley cut in glacial deposits which is now more than 80m below present sea level.

Thus, the floor of the North Sea, "terra incognita" only a few years ago, is today a fascinating area of geological research. It is not only a natural laboratory for the genesis and the occurrence of hydrocarbons but also an entrancing book of stratigraphy in which the sequence of pages reveals many discordances and a multitude of disharmonic structures as a consequence of the progressive uplift of domes of salt which was originally deposited 250My ago; domes which are still rising at the present day.

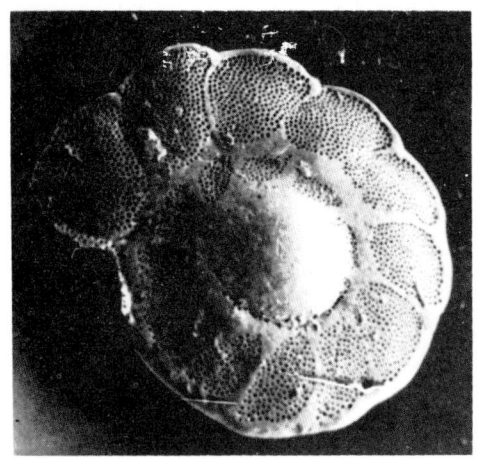

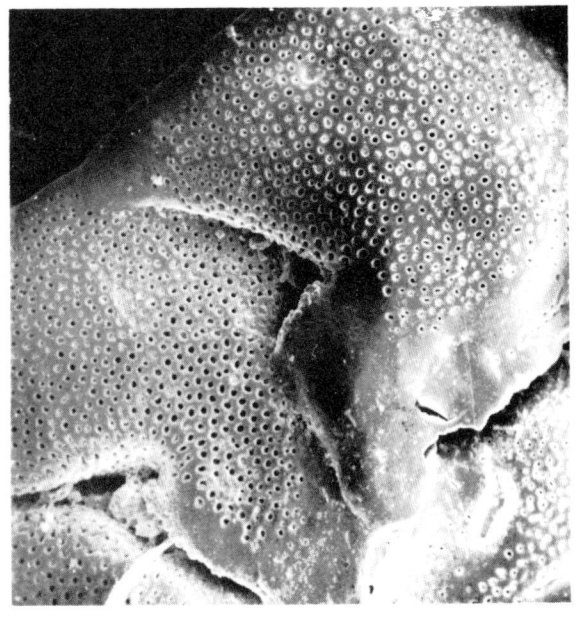

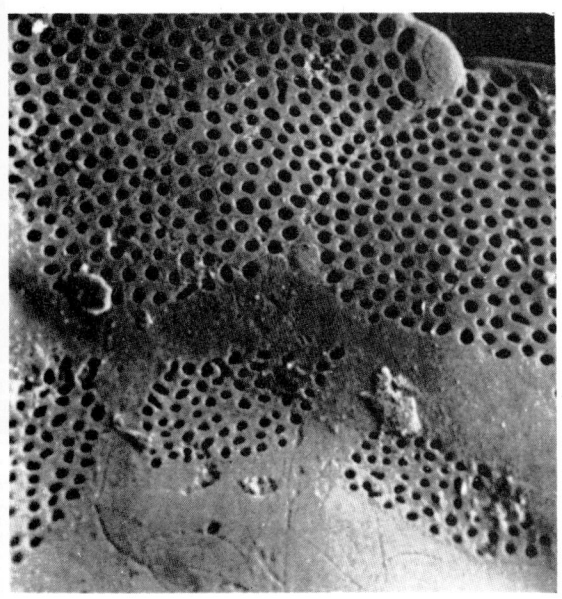

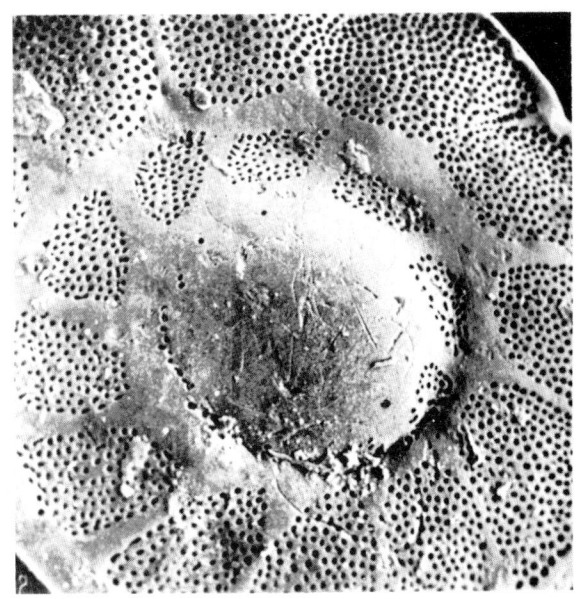

9.7 **Lutetian benthonic foraminiferid:** *Discorbis vesicularis* photographed by scanning electron microscope (Muséum, Lab. de Géologie) at several magnifications; top, left, x 40; bottom, right, x 100; top, right, x 150; bottom, left, x 200 (photo: Le Calvez).

The English Channel in the Cenozoic

10

Like the North Sea, the English Channel is *an epicontinental sea, the geological history of which is closely linked to that of the neighbouring land.* Between the discovery of nummulites south of the Scilly Isles in 1857 and the publication of the geological maps of the English Channel by the B.R.G.M. and I.G.S., a great deal has been accomplished.

The post-Hercynian history of the Channel can be considered in relation to two areas (LARSONNEUR, 1972), separated by a line from Caen to Portland. The eastern part is closely associated with the evolution of the Anglo-Paris basin, while the western part includes the south west of England (Cornubia) and Armorica and, since the Trias, has been a zone of subsidence.

The late Cretaceous sea had isolated Brittany as an island at the same time as it had connected the northern basins with the developing Atlantic. In the Maastrichtian the sea remained only in the western Channel, the Cotentin (limestones with *Baculites anceps*) and in the northern and central parts of the North Sea (Fig. 10.1).

It is probable that following this regression, the Dano-Montian transgression returned by the same route and flooded into the Paris region and the Mons basin. In the western Channel, *the Danian succeeds the Maastrichtian without a break* (CURRY et al., 1970) and occurs as biocalcarenites, about 120m thick, resembling those of Denmark and Belgium and like them contains *Globigerina pseudobulloides*. Despite these affinities, the two areas were apparently separated by the Pas-de-Calais isthmus which linked the Wealden-Artois anticline with the London-Brabant dome, and which, for the greater part of Tertiary time, acted as a barrier between the English Channel and the North Sea.

This barrier was probably breached from the Thanetian to the early Lutetian, though the strait was likely to have been very shallow. In the lower Ypresian however, the barrier separated the Ypres clays of Belgium from the Sparnacian lagoons of the Paris Basin.

In the Thanetian, as has been seen (p. 61), *the Paris basin was an extension of the North Sea.* Although grey, sandy clays of Thanetian age have been found in the western Channel, a direct link with the Paris Basin seems improbable. Moreover, in England, Thanet Beds are known only in the London basin and were not deposited in the Hampshire basin.

By contrast, in the Cuisian (Fig. 3.11, p. 66), the similarity of facies and fauna across the Channel (Isle of Wight and the Dieppe coast), the discovery of the Varengeville formation offshore from Dieppe and in the western Channel and the appearance in the Anglo-Franco-Belgian basin of foraminiferids (especially *Nummulites planulatus* and *Cuvillierina eocenica* all favour the suggestion that *the English Channel was open* at this time.

These conditions persisted into the lower Lutetian (Fig. 3.13, p. 68) although in the Paris Basin there was a brief emergence at that period. In the eastern part of the English Channel limestones with *Nummulites laevigatus* are well represented. Though it is certain that this nummulite and its associated tropical fauna entered the area from the west, it is surprisingly rare in the western Channel where mid-Lutetian glauconitic sandstones with *N.* cf. *aturicus* predominate.

In the middle Lutetian, the palaeogeography of the eastern Channel was changed by the renewed *uplift of the Weald-Artois anti-clinorium*, which separated the southern part of the North Sea from the Atlantic. These seas then remained separated until the opening of the Straits of Dover in the Holocene. The great distance between the western Channel and the eastern limits of the Paris Basin explains the precarious position of the Parisian cuvette and the tendency for it to be cut off in the late Lutetian (see p. 68).

Only one marine episode interrupted the inexorable approach to isolation of the Paris Basin. This was the *Biarritzian episode in the late Lutetian.* Although this transgression was short-lived, it was widespread, extending not only into the Paris Basin, but into the gulf of the lower Loire, into the Cotentin and into Hampshire (cf. Fig. 3.14 and p. 98). The late Lutetian is represented in the western Channel by biocalcarenites and calcareous sandstones containing *Orbitolites complanatus, Alveolina elongata, Fabiana cassis* and *Linderina brugesi.* This microfauna can also be found in the corresponding horizons in the Paris and Hampshire basins. However, *Nummulites brongniarti,* which lived in the gulf of the lower

Loire and on the Atlantic shores of Brittany, is not found in the western Channel.

It was in the late Lutetian that *Nummulites variolarius* became trapped in the Paris Basin where it became very abundant in the Auversian. At the same time *N. prestwichianus* and *N. rectus* were evolving in the Hampshire basin and *N. wemmelensis* appeared in the Belgian basin. In the eastern Channel, offshore from Fécamp, marls with *N. rectus* are found resting on sands with *N. variolarius* as a result of the overlapping of the English and Parisian Auversian facies (AUFFRET, *et al.,* 1975). The recent discovery of the Calcaire de St. Ouen (cf. p. 73) in the central Channel supports the concept of a closed sea in the Paris Basin during the Marinesian (Fig. 3.26, p. 77) which became an evaporitic lake in the Ludian. The sea persisted, however, at this time (Bartonian) in the western Channel and the two nummulites of Hampshire (*N. prestwichianus* and *N. rectus*) occur in calcareous sandstones. A

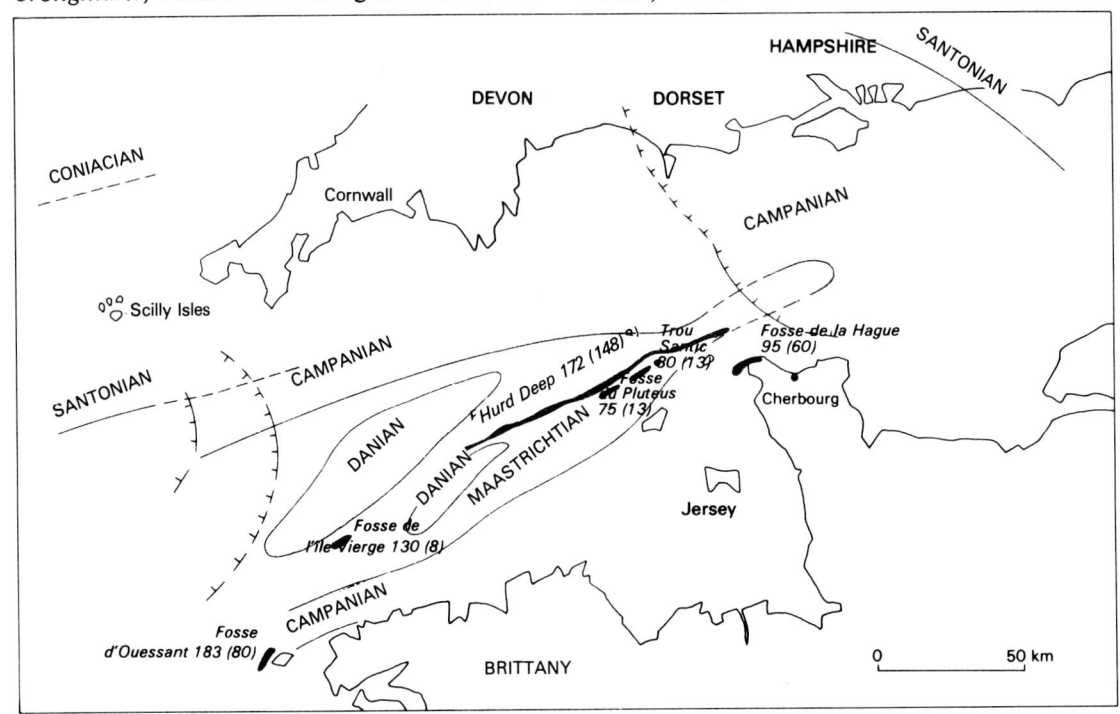

10.1 Geological map of the beds below the Tertiary unconformity of the western English Channel. The Danian has been included in the Mesozoic on which it is concordant (after Curry *et al,* 1971). Major deeps on the sea-bottom are indicated in (solid) green. Their depths are shown in metres; the first figure is the depth below present sea level; the second figure (in brackets) shows the depth below the general level of the neighbouring sea-floor.

southern species, *N. fabianii* is also present, but, like *N.* cf. *aturicus* in the mid-Lutetian and *N. brongniarti* in the late Lutetian, it did not penetrate very far to the east.

In the Stampian, the Channel was, for the last time, the route by which a transgression entered the Paris Basin, carrying with it *Archiacina armorica* (p. 77). This transgression had little effect, however, on the lagoons of the Hampshire Basin. There appears to have been some uplift in the northern part of the Anglo-Paris Basin at this time and it is not known how far the Stampian gulf extended into the eastern Channel. If it did so, it could only have been there for a very short time, because, in the late Stampian, the sea retreated southwards from the Paris Basin through the Loire channel and the English Channel became lagoonal (Fig. 3.34, p. 84).

In the Miocene, a new transgression, starting in the Aquitanian (Fig. 6.1, p. 154) was confined to the western part of the Channel. It

deposited 120m of the fine sands with abundant *Globigerina* (*G. praebulloides*) which rest discordantly on the Eocene. The transgression continued into the Helvetian (*G. bulloides*) ringing round Brittany, reaching the Cherbourg peninsula and extending south-eastwards to Blois, but it failed to reach the eastern Channel or the Paris Basin.

In the Pliocene (Redonian) a last marine invasion increased the area of the Miocene sea (Fig. 6.6, p. 158). This time the sea reached the Normandy coast at Fécamp where it deposited 30m of sands and clays. Although there is nothing to show that the Straits of Dover (Pas de Calais) were yet open, LARSONNEUR (1972) believes that the Pliocene gulf was the beginning of the eastern Channel. It is much younger therefore, than the western Channel, the outline of which can be seen in the early Mesozoic.

The opening of the Straits of Dover is certainly the most recent event in the geological

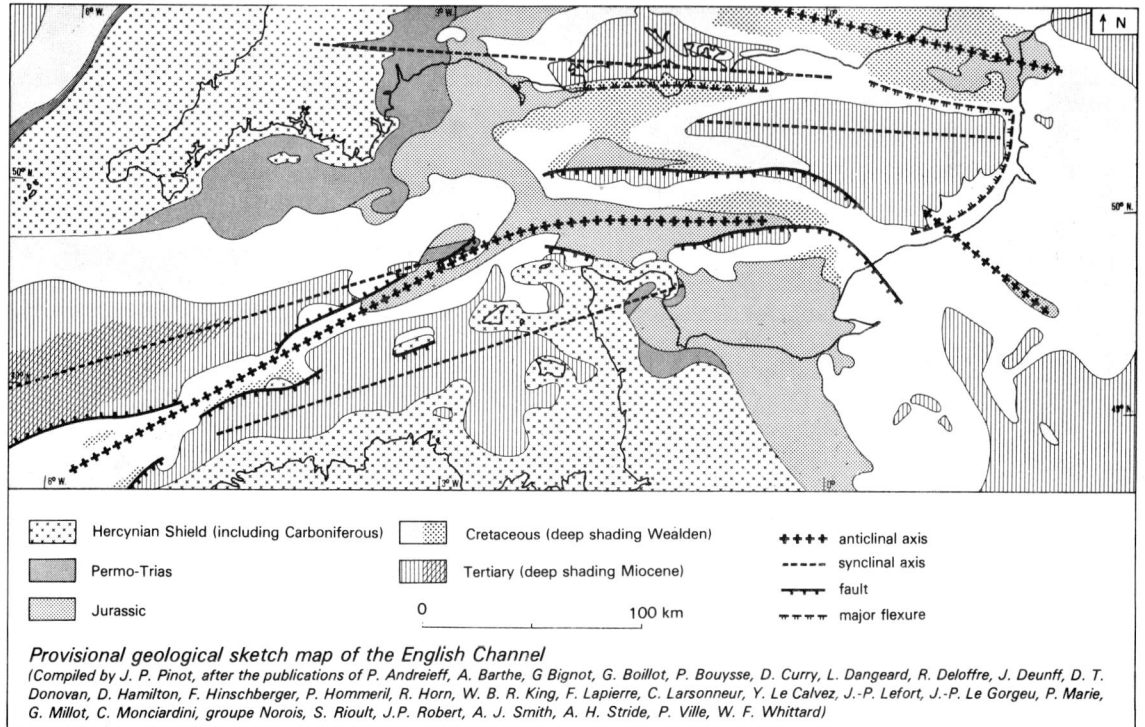

Provisional geological sketch map of the English Channel
(Compiled by J. P. Pinot, after the publications of P. Andreieff, A. Barthe, G Bignot, G. Boillot, P. Bouysse, D. Curry, L. Dangeard, R. Deloffre, J. Deunff, D. T. Donovan, D. Hamilton, F. Hinschberger, P. Hommeril, R. Horn, W. B. R. King, F. Lapierre, C. Larsonneur, Y. Le Calvez, J.-P. Lefort, J.-P. Le Gorgeu, P. Marie, G. Millot, C. Monciardini, groupe Norois, S. Rioult, J.P. Robert, A. J. Smith, A. H. Stride, P. Ville, W. F. Whittard)

10.2 **Provisional geological map of the English Channel** (after Pinot, 1971) (reproduced by pemission of the *Encyclopedia Universalis*).

history of the Channel. Carbon-14 dating of peat deposits indicates that these straits were formed less than 9000 years ago while the exchange of sediment between the Channel and the North Sea basin became effective only 2500 years ago. It is possible that the straits had become functional during the Riss-Würm interglacial (c. 80 to 100,000 years B.P.) at the time of the Eemian transgression when the Ostend beds with *Corbicula fluminalis* were deposited. Nevertheless, there is no reason to suppose that this was a permanent opening.

During the greater part of the Quaternary, the Thames and the Rhine discharged to the north, while the Somme and the Seine flowed to the west. However, during the formation of the last ice sheets (the Riss and the Würm) which blocked the North Sea, it is probable that at least part of the water of these rivers was diverted southwards to the Channel across the Pas-de-Calais isthmus into the Channel area, which was still largely emergent. On the other hand the pre-Loire and the pre-Seine rivers had flowed into and across the floor of the Channel since the Miocene. However, the fossil valleys of these rivers are often masked by accumulations of Quaternary sands and gravels.

When sea level fell during the Quaternary, karstic phenomena were developed on the emergent sea bed. These included trenches or "deeps" such as the "Santic Hole" near Roscoff and the Fosse de la Hague on the Cotentin coast (Fig. 10.1) (BOILLOT, 1963). Some of these trenches are of considerable size. For example, the Hurd Deep, a long, narrow depression in the centre of the western Channel, is cut to over 200m below the sea floor which itself is 172m maximum below sea level. The origin of these deeps is, however, controversial. They may be of karstic origin, as suggested above, or they may be the result of fluvial erosion or of tidal erosion of estuarine channels during the interglacial periods. In connection with this latter hypothesis, seismic exploration has shown the existence north of the Cotentin, of a series of valleys more than 200m deep, though they are now filled with alluvium.

It should be noted that the Hurd Deep has a NE-SW orientation, which is almost entirely influenced by the structure of the sea bed itself and is related to the prolongation of the Seine valley disturbance (Fig. 10.2). Off Fécamp this

structure swings round from an approximately WNW-ESE direction until it links up with the structural axis of the western Channel. Its track is marked by *negative gravity anomalies and positive magnetic anomalies*. These are believed to correspond to the separation of the basement into two plates by a rift (WEBER (1973)) which became abortive. This rift allowed basic rocks to rise (giving the positive magnetic anomaly) and was filled at the end of the Precambrian by Brioverian detrital sediments penetrated by granite batholiths (resulting in negative gravity anomalies) like those of the Cotentin and northern Brittany.

This rift in the earth's crust was probably reactivated in the Miocene and the gulf of the western Channel is simply the prolongation of the fluvial channel in which the Lozère sands were deposited. It extends 60km southwestwards beyond the edge of the continental shelf of the Western Approaches. According to BOILLOT *et al.* (1972) and LE PICHON *et al.* (1971) the formation of the Channel graben was the consequence of crustal phenomena, which also caused the opening of the Bay of Biscay.

There is thus a *marked contrast between the structure of the western and central part of the Channel which is very ancient and that of the eastern Channel which is little more than Plio-Quaternary in age.*

The structure of the western Channel is responsible for its continuing subsidence, which is, however, much less than that of the North Sea (600m of Tertiary in the Channel compared with 3500m in the North Sea). However, there is a marked contrast between the areas in another way. During the Eocene, the Channel was affected by a swelling of the earth's crust, due to a rise in the mantle (giving a positive gravity anomaly) which results in the appearance of an anticlinal zone with a core of Jurassic or lower Cretaceous beds which divides the Tertiary basin into two parts. A slight discordance is perceptible between the Danian and the Eocene and another, more marked, occurs between the Eocene and the Miocene. Folds and faults are developed in this interval; some at the end of the Eocene and the others in the Oligo-Miocene. A good example of this folding is seen in the monoclinal structure of the Isle of Wight where the Chalk and Eocene beds have been turned to a vertical position

(Fig. 10.3). In the western Channel, faults have two principal directions: NW-SE in prolongation of those of Devon and WSW-ENE, parallel to the axis of the Channel rift. They become west to east between the Cotentin and the Isle of Portland over a N-S dome-like structure, like that of the Straits of Dover, against which transgression in the Thanetian, Bartonian and the Miocene died out. The faults fragmented the sea bed into horsts and grabens in the Oligocene as in other regions of western Europe. Further east, the faults swing to the south-east to link up with those of the coast near Dieppe (the Fécamp fault). There is however no indication of any submarine prolongation of the Bray anticline.

Thus, although fragmentary, our geological knowledge of the Channel provides the key to the palaeographical evolution of the adjacent regions and reveals a close structural unity between the Paris Basin and the western borders of France. From this point of view it would be reasonable to call it the French Channel rather than the English Channel.

10.3 **The Bagshot Sands (Cuisian) of Alum Bay, Isle of Wight upturned to the vertical by an east-west fold. The younger beds are to the left hand side of the picture** (photo: Pomerol).

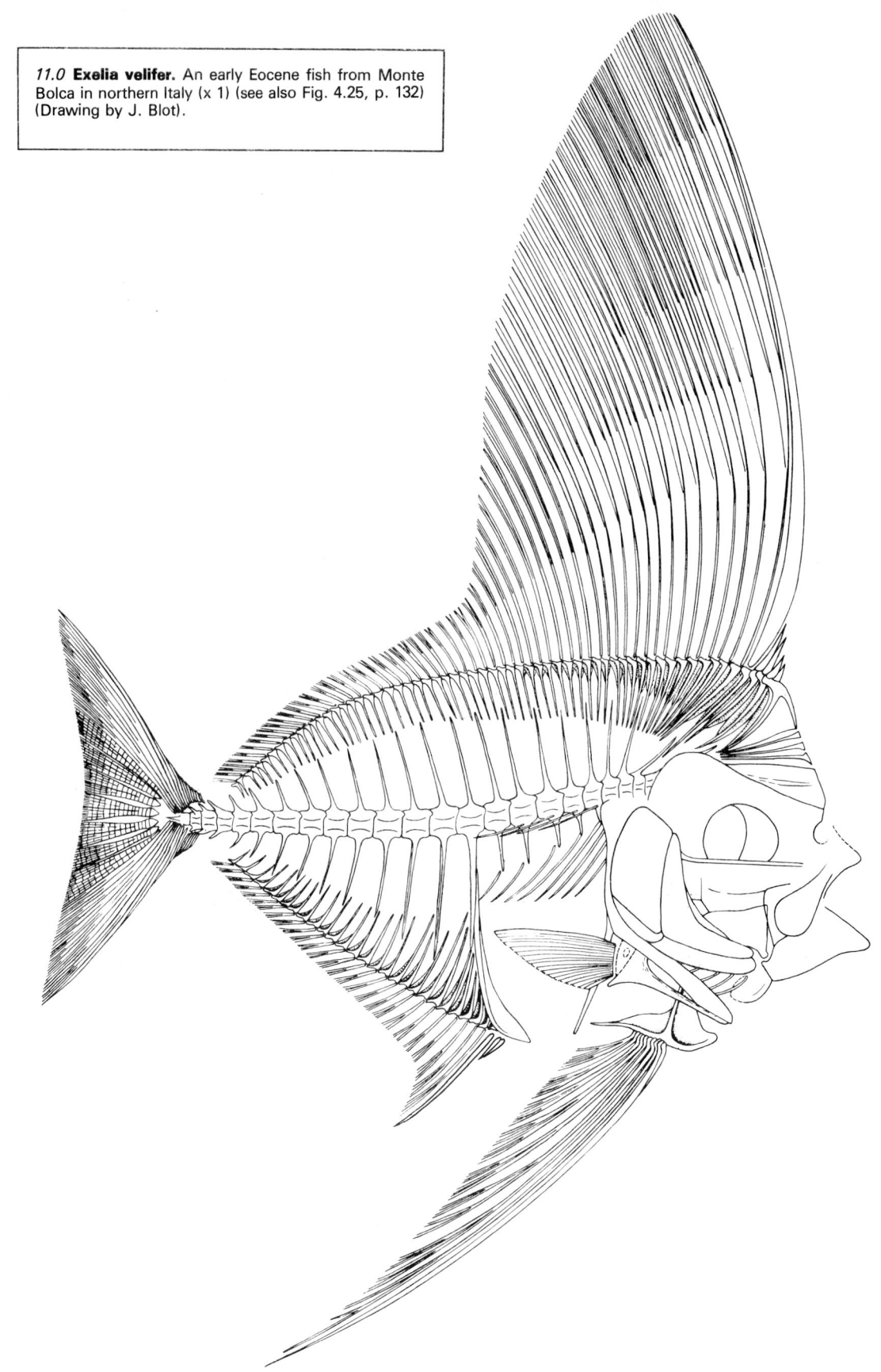

11.0 **Exelia velifer.** An early Eocene fish from Monte Bolca in northern Italy (x 1) (see also Fig. 4.25, p. 132) (Drawing by J. Blot).

World Tertiary | **11**

In this brief review of the stratigraphy and palaeogeographic evolution of the World's continents, the succession of phenomena during the Tertiary will be discussed. The Quaternary will be considered in Chapter 12.

I. – NORTH AMERICA

The North American continent was united to South America at the beginning of the Cenozoic, but the two were separated early in the Palaeocene by the development of a Strait further south than the present Panama isthmus, in the region of Venezuela and Colombia. By contrast, the northern part of North America was linked to Europe by way of Greenland and there was no communication between the Atlantic and Arctic Oceas until the Oligocene. The outline of North America during the Cenozoic was very similar to its present form (Fig. 11.1). The continent was never completely submerged by major marine transgressions. Indeed, the only transgressions to affect it were restricted to the eastern, south-eastern and western margins. On the east, the Atlantic coastal plain area was separated from the Gulf of Mexico coastal plain by the important Florida-Bahamas carbonate bank. On the western margin, a great thickness of sediment was deposited in the western coastal basin from which subsequent tectonic activity has formed the Coast Ranges of the Pacific margin.

Two major orogenic phases and one tectogenic phase affected North America in the Cenozoic. The first is the **Laramide orogeny** which uplifted the Rocky Mountains. This began in the late Cretaceous, continued in the Palaeocene and ended in the Eocene. During this period the products of erosion accumulated in vast continental basins east and west of the main chains, in which the remains of many mammals were entombed. Peneplanation of the Rocky Mountains ceased at the end of the Oligocene when their elevation was 1000 to 1500m.

In the Miocene, a phase of tectogenesis which continued into the Quaternary and was accompanied by the *extrusion of basalts*, affected the whole of the North American continent. These movements brought about the rejuvenation of the Rocky Mountains, where the Oligocene erosion surface is now at 4000m. In the Sierra Nevada, great faults brought the granitic basement to the surface. These movements also led to the rejuvenation of the Appalachians. The river systems (e.g.; the Mississippi) thus initiated caused active erosion and carried huge volumes of sand and clay to the sea. Not only were the mountains uplifted, but also the sedimentary basins, where the rivers cut deep canyons such as that of the Colorado river (more than 2000m deep) (Fig. 11.2).

11.1 Palaeogeography of North America in the Cenozoic. 1 – marine formations; 2 – continental formations; 3 – volcanic formations (Fig. 11.3 shows the continental basins of the western United States in more detail) (after Kummel, 1961).

11.2 The Grand Canyon of the Colorado River, showing a continuous section of over 2500m from the Precambrian to the Jurassic. The uncomformity between the Precambrian and the Cambrian is visible at the bottom right of the picture (photo: Pomerol).

In the middle Miocene, the **orogenic phase of the Coast Ranges** began and caused the folding of many thousands of metres of sediment. These movements still continue at the present day.

We shall describe first the rocks of continental facies, which permit correlation with the European stratigraphic subdivisions, and then the marine sequences.

1. The Continental Sequences

These are very thick deposits of clays, sands and sandstones, with intercalations of volcanic ashes and lavas, which accumulated in the basins bordering the Rocky Mountains (Fig. 11.4). On the west, the main basins are those of the Columbia and Colorado rivers and on the east, there are the Bighorn, Green River, Wasatch and San Juan river basins. All these basins have a rich vertebrate fauna, including many mammalian fossils from which it has been possible to reconstruct evolutionary series such as that of the *Equidae* (See Fig. 5.9, p. 146).

The Palaeocene sequence of North America is more complete than that of Europe (Table II, p. 22). The Puerco and Torrejon beds of the San Juan basin contain multituberculates (*Neoplagiaulax*), marsupials (*Thylacodon*), creodonts, primitive carnivores such as *Arctocyon*, lemurs (*Plesiadapis*) and insectivores. In France, the Thanetian fauna of Cernay is of comparable age (see p. 34).

The Eocene fauna of the Wasatch basin marks an important evolutionary step, which can also be seen in the Sparnacian of the Paris basin. The creodonts and the condylarths (*Phenacodus*) persist and the rodents (*Paramys*), artiodactyls and perissodactyls (*Hyracotherium = Eohippus*) appear. These are all known from the Lower Eocene of Europe.

The Middle Eocene (Bridgerian) fauna includes a tapir (*Lophiodon*), *Palaeotherium*, and *Orohippus* which succeeds *Hyracotherium*. True carnivores appear in the Upper Eocene (Uintan = Ludian) together with large forms of *Palaeotherium* and *Lophiodon* and primitive camels. The lemurs and *Palaeotherium* disappeared in the Oligocene of the White River, where the horses are represented by *Mesohippus*. The first hornless rhinoceros

(*Aceratherium*) appeared at this horizon. This fauna was able to reach Europe across the North Atlantic isthmus until the Oligocene. In the Neogene, migration across Asia to Europe became possible following the closure of the Ural sea and in the Miocene the link with South America was re-established. This enabled the *Equidae*, mastodons, tapirs and llamas to migrate southwards and allowed *Megatherium*, the opossum and the armadillo to invade North America.

2. The Marine Sequences

Three very different regions deserve attention. These are the **subsiding coastal plains** of the east coast and in particular, that of the Gulf of Mexico; the **Palaeogene geosyncline** of the Coast Ranges; and, finally, the **Florida-Bahamas plateau.**

The latter area is one in which a very considerable thickness of limestone has been built by corals, algae and molluscs and includes consolidated dunes composed of oolites. Two thousand metres of beds were deposited in the Cretaceous, 600m in the Tertiary and deposition has continued to the present day. The emergent parts of this platform, the Bahamas and Florida, are only just above sea level and a very slight rise of sea level would lead to their complete submergence. On the other hand, a fall of only 20m would expose an enormous land area. This area, with its uncertain limits between land and sea, its lagoons and its islands, is an *excellent model of limestone formation in the Paris Basin in the Jurassic and in the Palaeogene* (Figs. 11.5, 11.6, 11.7, 11.8).

A section across the **coastal plain of the Gulf of Mexico** (Fig. 11.9 and 11.10) and the gulf itself shows the accumulation of a tremendous thickness (some 13,500m) of detrital sediments, which still continues in the delta of the Mississippi river. Complementary to this subsidence is a system of flexures parallel to the coast, between Houston and the sea.

The principal type areas of the American Eocene marine formations are in Texas and Alabama. These include the Wilcox formation (Texas) of early Eocene age, which is essentially sandy, and the more calcareous Claiborne Formation (Alabama) of middle Eocene age,

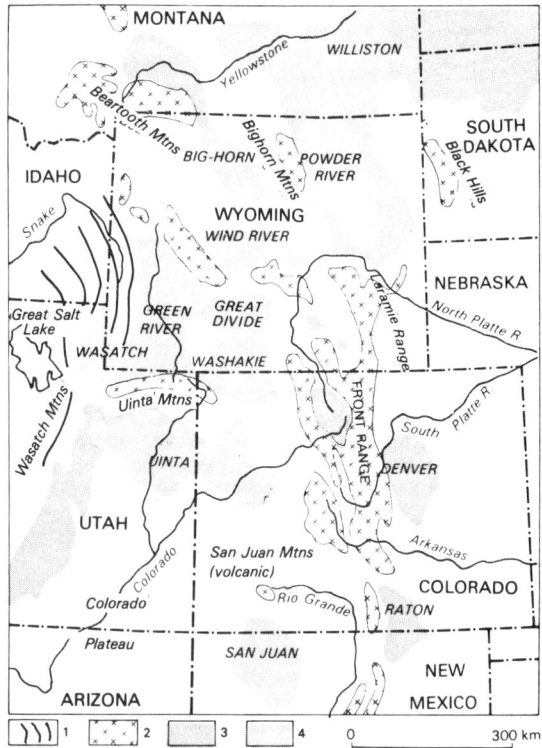

11.3 **Sketch map of the western United States (central Rocky Mountains region) showing the distribution of the principal continental basins.** 1 – Folds and faults; 2 – Precambrian; 3 – Palaeocene basins; 4 – Palaeocene basins including Eocene beds (after Kay and Colbert, 1965).

11.4 **Plio-Quaternary volcanism in Yellowstone Park, Wyoming.** The alternation of rhyolites and softer tuffs is responsible for the formation of the waterfall (Lower Falls of Yellowstone Canyon) (photo: Pomerol).

11.5 The Great Bahama Bank near Andros Island: banks of calcite mud (micrite) and of ooliths with mangroves. The boundaries are not static, but fluctuate between the land and the sea (photo: Pomerol).

11.6 Mangroves and water plants (submerged) on the coast of Florida Bay (photo: Pomerol).

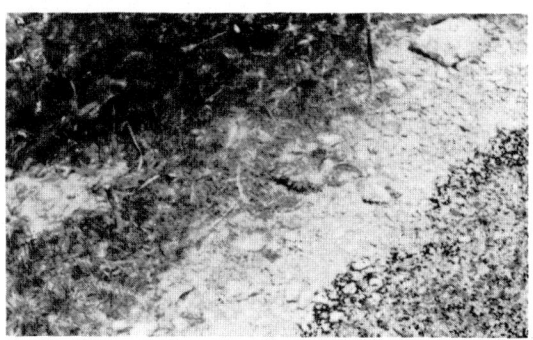

11.7 Detail of water plants (top, left), calcite mud (micrite) and Cerithium shell bank (bottom, right) on the coast of Florida Bay (photo: Pomerol).

the latter containing *Venericardia densata*, related to *V. planicosta* of the Lutetian. The upper Eocene is represented by the Jackson formation in Texas and contains the first *Lepidocyclina*. This genus occurs earlier in North America than in Europe, where it does not appear until the Miocene. By contrast, nummulites are almost entirely absent from North America, a few small forms occurring only in the Oligocene Vicksburg formation on the banks of the Mississippi. Correlation between the marine formations of North America and Europe is thus more difficult than between the continental sequences, and it is necessary to go into the Caribbean area to find close links between the Americas and Europe.

Correlation between the marine beds of the Gulf of Mexico and California is equally difficult. Despite the separation of North and South America during much of the Tertiary period, the connection between the Atlantic and the Pacific was far to the south and these oceans had different climates and faunas.

The Coast Range Province of the Pacific coast has about 8000m of geosynclinal sediments which include sands, clays and limestones (cf. Table III, p. 24, and Table X, p. 138). In the Eocene, the beds are largely detrital and often show a flysch type of rhythmic sedimentation. Nevertheless, there are occasional intercalations of carbonaceous horizons and hydrocarbons. In the Oligocene, a thick series of red sandy clays and conglomerates was deposited at the foot of the Sierra Nevada in south California.

The Palaeogene beds of the Coast Ranges are much less fossiliferous than those of the

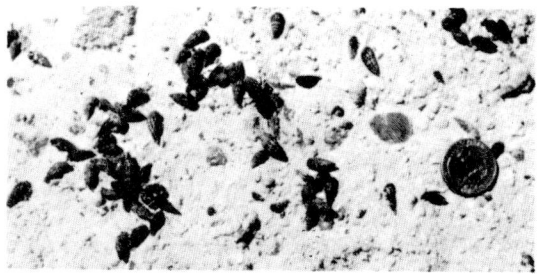

11.8 Cerithium in living position on the coast of Florida Bay (when fossilized the shells lose their colours) (photo: Pomerol).

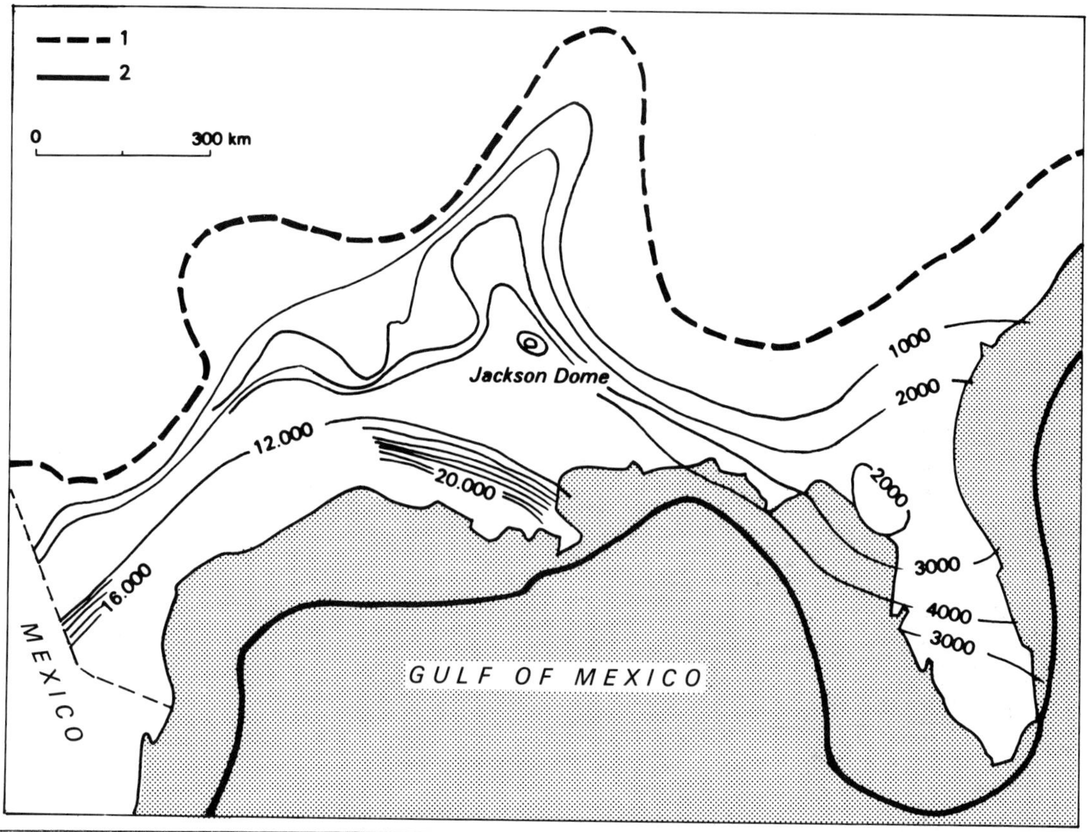

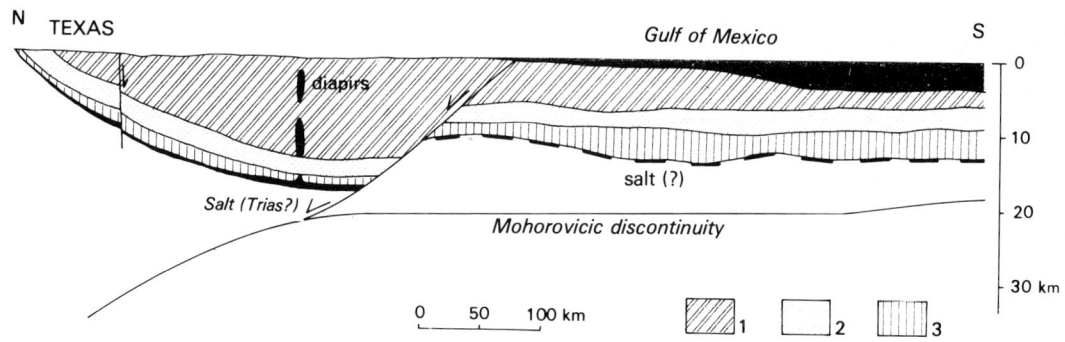

11.9 Tertiary subsidence in Texas and Louisiana around the Gulf of Mexico (after Murray and Toulmin Jnr. *in* Kay and Colbert, 1965). 1 – Tertiary shoreline at limit of transgression; 2 – Approximate position of outer edge of present day continental shelf. The thickness of the Tertiary beds is shown in feet.

11.10 North-south section from Texas to the Gulf of Mexico. The salt domes (diapirs) have acted as oil traps as in the North Sea (see Chapter 9, p. 201) (based on Worzel *et al* in Kay and Colbert, 1965). 1 – Tertiary; 2 – Cretaceous; 3 – Jurassic.

11.11 **Glyptodon typus,** an Edentate from the Pliocene and Quaternary of South America. This was a giant armadillo (4m long) with an enormous carapace and a short flexible tail covered with small plates. Its molar teeth had long roots like those of rodents. Its carapace served prehistoric man as a shelter (photo: Serrette).

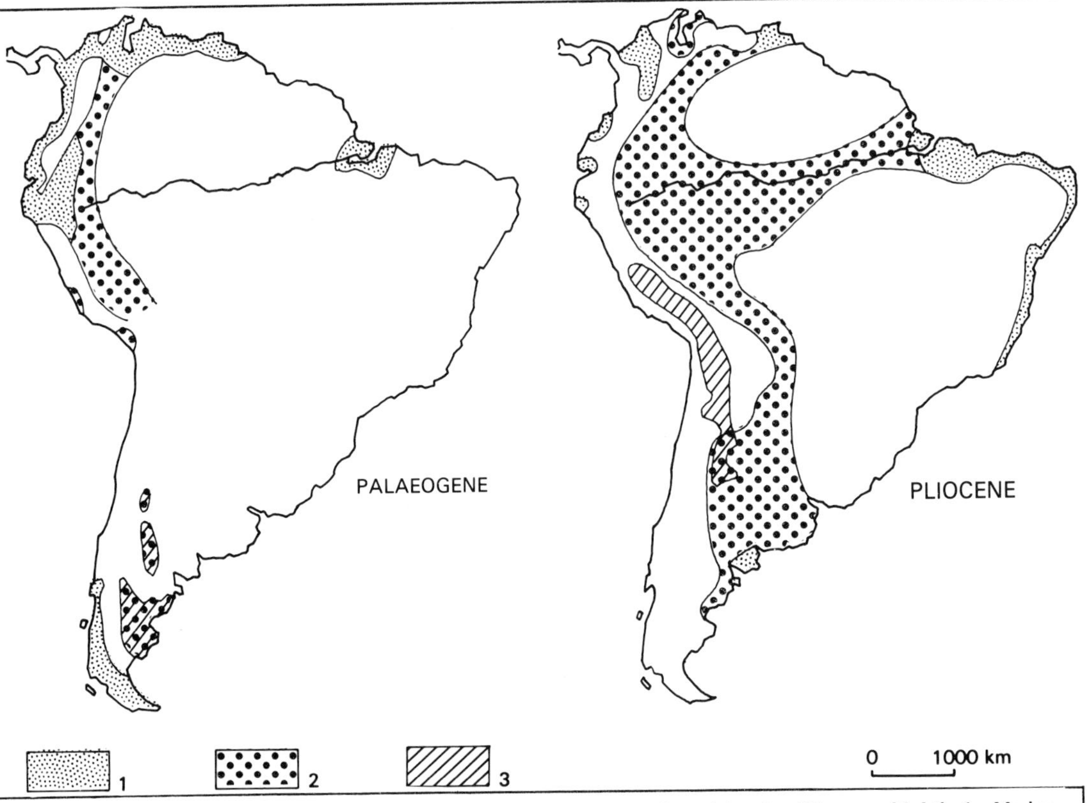

PALAEOGENE

PLIOCENE

0   1000 km

1   2   3

11.12 **Palaeogeography of South America in the Palaeogene (left) and in the Pliocene (right).** 1 – Marine formations; 2 – continental formations; 3 – volcanic formations (after Harrington *in* Kummel).

European basins. There are, however, species of *Turritella* similar to those of the European Cuisian and species of *Venericardia* comparable to those of the European Lutetian. Although detailed correlation is still uncertain, a broad outline can be based on the foraminiferids and on radiometric ages.

Miocene volcanism favoured the formation of cherts and diatomites by providing a source of silica. These occur in beds more than 1000m in thickness. *The Miocene was also a period of very active orogenesis* and erosion of the folded mountain chains led to the accumulation, in the Pliocene, of great thicknesses of sediment in elongated subsiding basins between the folds. In the Los Angeles basin, for example, 5000m of sediment were formed. These Pliocene deposits and even the Pleistocene beds are frequently folded and intense activity associated with the San Andreas Fault (Fig. 1.7, p. 20) is evidence of the continuing tectonic activity.

A cooling of the climate is clearly perceptible in the Pliocene. This is well shown by the foraminiferids, which are very sensitive to variations in temperature (*Globigerina pachyderma* and *G. bulloides*), and on the basis of these BANDY (1968) has suggested the occurrence of a glacial episode in the Pliocene about 3 million years ago and another about 1.8 million years ago.

The Tertiary beds of North America contain great mineral wealth. These include the major oilfields of the Gulf Coast in Texas and Louisiana, where the oil accumulated in Miocene rocks against diapirs of Triassic salt, and those of California where most of the oil is obtained from Miocene rocks, with the remainder from the Pliocene. The continental basins contain lignites and bitumen, and gold-bearing alluvial placers occur in the far West. Magmatic intrusions emplaced during the Cenozoic in the Rocky Mountains were accompanied by mineralised zones in which copper (Utah, Arizona, New Mexico) and gold and silver (Utah, Nevada) deposits occur. The Miocene diatomites of California are also actively exploited.

II. – SOUTH AMERICA

The South American continent also took on the broad outline of its present shape in the Cenozoic. At the time of the **Andean-Laramide orogeny** at the end of the Cretaceous and the beginning of the Palaeocene it had a temporary connection with North America, which explains the similarity between the floras of California and those of South America. This link also allowed the migration to South America of two notoungulates, *Artostylops*, known in Wyoming, and *Palaeostylops*, also occurring in Mongolia. Marsupials and edentates also made the crossing between the continents and persisted in South America throughout the Tertiary, though they then disappear from the other continents except Australia. Immigrants with African affinities, including platyrrhine monkeys and hystricomorph rodents, arrived in the late Eocene. According to HOFFSTETTER (1976) they crossed east to west by the equatorial current when that ocean was much less than its present width.

The edentates of South America probably showed the maximum diversifications of form, ranging from the anteaters to the armadillos (Fig. 11.11). Following the re-establishment of the link with North America after the Pliocene earth movements, which brought about a general uplift of the continents, a giant armadillo, *Glyptodon* (Fig. 11.11) migrated to the northern continent together with species of seven other groups, including *Megatherium*, a giant sloth, described by CUVIER (Fig. 12.11, p 232) in 1825. Fourteen other mammal families crossed the Panama isthmus from north to south, led by a great carnivore with dagger-like fangs, which ravaged the autochtonous fauna. By contrast, the elephants remained in North America, unable or unwilling to cross the "heat barrier" of the Panama isthmus.

The sea broke out of the Caribbean basin and penetrated to the east side of the Andean cordillera where it formed a north to south gulf, the Peruvian-Ecuadorean gulf (Fig. 11.12). This gulf remained in existence until the Oligocene and in it were accumulated the rich oil deposits of Peru. The sea withdrew in the Miocene at the same time as the Andes underwent further folding. This was followed in the Pliocene by a general uplift of the area. The Andes reached their present altitude, while major faults formed deep grabens in the interior, such as that of Lake Titicaca on the Bolivian-Peruvian border. As in the case of the Rockies, an enormous quantity of detritus containing a rich vertebrate fauna accumulated on the eastern side of the Andes.

As a whole, the Atlantic border of the South American continent remained stable, being invaded only by minor transgressions into the gulfs of Patagonia, the River Plate, Buenos Aires, Bahia and the Amazon, mostly in the Palaeocene and the Eocene.

Further north, the area of sedimentation of the Caribbean region is of great stratigraphic, tectonic and economic importance for it contains one of the world's richest oil deposits. Tectonic instability, volcanism and orogenesis characterise the northern part of Colombia and Venezuela, the island of Trinidad and the Greater Antilles. These areas formed a largely open sea between the Pacific and the Atlantic. Coastal cordilleras were built and subsequently destroyed during the Palaeogene, while the present outline was shaped in the Mio-Pliocene.

The stratigraphical interest of these formations lies in the fact that they contain a microfauna similar to that of southern Europe, although the large foraminiferids and especially the nummulites are different. The planktonic organisms have enabled good zonal sequences to be established (see Fig. 2.21, p. 46 and Table III, p. 24) and this biozonation is all the more easily transferable from the Caribbean to the Tethys because climatic conditions in the two areas were similar. These two seas were then much closer together than they are today, of course (see Fig. 1.3, p. 16). This biozonation is not, however, readily applied to northern seas, such as the North Sea, because planktonic organisms, like all others, are influenced by ecological factors, especially temperature.

AFRICA

The geological history of North Africa is inseparable from that of the Tethys and has been described in Chapter 8. However, the Palaeogene transgressions advanced far across the Sahara desert. It is probable, indeed, that in the Palaeocene, a trans-Saharan connection was established with the Gulf of Guinea (Fig. 11.13). On the east coast of Africa and on the west coast of Madagascar, Palaeocene basins are well-developed.

Africa, to which was welded the Arabian peninsula, remained isolated during the Cenozoic, except in the late Miocene when the formation of evaporites accompanied the almost complete desiccation of the Mediter- ranean basin. *Africa then became joined to Europe at its eastern and western extremities.* The Mediterranean re-opened through the Straits of Gibraltar by faulting. Its eastern end, however, remained closed, despite the presence of the Red Sea rift, which was blocked at its northern end by the Suez isthmus. This land bridge allowed the passage of horses and cattle into Africa in the Villafranchian.

The *great rift valleys of Africa*, bordered by high and often still active volcanoes, *were formed during the Pliocene.* The floors of the rifts are now occupied, in part, by elongated lakes (Albert, Edward, Kivu, Tanganyika, Nyassa). The rift divides into two parts near

Lake Victoria and in Ethiopia joins up with the Red Sea rift and that of the Gulf of Aden. This "triple junction" is situated in the Afar, a region drained by the Awash river. Geophysical evidence suggests that this is the point at which the continent will begin to break up and a new ocean will be initiated to continue the long term tendency towards the disintegration of Gondwanaland.

In addition to this "earth-shattering event" (for faults have always played an important role in the history of this stable continent), gulfs have been formed on the western and eastern coasts. These include the Gulf of Senegal and that of Nigeria which penetrated far towards the Sahara. Further south were the small gulfs of Cameroon, Gabon, Congo, Angola and South West Africa. The maritime fringe extending from Arabia to Mozambique on the east coast contains the classic fauna of the Mediterranean with *Nummulites* and *Lepidocyclina*. This fauna is also present in the small gulf at Port Elizabeth in the extreme south of Africa. Similar formations occur on the west coast of Madagascar which suggests that this was an island throughout the Cenozoic. This proposition is, however, contradicted by evidence of the migration of lemurs to Madagascar in the Oligocene and of the pigmy hippopotamus in the Pleistocene. The resulting enigma remains unresolved.

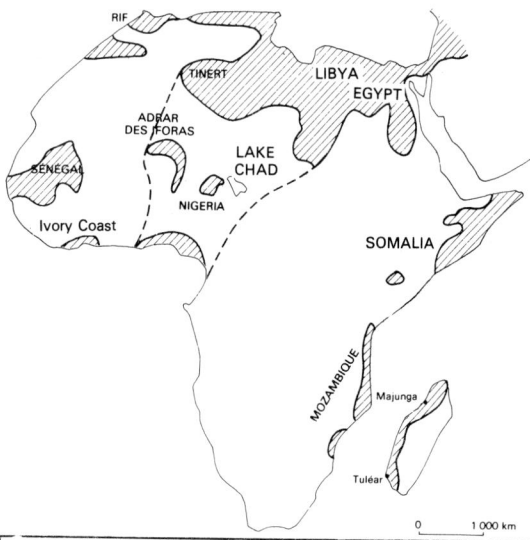

11.13 **Palaeocene seas on the African craton.** Borehole evidence indicates that the extent of the Palaeocene beds is much greater than had been supposed. It is likely that the sea spread in a continuous belt across Africa from the Gulf of Guinea to Libya. Most of the Palaeocene formations contain *Ranikothalia*, sometimes referred to as a cordate nummulite, although according to Blondeau, it is not on the phylogenetic line of the nummulites. The coastal basins extend offshore and are now being explored for oil. As far as is known, most of the coastal basins persisted throughout the Palaeogene.

IV. – ANTARCTICA

Marine Cenozoic deposits with molluscs, brachiopods, nautiloids, fish and penguins are present, as is the fossil flora containing *Nothofagus*, which is known throughout the southern hemisphere. Volcanism, faulting and orogenesis all occurred in the Cenozoic especially in the Trans-Antarctic mountains.

V. – AUSTRALIA, NEW ZEALAND AND NEW CALEDONIA

Australia was virtually untouched by marine transgressions in the Palaeogene. In the Neogene, however, gulfs penetrated widely along the south coast and less extensively across the west coast. Tectonic activity accompanied by faulting and volcanic activity temporarily caused the sea to withdraw at the Mio-Pliocene boundary. The sea returned briefly but left the Australian continent completely in the Pleistocene.

By contrast, the great interior lakes of southern Australia were formed in the Palaeogene. Lake Eyre, now usually dried up, is the last remnant of these lakes. The earliest known remains of a mammal, a kangaroo-rat *Wyniadia bassiana*, occur in the Oligo-Miocene of Tasmania. A fossil fauna consisting of monotremes and marsupials dates from the Plio-Pleistocene.

In New Zealand, on the other hand, peneplanation at the end of the Cretaceous was followed by almost complete submergence by the Palaeogene transgressions. The sediments were sometimes detrital, sometimes calcareous.

They were strongly folded in the Neogene when volcanism occurred. Following an almost total emergence in the Miocene when the two islands were united, a new marine invasion, more restricted than the first, again separated the islands in the Pliocene.

New Caledonia had an eventful history. In the early Eocene, transgression was followed by folding. A new invasion occurred in the late Eocene, when flysch with nummulites was formed, with intercalations of submarine basalt lava flows. Final emergence took place in the Oligocene, with the intrusion of peridotites, now superficially altered to nickel- and chrome-bearing serpentinites, which are found in conglomeratic Miocene littoral deposits. The Tertiary beds of these islands contain a series of microfaunas which enable correlation to be made with the Tethyan province.

VI. – ASIA

This enormous continent was separated from Europe during the whole of the Palaeogene by the Ural Sea which retreated at the end of the Oligocene to allow land migrations between America and Europe across Asia.

In the Middle East, an arm of the sea which passed north of Arabia to link Syria with Iran was the site of almost continuous sedimentation from the Permian to the Pliocene. Important fold movements then built the Zagros mountains and united Arabia with Asia. The Iranian plateau was a subsiding area and was progressively filled with sediment during the Palaeogene. To the north, these beds were subsequently folded to form the Elburz mountains which today border the southern side of the Caspian Sea.

The principal oilfields of southern Iran are in the Asmari limestone, of Oligo-Miocene age. Tilted Quaternary terraces show that in this country the crust is still unstable, due to tectonic activity.

Further east, **an arm of the Tethys separated peninsular India from Asia.** Thick clastic sediments and limestones with abundant nummulites were deposited. These beds are now found in Baluchistan, western Pakistan, Afghanistan and Kashmir.

The "welding" of India to the Asian continent occurred in the Oligocene and the

paroxysmal orogeny which uplifted the Himalayas occurred in the Miocene. At the same time, the Indo-Gangetic foredeep was filled with molasse. The great basalt extrusions of the Deccan, some 2000m thick, had been formed at the end of the Cretaceous. At that time the Indian peninsula lay much further to the south and was passing to the east of an emissive centre which is still active (Fig. 1.3, p. 16.)

Finally the Tethyan geosyncline reached the Far East and skirted the southern edge of the emergent Indonesian platform to which New Guinea and Australia were still attached. The Cenozoic transgression broke up Indonesia into a series of parallel ridges and trenches subjected to folding and volcanic activity from the Pliocene to the present time. However, in the Philippines, the only marine invasion of any importance occurred in the Miocene.

At the beginning of the Cenozoic, Japan was a continental region linked with Korea and Manchuria which was bordered on its eastern side by a geosyncline in which 7000m of Palaeogene sediments were deposited. Intense folding was accompanied by subsidence during the Miocene, and Japan became a string of islands except in the south. The sea retreated during the Pliocene and as a result Japan became linked to the mainland of Asia during the glacial episodes of the Pleistocene.

The Quaternary | 12

The term "Quaternary" was proposed by DESNOYERS in 1829, nineteen years after BRONGNIART had applied the term "Tertiary" to the formations overlying the Chalk in the Paris Basin. In his new term Desnoyers included those marine beds which are younger than those of the Seine basin and in so doing, included the whole of the Neogene! A year later, Marcel DE SERRES (1830) restricted the term "Quaternary" to the "Diluvium" which had been proposed by BUCKLAND in 1823 to describe the beds formed by the "Deluge". These were supposed to be pebble beds, overlain by peats and loams, the Alluvium. He stated that the Diluvium contained human remains, which, in the absence of a precise chronology, was not incompatible with events recorded in the Bible.

I. – SUB-DIVISIONS AND LOWER LIMIT

LYELL, in 1833, divided the last period of the Tertiary, the Pliocene, into a lower or older Pliocene and an upper or newer Pliocene. This latter epoch contains more than 70% of forms which are still living and for it LYELL proposed the term **Pleistocene** in 1839. The term **Holocene**, the second epoch of the Quaternary, was proposed by GERVAIS in 1867. It begins in 10,000 B.P.[1], and contains only species still living. The relationship between the Pleistocene and the glaciations was glimpsed in 1838 by AGASSIZ and established in 1846 by FORBES, who regarded the Pleistocene of LYELL as the "Glacial Epoch".

Thus the concept of a Quaternary or Pleistocene, formed of recent sediments and contemporaneous with the *appearance of Man and a period of cold climate*, was forged towards the middle of the last century. It was an intuitive notion and there was no question of assigning rigorous limits to the Quaternary.

The problem of the limits of the Quaternary was not eased by the creation of the Villafranchian and Calabrian stages. The **Villafranchian** was proposed by PARETO in 1856 to include a sequence of sands and continental marls at Villafranca d'Asti, near Turin in the northern Apennines. These beds have yielded remains of elephants, horses and cattle whose first appearance is considered by HAUG (1911) to mark the beginning of the Quaternary (see Fig. 4.21, p. 129).

The Calabrian stage was proposed by GIGNOUX in 1913 for the marine deposits of Calabria, in the toe of Italy, which yield a fauna indicating colder conditions than those of the Astian. These include the bivalve *Arctica islandica* (Fig. 12.2) and the foraminiferid *Hyalinea balthica*. However, he placed this stage at the top of the Pliocene until the International Geological Congress held in London (in 1948), when he agreed that it should be referred to the base of the Pleistocene.

Analysis of works published over more than a century reveals that four independent criteria have been used at various times as a basis for

---

1. All dates in the Quaternary are given in years "Before the Present" (= B.P.). By convention, the Present is defined as 1950 A.D.

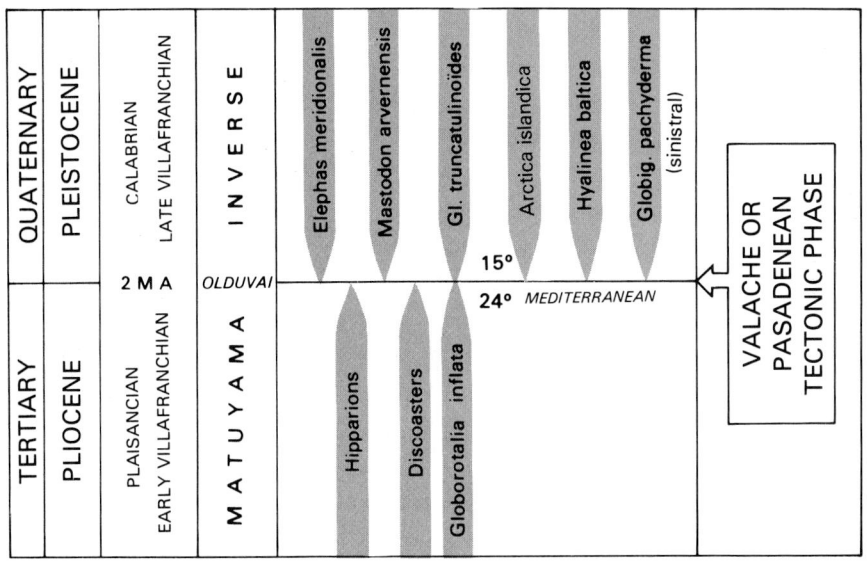

12.1 Some events marking the Tertiary-Quaternary boundary. Neither the Hominids nor glaciations are represented on this chart (see also Table XIII, p.   ). The Olduvai event was a short period of normal magnetic polarity in the long period of the Matuyama reversal and occurred about 1.9Ma. Some vertebrate palaeontologists would, however, place the Tertiary-Quaternary boundary at about 2.7Ma.

the existence of a Quaternary system. These are based on:

anthropology: the appearance of Man;
glaciology: the first glaciation;
vertebrate palaeontology; the appearance of *Elephas, Bos, Equus*;
invertebrate palaeontology: the appearance of a cold fauna of molluscs and foraminiferids.

Were these events synchronous? It will be shown in detail later, and stated briefly here, that they were not, for the following reasons:

the Australopithecines, the first hominids, existed at least about 4 million years ago, in the middle Pliocene;

the first glaciations probably date from 9Ma, at the end of the Miocene;

the lower Villafranchian dates from 3.5 to 4Ma, which puts it in the Pliocene;

the appearance of cold faunas in marine deposits was not simultaneous in all latitudes.

How then should the Pliocene-Pleistocene (Tertiary-Quaternary) boundary be defined? At the present time, it seems that the most suitable position for this limit is **at the base of the Calabrian,** as redefined by SELLI (1967) at Castella, in Calabria. The boundary then coincides with *the arrival of a colder fauna in the Mediterranean,* including *Arctica islandica* and *Hyalinea balthica,* and corresponds to a reduction in summer surface water temperature from 24°C to 15°C (based on $0^{16}/0^{18}$ measurements). This lowering of temperature is also shown by a predominance in the Mediterranean of sinistral over dextral forms of certain species of *Globigerina* such as *G. pachyderma.*

The base of the Calabrian is placed at about 2Ma, the date of a brief return to normal polarity of the Earth's magnetic field (the "Olduvai event") within the long period of the Matuyama inversion (Table XIII). In oceanic sediments, the discoasters disappear from the nannoplankton (Fig. 2.24, p. 51) and *Globorotalia truncatulinoides* appears among the planktonic foraminiferids. At about this time, sedimentation on the continental shelf of the Antarctic ocean ceased. This was associated with an increase in current velocity and with the thickening of the pack-ice.

12.2 **Arctica ("Cyprina") islandica.** A northern lamellibranch which appeared in the Mediterranean Sea at the beginning of the Calabrian.

This boundary divides the Villafranchian into an upper and lower part which straddle the Plio-Pleistocene boundary. It undoubtedy places the appearance of Man, elephants, horses and cattle, as well as the first glaciations, within the Pliocene. However, for the sake of clarity, the observations which form the historic foundations of the Quaternary system will be included in this chapter, and we shall examine in turn continental features, marine episodes, and finally Man and his works.

II. – CONTINENTAL FEATURES

With the exception of the development of Man and his industries, the dominant continental features of the Quaternary were the glaciations.

1. Glaciations (Table XIII)

Historically, the classical terminology of the European glaciations was established by PENCK (1905) in the Bavarian Alps. He recognised four glaciations which he named, from oldest to youngest, *Günz, Mindel, Riss* and *Würm*. The names are those of tributaries of the Danube (Donau) which now gives its name to a glaciation earlier than the Günz. An even earlier glaciation is named after another Bavarian river, the Biber, though this name is generally followed by a question mark.

There has been difficulty in establishing correlations between these Alpine glaciations and those of the ice sheets of northern Europe (Fig. 12.3) and North America (Fig. 12.4). The only correlations known with certainty are those of the two most recent glaciations:

Würm-Vistula-Wisconsin (Fig. 12.6)

Riss-Saale[1]-Illinois

These correspond respectively to the internal morianes (Würm) such as occur at Rives, west

_____

1. A Polish river.

Table XIII
Pliocene and Quaternary correlation
(see Table XIV for the late Quaternary)

Ages	Magnetic Stratigraphy	Marine stages	Marine faunas	Climatic and glacial sequences	Land faunas	Human activities	Hominids
HOLOCENE		Dunkerquian	Thousands of years 0 / 1	PRESENT SUB-ATLANTIC		IRON (Age of Metals)	Neanthropus *Homo sapiens*
			2 / 3	SUB-BOREAL		BRONZE	
		Flandrian = Versilian	4 / 5	ATLANTIC (climatic optimum)	Modern faunas	COPPER	
			6 / 7 / 8 / 9	BOREAL		**Neolithic**	
			10 / 15		Reindeer	**Mesolithic**	
10000 years						**Magdalenian** (Late Palaeolithic)	
	BRUNHES normal	Grimaldian regression −100m		WÜRM I to IV = VISTULA = WISCONSIN with 3 interstadials	Mammoth – Giant Elk (Megaceros) Woolly Rhinoceros (Rh. tichorhinus) Reindeer – Musk-ox (Ovibos moschatus) Lemming	**Solutrean** / **Aurignacian** / **Perigordian**	1st evidence of art
		Tyrrhenian II = Eemian = Normannian	80 millions of years 0.1	RISS-WÜRM	Aurochs – Giant Elk – Panther Wolf – Horse Rhinoceros mercki	**Mousterian** 1st Burials (Middle Palaeolithic)	Palaeanthropus *Homo neanderthalensis*
100000 years				RISS = SAALE = ILLINOIS	Reindeer (Rangifer tarandus) Mammoth (El. primigenius)	Levalloisian industry	
		Tyrrhenian I = Holstein sea (warm fauna)	*Strombus bubonius Conus guinaicus Mytilus senegalensis* 0.2	MINDEL-RISS	Ox – deer – lion – panther hyena Equus caballus	Acheulian	(Pithecanthropus, Atlanthropus, Sinanthropus)
			0.3	MINDEL = ELSTER = KANSAS and interstadials	Hyena – wolf – lynx – fox bear	Clactonian industry / Invention of fire (Early Palaeolithic)	
0.5 MA					Rhinoceros tichorhinus Elephas antiquus and Equus stenonis	Abbevillian (Chellean)	Archanthropus
0,7			0.6	GÜNZ-MINDEL	Epi-Villafranchian = Cromerian (Saint-Prest)	1st biface tools	
Jaramillo **1 MA**		Sicilian (cold fauna)		GÜNZ and interstadials	Rhinoceros etruscus Rh. mercki, Elephas antiquus and Elephas meridionalis Equus stenonis Hippopotamus major	Pebble culture	
Gilsa 1.6	MATUYAMA inversion			DONAU-GÜNZ	Tiglian		
		Calabrian (cold fauna)	*Arctica islandica Hyalinea balthica* *Globorotalia truncatulinoides*	DONAU?	Extinction of Mastodon and Deinotherium Late Villafranchian = Pretiglian (Saint-Vallier) Elephas meridionalis Mastodon arvernensis		(Australopithecus) *Homo habilis*
Olduvai 1.9 MA			Extinction of Discoasters *Gl. inflata*		Extinction of Hipparion Equus stenonis Mastodon borsoni		Australopithecines
2.5	GAUSS normal	Plaisancian		BIBER?			
3 MA **Mammoth**		Astian	*Globorotalia crassaformis*		Early Villafranchian = Reuverian (lower Perrier)	1st stone tools	
3,3	GILBERT inversion		*Gl. exilis*		Appearance of horses ox – elephant – deer		
4 MA		Tabianian / Zanclean	*Globorotalia margaritae*		Ruscinian		
5.3 MA **MIOCENE**	Period V normal	Messinian			Turolian = Pikermian		

(Left margin vertical labels: **PLEISTOCENE**, **PLIOCENE**)

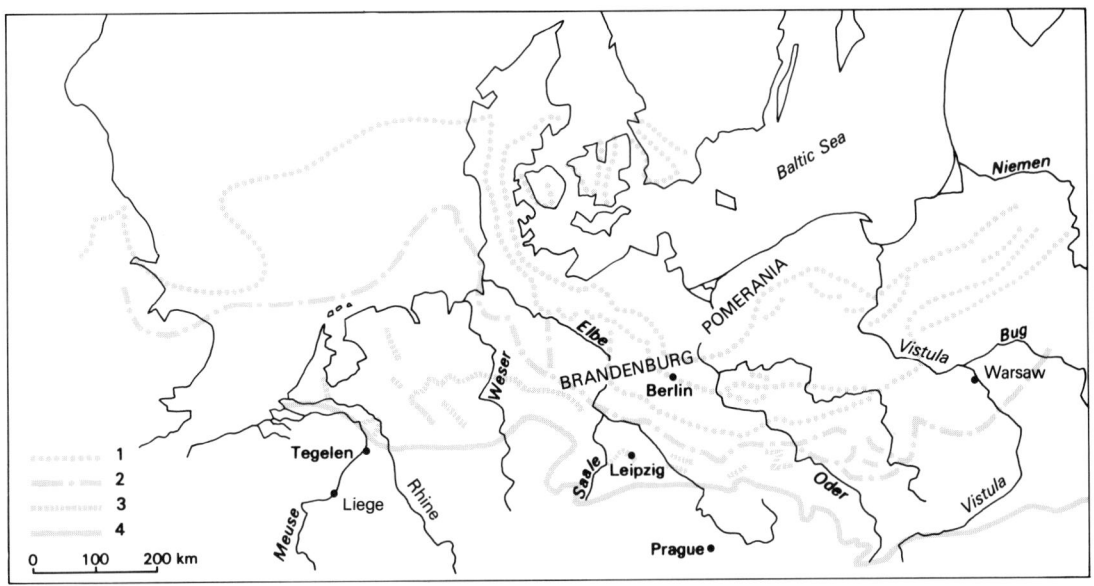

12.3 **Some retreat stages of the ice-sheet of Northern Europe.** 1 – moraines of the Vistula (Würm) with successive positions: Brandenburg, Poznan and Pomerania; 2 – moraines of the Warthe (internal moraines of the Riss); 3 – moraine of the Saale (external moraines of the Riss); 4 – southern limit of erratic blocks (after Woldstedt and St.-Clair Dujon, original).

of Grenoble; and to the more widespread external moraines (Riss), which reach Lyons and cover the region of Dombes (Fig. 12.5). The relationship between the Mindel, Elster[1] and Kansas glaciations is less certain and that between the Günz and the Nebraskan still more uncertain.

It is in fact extremely difficult to decipher all the climatic oscillations on the continents during the Quaternary, since each glacial advance largely destroyed the evidence for the preceding glaciation. Thus, in the Alps, it is often only possible to distinguish two stages of down-cutting of valleys and two morainic episodes. These are the Riss glaciation with its hanging valleys and external moraines, and the Würm glaciation corresponding to the last stage of down-cutting and its internal moraines.

Despite the difficulties of deciphering the evidence for climatic fluctuations on the continents, considerable progress has been made in the use of deep-sea sediments for this

12.4 **The North American ice-sheet** (after R. Flint, 1971) (see also Fig. 12.6).

1. A Polish river.

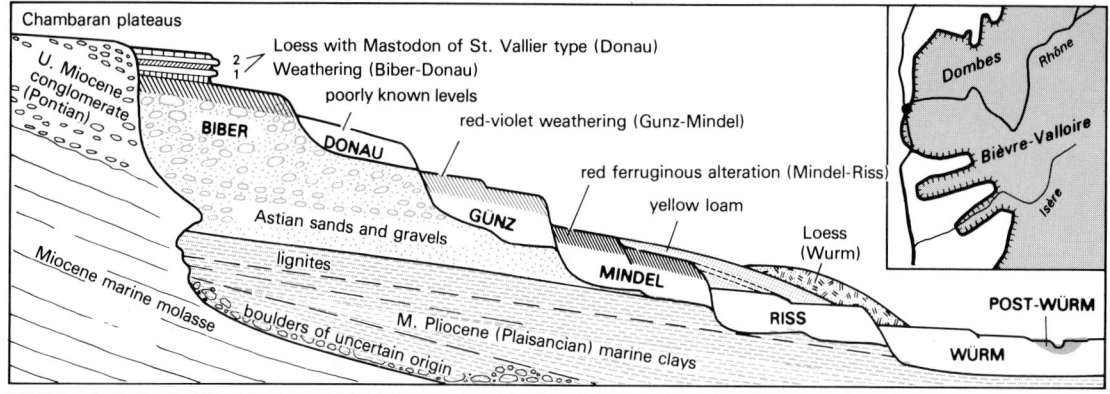

12.5 The fluvioglacial deposits of the Bièvre-Valloire (Bas-Dauphiné). Transverse section showing the sequence of Pliocene and Quaternary deposits laid down in a valley cut in late Miocene-Pliocene (Pontian) times (after F. Bourdier (1959), modified).

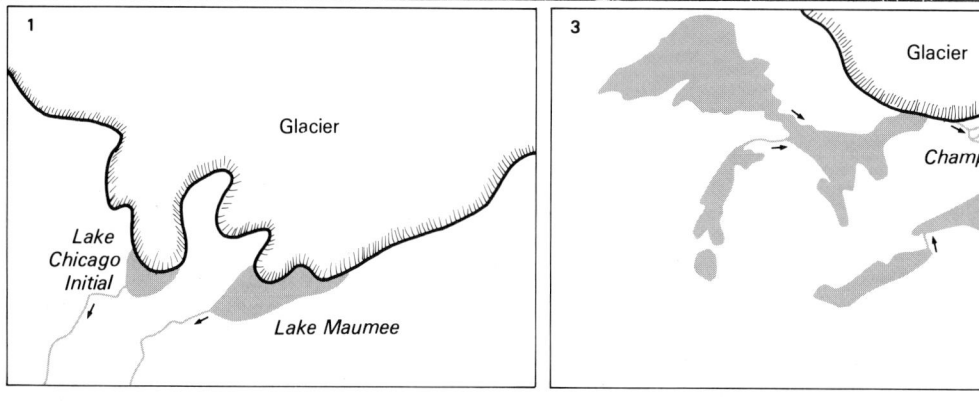

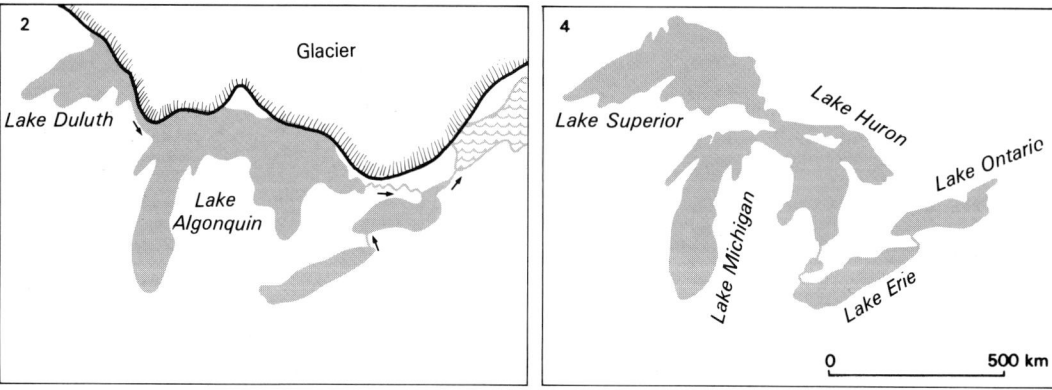

12.6 Some stages in the retreat of the North American Würm (Wisconsin) ice-sheet (after Kay and Colbert, 1965).
1 – The initiation of the Great Lakes – overflow occurred towards the south.
2 – The beginning of the retreat of the ice fronts – the direction of overflow changed and in some areas was reversed. The sea entered the estuary of the St. Lawrence river (the Champlain sea).
3 – Main period of retreat of the ice fronts – melting continued, sea level rose and the Champlain sea reached its maximum extent.
4 – Present day – with the removal of the great weight of ice, the continent has risen (isostatic readjustment) and the sea has retreated. A similar sequence of events occurred in the Baltic area as the Scandinavian ice-sheet melted.

purpose. So far, eight fluctuations in climate have been recognised in the last 700,000 years, that is to say, in the Brunhes normal magnetic epoch. These oscillations probably correspond to four stages of the Würm, two stages of the Riss and two stages of the Mindel glaciations.

It has already been pointed out that the first signs of climatic cooling appeared in the Miocene at about 9Ma and there is still uncertainty as to the age of the Antarctic ice sheet.

There is a continuing need to seek new elements of correlation between the sequence of glaciations on the one hand, and transgressions (marine and fluviatile terraces), periglacial deposits (loess), human industries and volcanic eruptions on the other.

2. Terraces

A simplistic view would suggest that each interglacial stage would correspond to a marine terrace at a constant level on every coast on the globe and that conversely a glaciation would cause a universal regression. In general, this is true and at the maximum of the Würm glaciation most coastlines stood at about 100m below present sea-level. There are, however, *many occasions when this simple concept of eustatic change requires modification*. Orogenic or epeirogenic deformation or isostatic readjustment may bring about local changes in the relative level of the land and the sea. It is also possible that episodes of high sea level are not concordant with interglacial periods, but with periods of reduction in volume of ice sheets (Fig. 12.6). This lack of concordance will influence the level of downstream terraces of rivers, where downcutting will occur during the growth of ice sheets, while at the same time the upstream terraces are prograding as the advancing glacier front releases great quantities of detrital material (Fig. 12.6, 12.7). It is also in the periods of growth that precipitation is greatest. In the Sahara these phases are called "pluvial". The water-courses then carried very great loads of sand, which was redistributed by the wind in travelling dunes in the interpluvial periods.

3. Loess

This consists of very fine grained (20μ)

continental material. Though generally clays or silts, loess deposits are sometimes calcareous. The constituent material was carried by the wind from the terminal moraines of the major glaciers (Fig. 12.8).

A chronology based on loess deposits has been used for a long time. The distinctive characteristics of these are their colour, structure and the man-made implements which they yield. In general, each loess deposit is separated from the next by a fossil soil formed in a more temperate period (an interstadial or interglacial period). In the Paris region, BOURDIER (1950) has established a chronological sequence based on loess, river terraces and human activities (Fig. 12.9 and 12.10).

4. Faunas and floras

The Villafranchian began in western Europe with the arrival of elephants, horses and cattle. The **lower Villafranchian or Reuverian** has yielded *Mastodon borsoni* and *Equus stenonis*, and also *Hipparion*, a survivor from the Miocene. This stage, it should be remembered, is still in the late Pliocene (Plaisancian-Astian, Table XIII).

The **early Villafranchian** is represented in France by the older fauna of Perrier in the Auvergne. The younger fauna of Perrier and that of Saint-Vallier in the Rhône valley contain *Mastodon arvernensis* and *Elephas meridionalis*, and mark the beginning of the Quaternary (**late Villafranchian or Praetiglian**). **The Tiglian** consists of continental beds intercalated in the

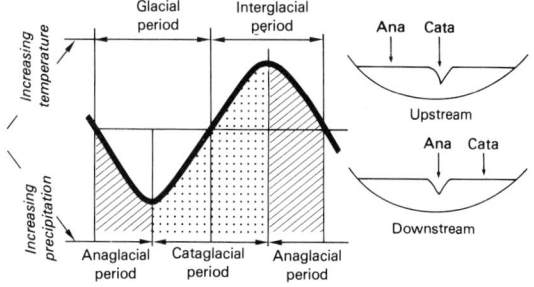

12.7 **The position of river terraces in relation to periods of glacial advance and retreat.** Marine terraces behave in a fashion similar to the terraces of the lower part of a river.

marine clays of Tegelen in Holland and contains mammalian remains and pollen characteristic of the temperate climate of the Donau-Günz interglacial stage. The upper part of the marine **Icenian** deposits of East Anglia is correlated with the top of the Tiglian (see p. 168).

The **Epi-Villafranchian,** found in the upper valley of the Eure (Saint-Prest), corresponds to the fluviatile **Cromer Forest Bed (Cromerian)** of England. The flora of these beds (from which *Tsuga* has disappeared) is typically Quaternary and characterises, in particular, the Günz-Mindel interglacial (Table XIII). *Elephas meridionalis* is still present and is joined during the warmer episodes by *Rhinoceros mercki, Hippopotamus major* and *Elephas antiquus*, with its oblique molars ("warm fauna"). *Mastodon* is now extinct (Fig. 12.11).

The rigorous climate of the Mindel glacial period caused the extinction of certain species (e.g. *E. meridionalis*) and led to the appearance of a "cold fauna" with *Elephas trogontheri* and *Rhinoceros tichorhinus* (the woolly rhinoceros). These were joined in the Riss glaciation by *Elephas primigenius* (mammoth), *Rangifer tarandus* (reindeer) and *Ursus spelaeus* (the great cave bear).

Mammoths have been preserved intact in the frozen soils of Siberia, mummified in Galicia in ozokerite, a natural paraffin wax, and figure in the cave drawings of early man in the Mousterian. The mammoth was smaller than its predecessor (*E. planifrons*), but larger than the modern elephant, and stood about 3.5m at the shoulder. It possessed formidable tusks developed from the upper incisors, some 4 to 5m in length. Its wool was short (about 2cm), but it also had a brown fur coat with hair 50 to 80cm long. This fur, together with a layer of

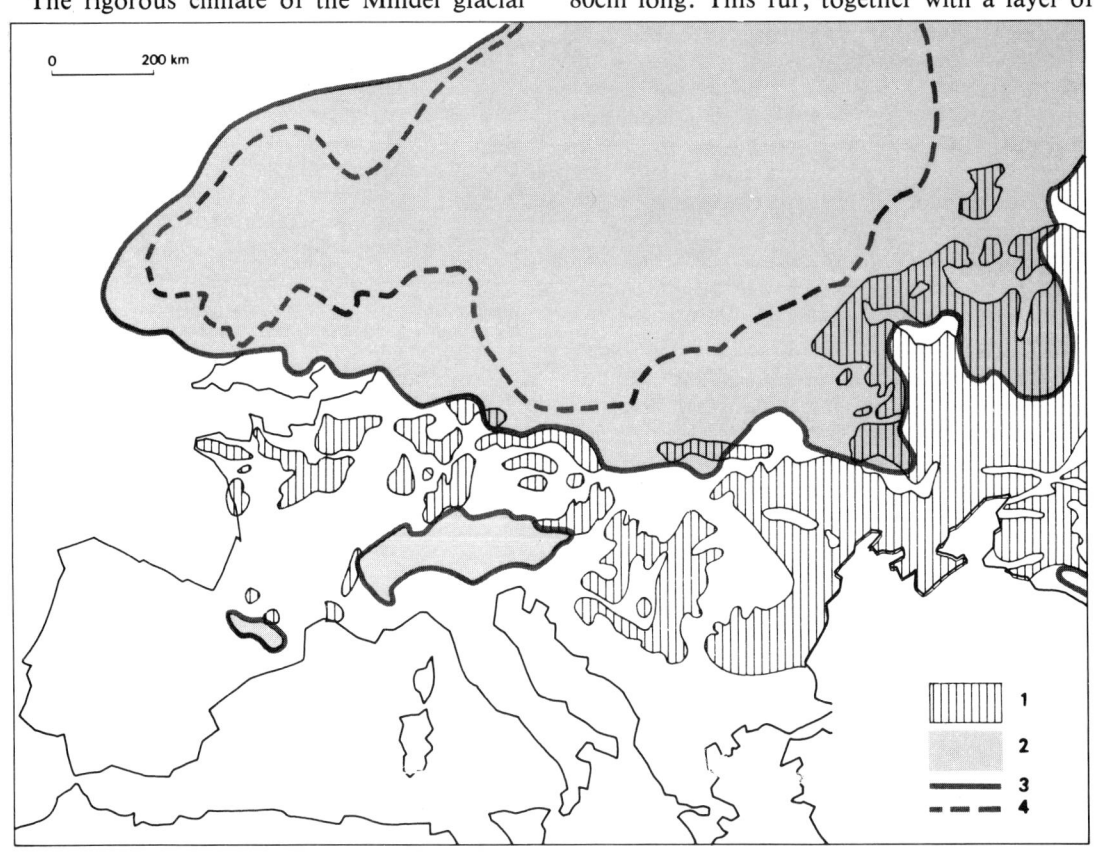

12.8 **Distribution of loess in the European periglacial area** (after Grahmann and Flint, simplified). 1 – loess; 2 – glaciers; 3 – external moraines; 4 – internal moraines (see Fig. 12.3).

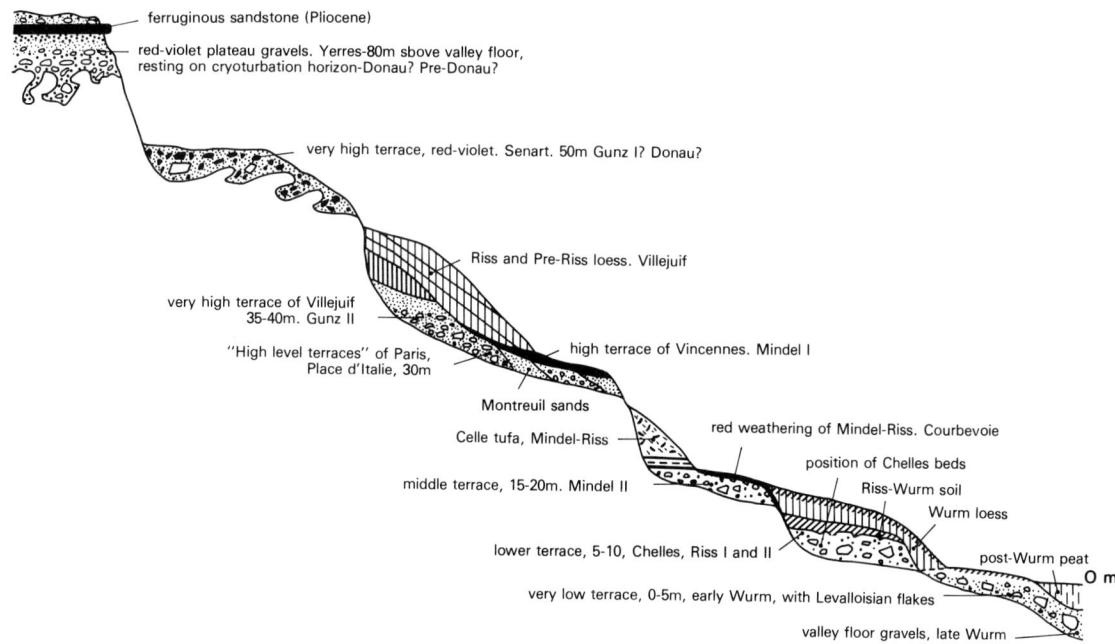

ferruginous sandstone (Pliocene)

red-violet plateau gravels. Yerres-80m sbove valley floor, resting on cryoturbation horizon-Donau? Pre-Donau?

very high terrace, red-violet. Senart. 50m Gunz I? Donau?

Riss and Pre-Riss loess. Villejuif

very high terrace of Villejuif 35-40m. Gunz II

"High level terraces" of Paris, Place d'Italie, 30m

high terrace of Vincennes. Mindel I

Montreuil sands

Celle tufa, Mindel-Riss

red weathering of Mindel-Riss. Courbevoie

position of Chelles beds

Riss-Wurm soil

middle terrace, 15-20m. Mindel II

Wurm loess

post-Wurm peat

lower terrace, 5-10, Chelles, Riss I and II

0 m

very low terrace, 0-5m, early Wurm, with Levalloisian flakes

valley floor gravels, late Wurm

12.9 Chronology of the Quaternary in the Paris region (after Bourdier, 1959).
The relative altitudes are indicated in relation to the stream level. The complex deposits of Chelles are, in reality, much younger than the "Chellean" (now called Abbevillian) which is believed to correspond to the Günz-Mindel interglacial stage (epi-Villafranchian or Cromerian, see Table XIII, p. 226).

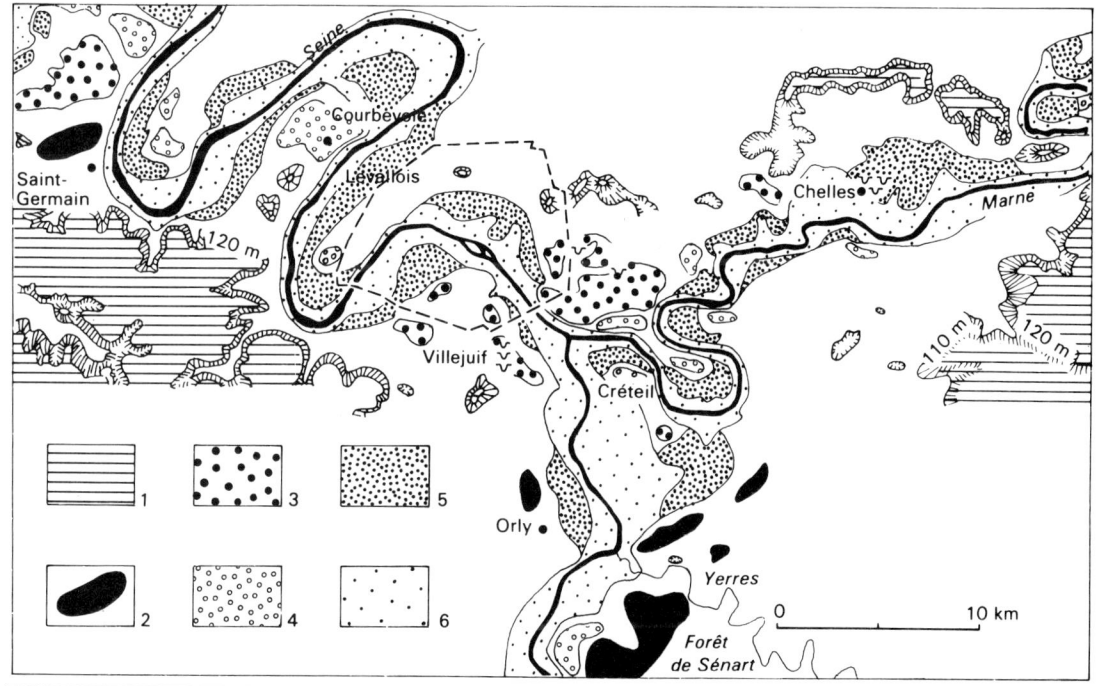

12.10 Quaternary surfaces in the Paris region. 1 — pre-Quaternary surface (+ 115m); 2 — very high level surfaces (50-80m); 3 — high level surfaces (30-40m); 4 — middle level surfaces (15-20m); low level surfaces (5-10m); 6 — modern alluvium (Holocene) (after Bourdier, 1959).

The Quaternary

Megatherium

1 m

Mammoth

Woolly Rhinoceros

1 m

Cave Bear

Sabre-toothed Tiger

Moa

1 m

Giant Deer

25 cm

Dodo

12.12 **Elephas meridionalis** (Villafranchian) was more than 4m in height and 6.80m in length. It had curved tusks and large molars with few, thick enamel ridges (photo: Serrette).

12.11 **Some extinct Quaternary animals** (after Augusta and Burian, Grassé, etc.). *Megatherium* from the Pleistocene of South America belonged to a large extinct group of South American Edentates (animals without incisors and canine teeth). This was Cuvier's "giant sloth" (5m in length). The development of the jugal apophysis (cheek bone) and the expansion of the lower jaw gave the face a massive appearance. The mammoth (*Elephas primigenius*) survived until the last glaciation. Frozen carcases have been found in the north of Russia and serological examination has shown that the Indian elephant is directly descended from the mammoth. It is not impossible that mammoths still lived in the Siberian tundra until comparatively recent times. The woolly rhinoceros (*Rhinoceros tichorhinus*) was still present in the Magdalenian and was widely represented in engravings by hunters of that period. As with the mammoth, its possession of a thick fleece is characteristic of a cold climate and distinguished it from the modern rhinoceroses. The cave bear (*Ursus spelaeus*) was very large. *Dinornis*, the great Moa of New Zealand, did not survive the occupation of the islands by the Maoris. *Aepyornis* from Madagascar was a similar bird and stood some 3m high. Its egg had a volume of about 10 litres. The giant deer (*Megaceros giganteus*), found in the Irish peat bogs and across all Europe, was remarkable for the size of its antlers. It lived before the last glacial epoch in the Eemian (see Table XIII, p. 226). The sabre-toothed tiger (*Machairodus* = *Smilodon*) had upper canine teeth so well-developed that the use of the term "sabre" is not inappropriate (see Fig. 5.11). This animal was widespread across Europe and America in the early Quaternary. The Dodo (*Didus ineptus*) survived in the Island of Mauritius until the 18th century. About the size of a turkey and unable to fly, the dodo was an easy prey for hunters and quickly became extinct. It is known essentially from descriptions and drawings. (Reproduced, by permission, from the *Encyclopedia Universalis*).

fat, 7cm thick, gave this animal considerable protection in cold climates; a protection which *E. meridionalis* (living in warm climates) lacked (Fig. 12.11 to 12.13).

During the marine transgression which accompanied the third interglacial, the isolated descendants of *E. antiquus* gave rise to the dwarf elephants of the Mediterranean islands of Sicily, Malta, Rhodes and Cyprus; an example of regressive evolution. The modern African elephant (*Loxodonta africana*) is also a descendant of *E. antiquus*, but the Asiatic (or Indian) elephant (*Elephas indicus*) is a descendant of the mammoth and has inherited its small ears and its molar teeth with their close packed ridges of enamel. *Hippopotamus major* was closely related to the modern *H. amphibius*.

The cold fauna persisted during the Würm glaciation and further new forms appeared. These included the extinct Giant Irish "Deer" (*Megaceros hibernicus*), which has been preserved in the peat bogs of Ireland. This animal was 2m at the shoulder and 3.5m across the antlers, which weighed 70kg. Also present was the musk-ox (*Ovibos moschatus*) which, in the Holocene, became restricted to the arctic regions. These great mammals were accompanied by *a great variety of rodents, which have been used for long-distance correlation and as indicators of climate*. They are, therefore, specially useful in palaeogeographic reconstruction.

The wild ox or Aurochs (*Bison priscus*), *Bos primigenius*, the ancestor of the domestic ox and *Equus caballus*, the modern horse, browsed in the steppe regions, with their milder climate. *Equus caballus* is the descendent of *Hipparion* of the Mio-Pliocene and *Equus stenonis* of the lower Villafranchian. The modern horse (*E. caballus*) originated in North America and reached Europe by way of Asia. It became extinct in North America during the time of the last glaciation, trapped between the cold of the north and the deserts of the south, which it was unable to cross and so to reach South America.

It is impossible, unfortunately, to build a rigorous chronology on the basis of these faunas because their distribution was dependent on latitude and even at the same latitude, mixed faunas existed. A similar difficulty occurs with pollen analyses which have general climatic significance, but only local stratigraphic value (Fig. 12.14). A number of floral groups indicate particular climatic regions as follows:

> *Dryas* (dwarf member of the Rosaceae), *Salix polaris* (willow), *Betula nana* (dwarf birch) of the modern tundra;
> *Pinus, Betula alba*, and grasses of the cold steppes;
> Spruce, oak, and beech of temperate forests;
> Box, vine and rhododendron of warmer climates.

The sub-divisions of the Holocene (Table XIV, p. 246) are based largely on pollen associations. These divisions are (from older to younger) as follows:

> Newer *Dryas* (the last cold phase);
> Pre-Boreal and Boreal (hazel, oak);
> Atlantic or climatic optimum (oak, spruce);
> Sub-Boreal, characterised by lower temperature and heavy precipitation (beech, alder);
> Sub-Atlantic, in which the forests were reduced by human activity.

5. Radiometric ages

The application of radioactive dating is at present hindered by the gap which exists between the potassium-argon method, restricted to ages greater than 1.5Ma, and the carbon-14 method, restricted to ages less than 50,000 B.P. Recent research suggests two partial solutions. The uranium-234 – thorium-230 method, with a half-life of 250,000 years, has enabled the Tyrrhenian beach deposits with *Strombus* of the cave of the Madonna dell'Arma in Italy to be dated at 85,000 ± 5,000 years. Since this beach deposit underlies a bed containing Mousterian implements, the beginnings of the Würm Glaciation can be dated to 90,000 to 80,000 years (Table XIV). This method, in association with the thorium-230 (or ionium) – radium-226 method (with a half-life of 75,200 years) when applied to coral reefs, has allowed the dating of two marine terraces at 120,000 B.P. ± 12,000 and 80,000 B.P. ± 8,000, respectively correspondingly to the Riss-Würm interglacial, and a

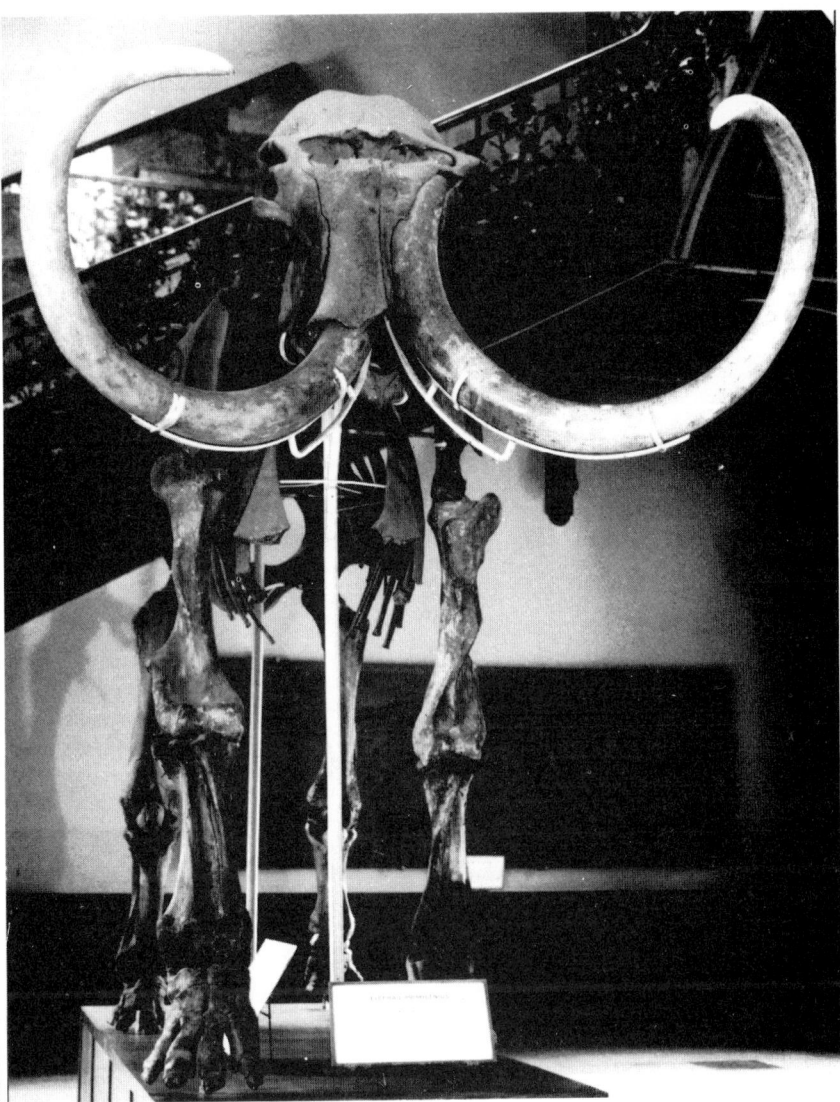

12.13 **Mammoth** (*Elephas primigenius*), though slightly smaller than *Elephas meridionalis* (about 3.5m high) had a relatively larger head, with two very large tusks curving upwards, up to 4m long. The molar teeth of this animal, which have many thin enamel ridges, are often found in the periglacial loess and the alluvium of European rivers. It became extinct in the middle Palaeolithic.

third at 200,000 B.P. ± 20,000 corresponding to the Mindel-Riss interglacial. Of these three levels, two are at about present sea level and that of age 120,000 B.P. is slightly above.

The use of the carbon-14 method (half-life: 5,730 years) has led to a reappraisal of the volcanic episodes of Mont-Dore, formerly believed to be of Mio-Pliocene age. A great number of the lavas and volcanic bombs have proved to be Villafranchian (BROUSSE *et al.*, 1969) and the volcanism of the Puy chain occurred between 35,000 and 4,000 B.P.[1], and even 900 B.P. for ash in the swimming pool at Clermont Ferrand. Even if this last date is

ignored, there can be no doubt that Man witnessed volcanic activity in the Auvergne and it is not impossible that the region will experience renewed activity in the future (PETERLONGO, 1972).

6. Magnetic Stratigraphy

The Quaternary began in the period of the **Matuyama inversion** (Table XIII, p. 226) at the time of a brief return to normal polarity, the *Olduvai event*, which occurred at about 1.9Ma

1. These dates are based on the examination of wood from trees destroyed at the time of the eruptions. – Ed.

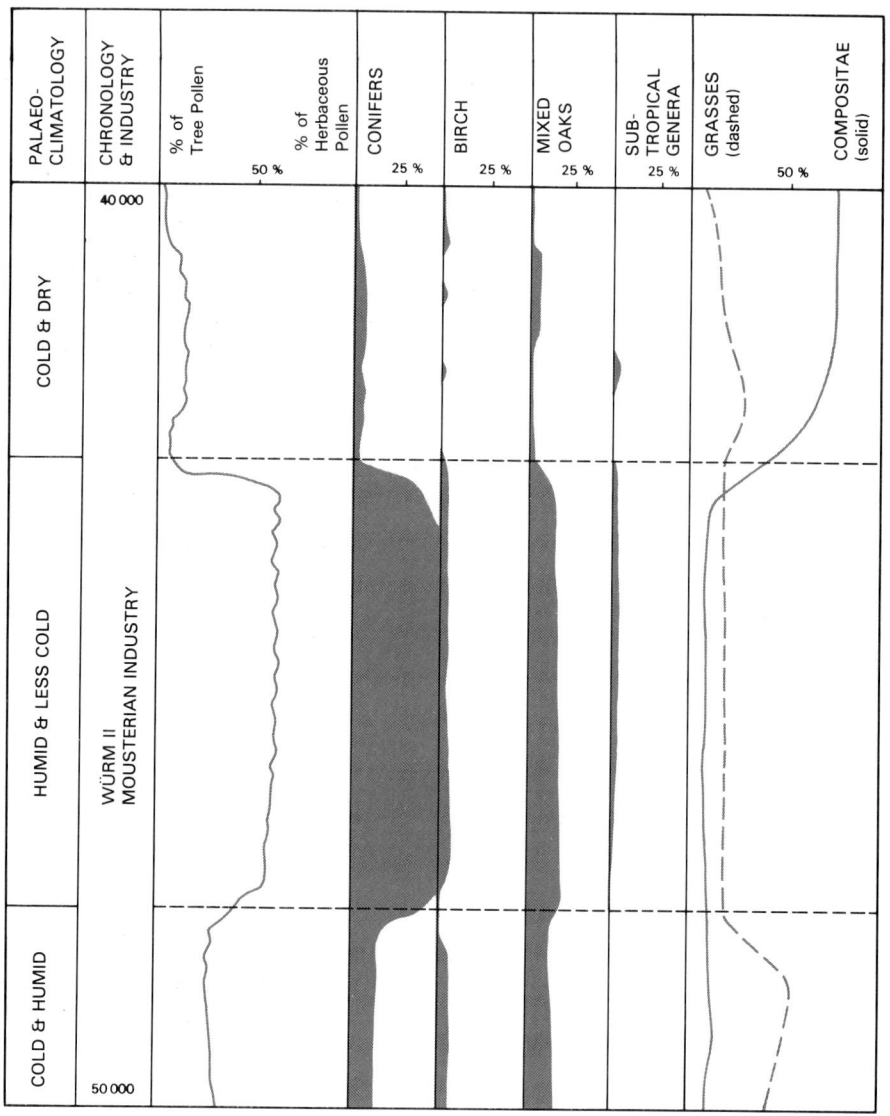

12.14 Simplified diagram showing climatic variations in the Mediterranean Languedoc area during the Würm II phase. Based on pollen analysis of Mousterian sediments from the Hortus cave (J. Renault-Miskovsky, 1967). The evidence from pollen analysis shows three distinct climatic phases:
1 – a phase characterised by pines (of woodland type), birch, mixed oaks and grasses indicating a cool, humid climate;
2 – a phase less cold and less humid during which conifers, oaks and birch co-existed. The grasses were less important and several warmth-loving genera appeared (olive and pistachio);
3 – a cold and dry phase during which trees were replaced by steppelands with *Compositae*.

and coincided with the base of the Calabrian. At about 0.7Ma, the Matuyama epoch was succeeded by the normal Brunhes epoch, with the magnetic poles in roughly their present positions. The reversals of magnetic polarity and the "events" which sub-divide the magnetic epochs form the basis of a chronological scale of special value in the

Pliocene and the older Quaternary where they complement (and often confirm) the absolute dating. An important feature of the magnetic method of dating lies in its applicability, not only to continental deposits, but also to marine rocks such as those situated on either side of the mid-Atlantic ridge (Fig. 12.15).

7. Other techniques of absolute dating: varves, tephrochronology, dendrochronology, lichenometry, historical records,

These techniques are concerned essentially with dating the deposits of the Holocene. The Swedish geologist DE GEER (1934) observed that thinly bedded, rhythmic lacustrine deposits or **varves** were deposited during the melting of the Scandinavian ice-sheets. Each varve consists of a light-coloured layer made up of detritus released by the melting of snow and ice in summer and a dark-coloured layer, with a pre-dominance of organic matter, formed in autumn and winter. The pairs of layers vary in thickness year by year and by selecting suitable layers (or, better, sequences of layers) it is possible to correlate the lower part of a younger section with the upper part of an older, adjacent, section. The detailed measurements that DE GEER and his students made enabled him to date the last advance of the Vistula glaciation (the Brandenburg moraine) at about 25,000 B.P., an age since confirmed by radiometric measurements.

The ashes thrown high into the atmosphere during volcanic eruptions often bury trees when they fall back to earth and dating of their wood by carbon-14 methods allows the determination of the date of the eruption. This technique of **tephrochronology** (tephros = cinder, ash) has revealed an important phase of reburial (by ash) in the Massif Central about 8300 B.P. It also led to the discovery in the Antarctic ice cap of ash from the eruption of Krakatoa, Indonesia in 1883 and from Katmai in Alaska in 1912.

The counting of rhythmic phenomena, such as varves, can be applied to the annual rings of trees, as seen in a felled trunk (**dendrochronology**). Thus by examining fallen trees or the framework of old wooden buildings, it is possible to cover a period of about a thousand years. From the annual rings of *Pinus*

ponderosa taken from the ruins of Wupatki in Arizona, SMILEY (1959) has calculated the eruption of Sunset Crater occurred between the summers of 1064 and 1065.

This is almost in the period of historical documentation, as is **lichenometry,** which is concerned with the rate of growth of lichens. In Alaska, the diameter of patches of *Rhizocarpon geographicum* increases with age. On a moraine dated at 1580, the average size of patches is 170mm, for moraines of 1650 it is 140mm, for one of 1830 it is 30mm and for a moraine exposed in 1937, the diameter is only 4mm.

Human chronicles are as old as the art of writing, but it is often difficult to decipher them and to assess their reliability. Nevertheless, documents often provide the best evidence of the date of geological events within historic time. The best account of the eruption of Vesuvius in 79 A.D. is that by Pliny the Younger.

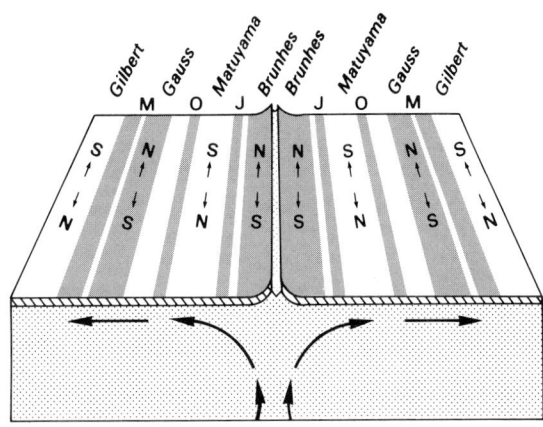

12.15 Plio-Pleistocene magnetic reversals on either side of the mid-Atlantic ridge. The particles of magnetite contained in the basalts of the oceanic crust record the following polarities: Brunhes normal epoch (present to 0.7Ma); Matuyama reversed epoch (0.7 to 2.5Ma) with the Jaramillo event at 1Ma and the Olduvai event at 1.9Ma; Gauss normal epoch (2.5 to 3.3Ma) with the Mammoth event at 3Ma; the Gilbert reversed epoch (3.3 to 5.3Ma). The symmetrical disposition of the bands in relation to the median ridge is one of the principal arguments in favour of the hypothesis of ocean floor spreading.

III. – MARINE EPISODES

1. Calabrian

The late Pliocene came to an end with a regressive phase in which the Astian sands were deposited. This was followed by the Calabrian transgression which deposited marls followed by sands, and brought the first cold fauna into the Mediterranean. This included *Arctica islandica, Chlamys islandica, Mya truncata, Pecten maximus* and *Panopea norvegica*. Amongst the foraminiferids were *Hyalinea balthica* and sinistral forms of *Globigerina pachyderma*. In the later part of the Calabrian (the Emilian) the boreal element of the fauna disappeared and a warmer fauna was re-established. As a result of recent tectonic activity the Calabrian deposits have been uplifted to 100m in the French Maritime Alps and to 500m in the Apennines.

The Calabrian began at about 2Ma and is generally correlated with the middle and upper Villafranchian. It is thought that it occurred at the time of a very early glaciation, the Donau or Danube glaciation. The disappearance of the cold fauna in the Emilian appears to be associated with the Donau-Günz interglacial (Table XIII, p. 226, and Fig. 7.7, p. 171).

2. Sicilian

A second cold episode followed, marked by the return of *Arctica islandica*, the disappearance of some Calabrian species and the development of new species. This was the Sicilian I stage, now found at an elevation of 80-100m in the Gulf of Palermo. As in the Calabrian, the sea became warmer again at the maximum extent of the transgression, with a "warm" fauna (Sicilian II). This second Quaternary sedimentary cycle was probably developed during the long period of the Günz glaciation and its interstadials.

3. Tyrrhenian

Following the post-Sicilian regression, which possibly correlates with the Mindel glaciation, the sea regained a level of + 30m and deposited sediments with a warm water fauna, known as the Senegalian fauna, which includes *Strombus bubonius, Mytilus senegalensis, Cardita senegalensis* and *Conus testudinarius*. These Tyrrhenian beds are shelly sands, calcarenites and partially cemented sands, in a molasse facies which in Italy are given the name "panchina". In the Maritime Alps, the Tyrrhenian beds occur at levels between 10 and 30m, whereas near Marseilles they are below sea level. It must be emphasised that, in tectonically active regions like the Mediterranean, beds of the same age may not occur at the same elevation and conversely, beds of similar elevation are not necessarily of the same age (see Fig. 8.14, p. 200).

This idea of the concordance between the level of marine terraces (or raised beaches) and the corresponding interglacial periods (see Fig. 12.7, p. 229), seductive though it may be, cannot be substantiated by recent oceanographic discoveries. There is, for example, a

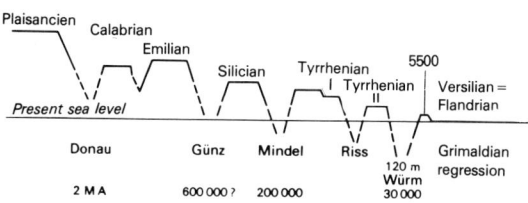

12.16 The succession of marine levels in the Quaternary and their correlation with the glaciations. The relative altitudes result from the combination of the effects of isostasy, subsidence and tectonic deformation. Some authors believe that the regression between Tyrrhenian I and Tyrrhenian II corresponds to a Würm interstadial. It is then necessary to shift the positions of the preceding glaciations one place to the left (after Blanc, 1942).

great regressive phase within the Tyrrhenian period, which separates it into two sub-stages: **the Tyrrhenian I (or Eotyrrhenian)** and the **Tyrrhenian II (or Neotyrrhenian).** The low levels of the latter, 2 to 8m above sea level probably correspond to the Riss-Würm inter-glacial[1].

In the North Sea basin during the Tyrrhenian I sub-stage, a gulf covered part of Denmark and reached as far as Hamburg depositing the Yoldia clay. This gulf is known as the Holstein sea (the Baltic Sea was not yet in existence). During the Tyrrhenian II sub-stage, a new transgression (known as the **Eemian** stage[2]) covered not only the margins of Holland but advanced across north Germany to eastern Prussia, foreshadowing the Baltic sea. In other areas, the Tyrrhenian II has received various regional names: the **Ouljian** stage, on the Atlantic coast of Morocco; the **Normannian** on the shores of the English Channel, where beds of varying age are found at elevations between 0m and 25m. Along the coasts of the Armorican massif and in Normandy, many deposits found very close to present sea level have been attributed to interstadials of the Riss and the Würm.

The Tyrrhenian II transgression is followed everywhere by the **Grimaldian regression** which is contemporaneous with the Würm glaciation. The sea level fell by at least 100m (Fig. 12.16) and the greater part of the continental shelf of north-west Europe (especially the floor of the English Channel and North Sea) was uncovered (see Fig. 1.4, p. 17).

For a third time (after the Calabrian and the Sicilian) a cold fauna entered the Mediter-ranean. Remnants of this fauna, which included *Arctica islandica* and *Panopea norvegica*, dated at 20,000 B.P., have been found on the eastern part of the Ligurian platform at the exceptional depth of 200m as a result of faulting.

This great regression, the only well-known one in the Pleistocene and probably the most important, has had for the Quaternary,

consequences analogous to those of the Messianian (or Pontian) regression of the late Miocene. In all the seas of the world, most coastal islands became re-attached to the Mainland. For example, the Bahamas formed a vast plateau on which the emergent coral reefs were destroyed and water-courses cut valleys to the edge of the continental slope. The Grimaldian regression culminated with the Würm II glaciation (about 45,000 B.P.) and was accompanied by cold water faunas which are found only in cores from deep water and never along the present day shoreline. The strong winds, which deposited loess on the continents, redistributed the sands uncovered by the retreat of the sea into immense dunes. Some of these became consolidated and are now found below present sea-level in the Mediterranean. On the Atlantic coast, these dunes have served as a source for sediment for modern sandy beaches.

4. Flandrian = Versilian

In the last stages of the Würm, a new transgression, with a cold fauna containing *Arctica islandica*, was initiated. However, the sea gradually became warmer and rose slowly to its present level. This is the Flandrian transgression of the northern seas. In the Mediterranean, the same transgression was named the Versilian (from Versilia, the alluvial coastal plain of eastern Liguria, near La Spezia (BLANC, 1966).

The Flandrian transgression proceeded in a step-wise manner during the later stages of the Würm. According to LARSONNEUR (1972), sea level in the English Channel was still 40m below present level in the Preboreal period (9000 B.P.). It then rose quickly (at more than 2cm per year) until the Boreal (8000 B.P.), at which time the Straits of Dover were opened. The rise in sea level continued, 10m below present sea level being attained in the Sub-Atlantic period, and the present level was reached about 2000 B.P. – at the beginning of the Christian era. On the coast of West Africa, this transgression attained its maximum about 5000 B.P. in the Nouakchottian stage.

A level, 3m above present sea level, has been recognised in several parts of the world, but has not been found in the English Channel. This

---

1. Authors are not in agreement on the correlation between the marine stages and the glaciations which is proposed here. Thus for some authors, the Calabrian lasts until the end of the Günz glaciation, the Sicilian corresponds to the Mindel-Riss interglacial, and the Tyrrhenian to the Riss-Würm interglacial as well as to the Würm glaciation. At the present time, it is not possible to be certain about any of these correlations.

2. From the Eem valley south of the Zuider Zee, now the Ijssel Meer.

level occurred at the time of the "climatic optimum" about 5000 B.P. Nevertheless, a small positive oscillation, the Dunkerquian, which took place at the beginning of the Christian era (the Sub-Atlantic) was responsible for depositing marine sediments just above the highest beach levels of the present day.

Certain flat-bottomed gulfs which are in communication with the open sea by way of a shallow sill record the oscillations of sea level with a precision of a few hundred years. The Baltic is such a gulf, being sometimes marine, sometimes lacustrine as sea level changed. However, the interpretation of these oscillations is made difficult in the case of epicontinental seas because of the isostatic readjustment which followed the break up of continental ice sheets.

IV. – MAN AND HIS ACTIVITIES

1. The Hominids

a) Oreopithecus and Ramapithecus: the development of man-like forms.

The evolution of the primates probably began in the Eocene, but the earliest recognised fossil with hominid affinities is *Propliopithecus* of early Oligocene age from the Fayum in Egypt. The primate line divided in the Miocene with one line ascending to the apes and the other to man (Fig. 12.17). This second line has a number of side branches of relatively short duration. The first of these led to *Oreopithecus* Late Miocene) of which several examples have been found in Tuscany during the past century. In 1957, an almost complete skeleton was discovered in the lignite beds of Grossetto. "Grossetto man" had a flat face and small canine teeth and was without a diastema (gap between front teeth and cheek teeth), a character which differentiates it from the apes and links it to Man. The pelvis was broad and basin-shaped and the strong lumbar vertebrae show that this creature could stand upright, a fundamental characteristic of the Hominidae.

The development of man-like (humanoid) forms is characterised by the acquisition of two features: a permanently upright stance, which frees the hands for the making and use of tools; an increase in the volume of the cranial cavity and a corresponding development of the brain, which led to an increase in intelligence and powers of thought. From this latter characteristic two important consequences followed though material proof was not forthcoming for a long time. The first was artistic ability, whose first manifestations did not appear until the end of the Aurignacian about 25,000 B.P. The second was language, although the use of writing was only discovered by the Sumerians in Mesopotamia about 6000 B.P.

It is now known that *Oreopithecus*, which was a tree dweller, lay on an evolutionary sideline. However, at the same time (the late Miocene) another animal, *Ramapithecus*, lived in India and Kenya. This also had a jaw with much reduced canine teeth and was clearly related to the Pliocene australopithecines which now appear to provide the oldest and most reliable marker on the path of Man's evolution.

b) Australopithecines and Homo habilis.

The Miocene *Ramapithecus* was only a first step and it was necessary to wait for the Pliocene australopithecines to be certain of recognising the true line of ascent of Man. The first of these was discovered in South Africa in 1925 and was given the name *Australopithecus* (Australo = southern, *not* Australian). The centre of research on the origin of Man subsequently shifted to Tanzania, then to Kenya and to Ethiopia (Fig. 12.18).

In 1959, LEAKEY discovered a skull of *Australopithecus*, which he called *Zinjanthropus*, in the now well-known Olduvai Gorge in Tanganyika (now Tanzania), east of Lake Victoria. These hominids were of small stature, about 1.0 to 1.3m tall and were bipedal. They had a flattened skull, a marked prognathism (a projecting jaw) and a strong supra-orbital ridge. The lower jaw was similar to that of modern man, but the teeth were rather stronger. The brain capacity of about

500cm³ was less than half that of modern man and shows that the upright stance and the use of the hands preceded the development of intelligence.

When first discovered by LEAKEY, *Australopithecus* was thought to be 500,000 to 600,000 years old and coincident with the beginning of the Quaternary as then under- stood. This and other discoveries profoundly disturbed the established concepts of pre- history. Potassium-argon dating of the oldest beds then known in the Olduvai Gorge gave an age of about 2Ma and in these beds the remains of a hominid much closer to man were found. This had a cranial capacity of 650 to 700cm³ associated with larger and more curved parietal

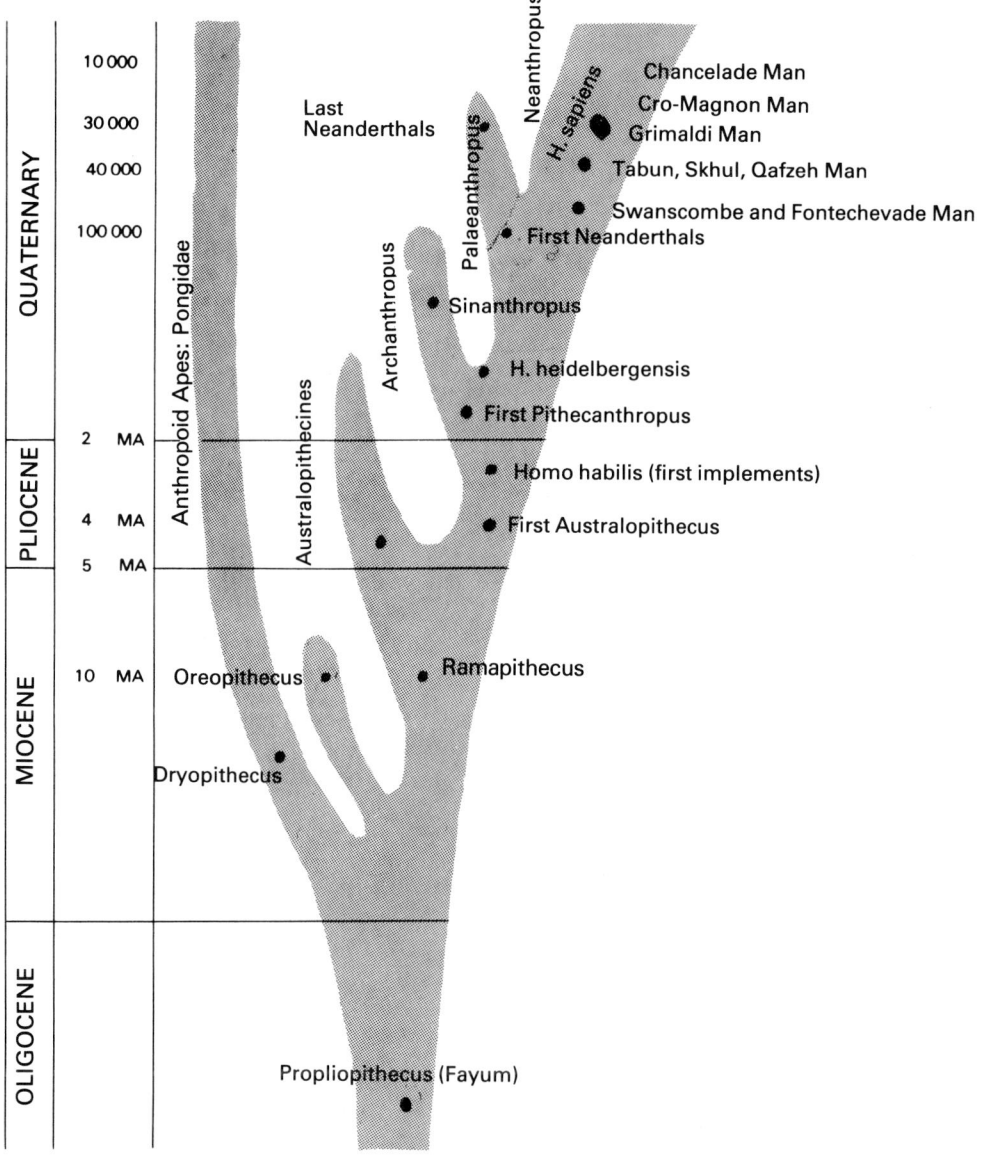

12.17 **The evolution of the Hominidae since the Oligocene** (see text for discussion).

bones. It had a hand capable of grasping, though less perfectly than that of modern man, and was fully adapted to a bipedal gait.

LEAKEY also discovered, in association with these remains, worked tools, made from pebbles ("Pebble culture") (Fig. 12.22) and evidence of shelters. He called this species *Homo habilis* since it showed a marked advance over *Australopithecus* both in its anatomy and in its ability to fabricate tools. Very soon after this COPPENS (1971) discovered remains similar to those of *Homo habilis* at a site 400km north of Lake Tchad and he named these *Tchadanthropus uxoris*. Thus, about two million years ago, two animals with man-like characteristics co-existed. Though both were bipedal, one was more advanced than the other, having learnt to make implements.

But these startling discoveries did not stop there. Further discoveries of australopithecines were made by LEAKEY east of Lake Rudolph in Kenya and by COPPENS in the Omo valley in Ethiopia (Fig. 12.18). The oldest of these remains occurred in beds nearly five million years old. It had, therefore, to be accepted that **the first hominids lived in the Pliocene** and that the beginning of the Quaternary could no longer be equated with the appearance of Man or Man-like creatures.

As the number of finds increased the problems of the origin of Man were clarified but also becoming more complex. They were simplified because the genera *Australopithecus* and *Zinjanthropus* were re-united into a single genus (*Australopithecus*), divided into two species: a massive form, *A. robustus* (formerly *Zinjanthropus*), and *A. africanus*, a more slender form. It is still not known, however, if *A. africanus* is the female of *A. robustus,* thus showing marked sexual dimorphism, or if it is the ancestor of *Homo habilis,,* whose ancestry cannot otherwise be traced beyond 2Ma.

c) **Archanthropoids: Pithecanthropus and Sinanthropus** (Fig. 12.19). This branch in the history of the hominids probably spanned more than a million years. This group had a low facial angle and was without a prominent chin. It had prominent eyebrow ridges and a pronounced bulge at the back of the skull (a "chignon"). The cranial capacity was close to

1000cm^3, considerably more than that of the australopithecines. Believed to have discovered fire, they lived about 500,000 years ago. The first remains were found in Java in 1891 and given the name *Pithecanthropus*, then more were found in China, south west of Peking, and named *Sinanthropus*. More recently in 1954, ARAMBOURG found evidence of this group at Ternifine in Oran (Algeria) and in Morocco.

These hominids, sometimes grouped under the name *Homo erectus*, used bifaced stone tools of Chellean and Acheulian type. A massive jaw, though without a chin, found at Mauer, 10km south east of Heidelberg, has been attributed to a closely related form *Homo heidelbergensis*. This is the oldest human find from Europe, though not the only member of the group. In 1971, DE LUMLEY discovered a skull 200,000 years old (the beginning of the Riss glaciation) in the Arago Caves, near the village of Tautavel, 19km north west of Perpignan. This skull was associated with 10,000 objects of early Tayacian and middle Acheulian age and a fauna including reindeer, elephant, horse and *Rhinoceros mercki*.

d) **Neanderthal Man and Modern Man.** From an undetermined date, probably near the end of

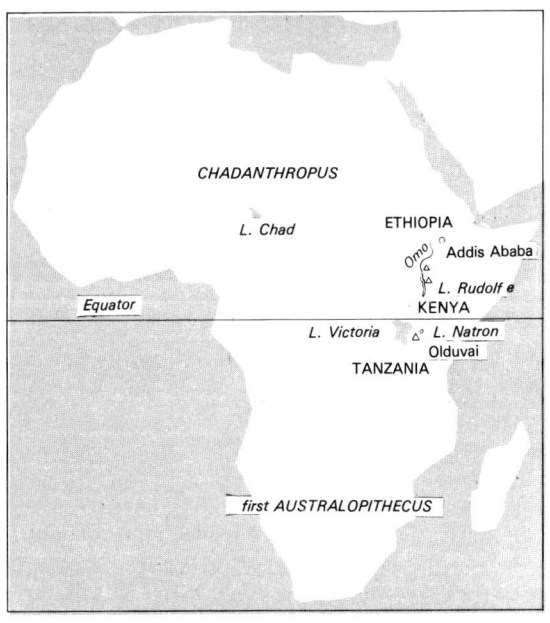

12.18 **The locations of the principal deposits containing the remains of Hominids in Africa.**

the Riss period, a little before 100,000 years ago, several groups of hominids, known as palaeanthropoids or Neanderthalers, lived in Europe. Their origin from the previous group is no longer in doubt. A parietal bone of an adolescent of an intermediate type has been found by DE LUMLEY (1969), in the Lazaret cave on the slopes of Mont-Boron at Nice, where the

youth had sheltered at the edge of a Mediterranean 20m lower than present sea level, about 130,000 years ago in the late Acheulian.

The Neanderthalers still had no chin, but the volume of their cranium was similar to that of modern man (1500cm^3). The frontal lobe was, however, smaller than that of *Homo sapiens* (36% compared to 43%) and the convolutions

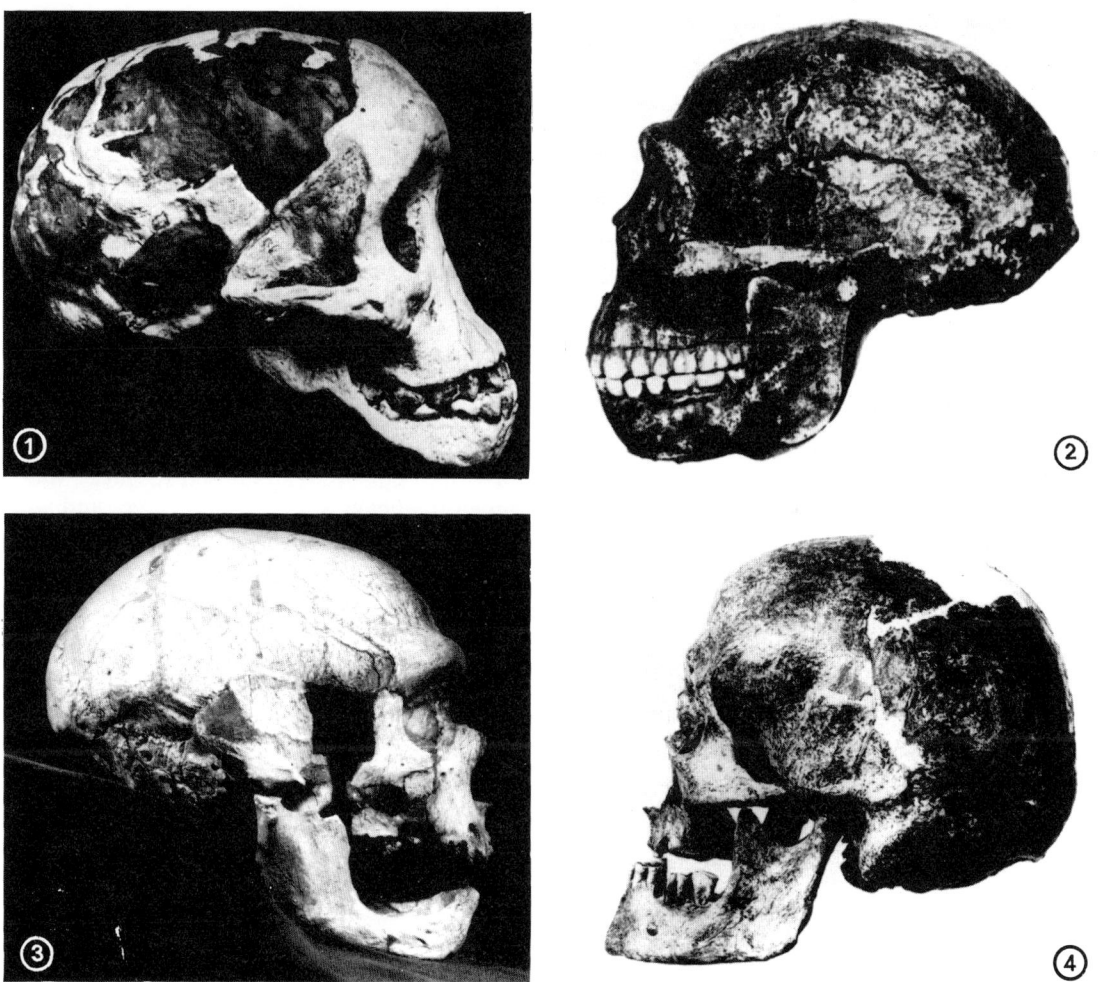

12.19 **Skulls of the principal types of Hominids:**
1 – *Australopithecus africanus* with simian characteristics and a cranium of small capacity.
2 – *Sinanthropus pekinensis* from Chou-Kou-Tien, near Peking; the brain volume increases; prognathism (i.e. having projecting jaws and low facial angle), absence of chin, brow-ridge or "visor" above eye sockets and post-orbital contraction; cranium elongated to the rear ("chignon").
3 – *Homo neanderthalensis* from Chapelle-aux-Saints (Corrèze); prognathism decreases but the visor and chignon persist; the post-orbital contraction is less marked (greater capacity of the brain); no chin.
4 – *Homo sapiens*, Chancelade (Dordogne); the cranium predominates over the face, the visor and chignon are absent, high forehead, prominent cheek bones, chin well developed (Musée de l'Homme, Paris).

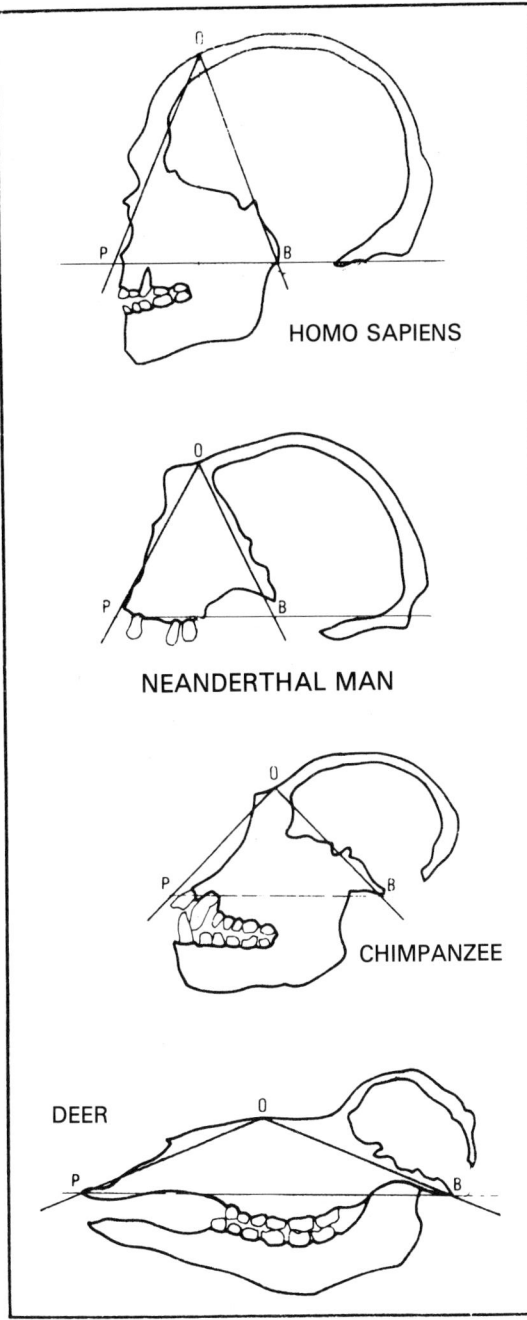

HOMO SAPIENS

NEANDERTHAL MAN

CHIMPANZEE

DEER

were less numerous and less complicated (Figs. 12.20 and 12.21).

Prognathism was dominant and the supraorbital ridges were conspicuous, but the "chignon" was much reduced. Their industry was Mousterian and they buried their dead in graves. One of the most typical is "Chapelle-aux-Saints Man" found near Brive (Corrèze), but perhaps the best known is *Neanderthal Man* found near Düsseldorf in the Westphalian Rhineland.

It may be noted that, while the earliest remains of man were discovered in Africa and Asia, the forms which immediately preceded modern Man are chiefly known from Europe.

In fact, it seems that from about 100,000 years ago, *two branches of the Hominidae have co-existed*. The forms previously called "*pre-sapiens*", such as *Swanscombe Man* from the south side of the Thames estuary and *Fontéchevade Man*, discovered in a cave on the bank of a tributary of the Charente, near Angoulême, lived about 100,000 years ago at the beginning of the Riss-Würm interglacial. They both possessed an upright forehead which lacked the heavy supra-orbital ridge, and were very little different from *Homo sapiens*, though this species is much more recent (30,000 B.P.). This group is called the neanthropoids. Without further evidence, it would seem unjustified to link these "cave-men" with modern man. However, the finding in Israel of *Tabun, Skühl,* and especially *Qafzeh Man* has helped to fill the gap. In Qafzeh VAN DER MEERSCH (1966) found twelve skeletons interred near the

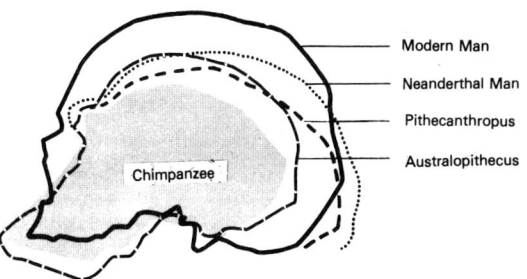

Modern Man
Neanderthal Man
Pithecanthropus
Australopithecus

Chimpanzee

12.20 **Profiles of skulls of mammals.** In the sequence from deer to man, the face shortens (the angle POB decreases) and the volume of the brain increases (after Cailleux).

12.21 **Comparative profiles of the skulls of the chimpanzee and the hominids.** Three groups are apparent: 1 – chimpanzee and Australopithecines; 2 – *Pithecanthropus* and Neanderthalers; 3 – *Homo sapiens*.

entrance to a cave which are of typical *sapiens* type and are about 40,000 years old.

Thus, when the Neanderthalers died out about 30,000 B.P., *Homo sapiens* had already existed for several tens of thousands of years and was practising, it seems, a more evolved Mousterian industry. This was succeeded by the Perigordian industry. The Aurignacian industry was practised by *Cro-Magnon Man* discovered at Les Eyzies in the Dordogne. *Grimaldi Man* (from Grimaldi near Menton, east of Nice) is slightly younger than *Cro-Magnon Man*, while *Chancelade Man* (from Chancelade, near Perigueux) is still younger. Hominids of the same age have been found in Africa at Olduvai and in northern America.

The first people to occupy the North American continent arrived during the Würm II period when man and reindeer were able to cross the Bering isthmus which formed a bridge as a result of the lowering of sea level during this glacial episode. Other people penetrated into Australia which was still joined to New Guinea.

This brief review of the evolution of the Hominidae leads to the following assertions:

– **Australopithecines appeared in the Pliocene at about 4Ma.**

– **The lower limit of the Quaternary cannot be defined on the basis of the appearance of Man.**

– **The branching pattern of evolution of the Hominidae has shown that several species have co-existed,** though the level of evolution in the branches may have differed at any particular time, as for example, between:

> *Australopithecus* and *Homo habilis*
> the Neanderthalers and *Homo sapiens*.

– **Certain stages of evolution have resulted from the convergence of characteristics in groups evolved independently** in different regions. This is found, for example, in:

> *Pithecanthropus* and its relatives of China, Indonesia, Ethiopia and Africa,
> the Neanderthalers of the various provinces of Ethiopia, Africa and Europe.

– *Homo sapiens* **is much older than had been supposed,** having appeared in the Riss-Würm interglacial, between 100,000 and 80,000 years ago.

2. Human industries and activities

Man had very soon learnt to work stone, but the use of other materials did not occur until about 6000 B.P., in the Neolithic period. At that time polished stone implements replaced the chipped or flaked tools that had been in use since the early Quaternary. Polished stone tools were still used by Australian aborigines up to a century ago and are still used by some tribes in Africa and Indonesia.

a) **Early Quaternary.** The earliest known implements made by Man, dated at 2.2Ma, were found by CAVAILLON (1971) in the Omo Valley in Ethiopia. These are slightly older than those used by *Homo habilis* at Olduvai at about 2Ma. These were **pebble cultures** and three main types of implement occur. The simplest tools were *choppers* made by removing two or three flakes from one end of a pebble, from one face only. By removing flakes from both faces a *chopping tool* was formed. By removing flakes from various directions polyhedral forms were made (Fig. 12.22). Choppers have been found in the Vallonnet cave at Roquebrune – Cap Martin, while polyhedral tools have been found in the Villafranchian terraces of the River Têt (Roussillon).

b) **Early Palaeolithic: Chellean- Abbevillian-Clactonian.** The deposits at Chelles, near Meaux (Seine-et-Marne) contained stone implements which became known as Chellean. However, the Abbé BREUIL (1939) proposed that the term Chellean should be replaced by the term Abbevillian (from Abbeville, Somme). This industry is characterised by bifacial tools with deep flake scars, having a yellowish-grey patina and a continuous sharp rim (BOURDIER, 1959). These are the typical *coup de poing* or hand-axes (Fig. 12.22). The flakes were also utilised, and have been given the name Clactonian (from Clacton-on-Sea, Essex), but to do this is to split up a homogeneous industry. The Abbevillian/Clactonian industry occurred in the Günz-Mindel interglacial and during the Mindel. This industry was the work of Heidelberg Man and his contemporaries, who also invented fire.

c) **Middle Palaeolithic** (Table XIV).

1) **Acheulian, Levalloisian and Tayacian;** Acheulian hand-axes (from St. Acheul,

The Quaternary

Table XIV
The correlation of the late Quaternary

Time		Ages	Industries		Man
Present 0	HOLOCENE – POST GLACIAL		Space / Atom		HISTORIC
25					
100		Subatlantic	Electricity / Internal combustion engine / Jesus Christ / La Tène		
2 000					
2 500			**Iron**		
3 000		**Sub-Boreal**	**Hallstatt**		
4 000			Bronze		
4 500			Copper		
5 000		**Atlantic** (climatic optimum)	**Neolithic**		*Megaliths (beginning of writing: Sumerian)*
6 500		**Boreal / Preboreal** / Late Dryas	Tardenoisian / **Mesolithic** / Azilian		
10 000	WÜRM (last glaciation)	Würm IV / Late Glacial — **Alleröd** Early Dryas	**Magdalenian**	UPPER PALAEOLITHIC	CHANCELADE
15 000		**Lascaux** Würm III-IV			SAPIENS
20 000			Solutrean	Perigordian / Aurignacian	*1st appearance of art*
25 000		**Würm III**			GRIMALDI / CRO-MAGNON
30 000					
35 000		**Würm II-III** inter-glacial	Mousterian		HOMO NEANDERTHALENSIS — HOMO
40 000		**Würm II**		MIDDLE PALAEOLITHIC	
50 000		**Würm I-II** inter-glacial	Levalloisian		*1st villages*
c. 80 000		**Würm I**			
		Riss-Würm			

246

Amiens) are lighter and thinner than earlier axes. Their edge is more regular and their shape more geometrical (almond, oval, triangle). Some elongated flakes, called "flake tools" appeared, as did the first scrapers. To obtain the best results, the tool-makers prepared a core of flint in such a way that the form of the flake to be struck off was accurately pre-determined. This is the *Levalloisian technique* (from Levallois-Perret, north west of Paris) which persisted into the succeeding period, the Mousterian.

Derived from the Abbevillian, the Acheulian began in the Mindel-Riss interglacial and persisted throughout the Riss glaciation, the Riss-Würm interglacial and into the Würm I phase, where it ended with the Micoquian (from La Micoque aux Eyzies, Dordogne), noted for its triangular hand-axes. Good correlation has been achieved between the industries of the Middle Palaeolithic and the fluviatile deposits of the terraces of the River Somme near Amiens.

Another culture developed in parallel with the Acheulian during the Riss glaciation. This was the Tayacian (from Tayac, Dordogne) which contains few flake tools, but has abundant arrow heads and scrapers. The last Acheulian men, the Neanderthalers and the first *Homo sapiens* appear to have lived in communities totally independent of each other and to have practised their industries in isolation. People using the Acheulian type of implement lived in the Lazaret cave at Nice about 130,000 B.P. where they had built a shelter in the cave entrance (DE LUMLEY, 1969) (Fig. 12.23).

2) **Mousterian:** The stone implements perfected and diversified by the Neanderthalers have been distinguished as Mousterian (from Le Moustier, Dordogne). Large hand-axes are rare, but skilfully worked flakes fashioned into scrapers, arrow-heads and points are abundant. For hunting they used the bolas, a weapon consisting of stone balls attached to a leather thong which was thrown so that it entangled the legs of wild animals. The people built shelters and divided up caves before moving into village groups in Würm I (80,000 to 55,000 B.P.). They practised burial customs and dug graves, such as that (3m deep) in which Chapelle aux

Saints Man was found.

d) **Late Palaeolithic:** Modern man (*Homo sapiens*), who already existed at the time of the Riss-Würm interglacial (100,000 to 80,000 B.P.), was alone in surviving the most recent glaciations, the Würm III and IV. Rapid progress in the development of tool-making continued through four stages: Perigordian, Aurignacian, Solutrean and Magdalenian.

1) **Perigordian:** This was derived from the Mousterian and is marked by the appearance of end-scrapers (grattoirs), side-scrapers (racloirs), gravers (burins), saw blades and a narrow pointed blade with a blunt back edge. This industry began at the beginning of the Würm II to Würm III interglacial phase (35,000 B.P.) and continued through the Würm III glacial phase.

2) **Aurignacian:** This culture developed in parallel with the Perigordian and Mousterian during the Würm III phase and is named after deposits at Aurignac in the upper Garonne valley. The implements are highly developed and include long thin blades, various gravers (burins), some of which were used to work bone or ivory, and notched tools which could be hafted. The first bone points (possibly for spears) made their appearance at the same time as the first sculptures, typified by the rounded figure of "Venus" (Figs. 12.25 and 12.26).

3) **Solutrean:** Towards the end of Würm III, the Perigordian and Aurignacian were replaced by the Solutrean industry (from Solutré, Saône-et-Loire). The workers in this industry made long, thin flint blades resembling laurel leaves, spear heads, thin scrapers and delicate notched points (pointes à cran). The Solutrean industry lasted until the end of the Würm III to Würm IV interglacial which was marked by the warm Lascaux episode. This culture was characterised by great inventiveness and was responsible for the introduction of sewing using bone needles and for the use of the bow and arrow (with bone-tipped arrows) for hunting.

4) **Magdalenian:** The Magdalenian civilisation (from the La Madeleine cave, Dordogne) spanned the period of the Würm IV glaciation (from about 18,000 to 10,000 B.P.) and was the last phase of the late Palaeolithic period. The working of bone and ivory reached a high degree of perfection with the production of

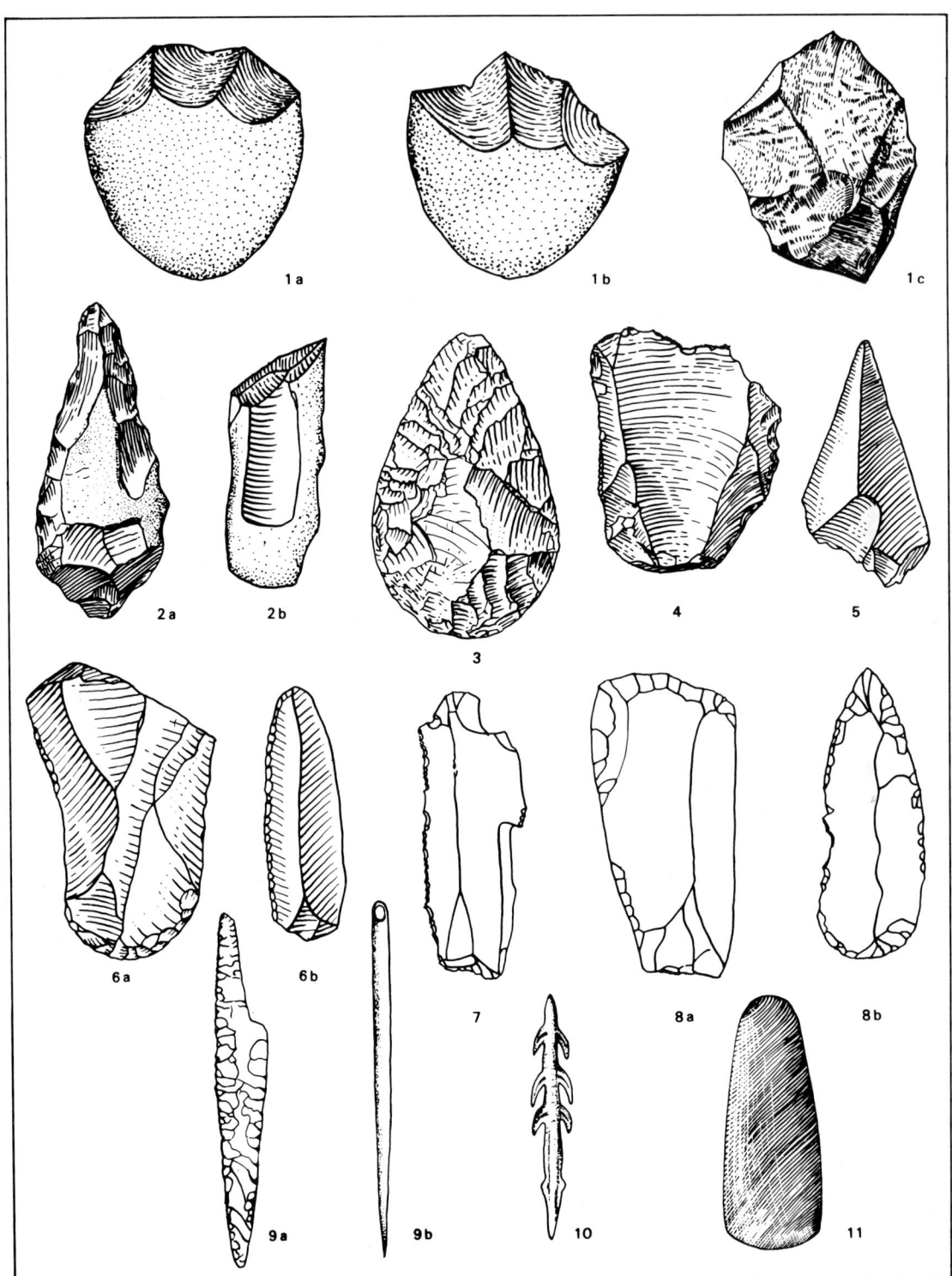

needles, fish-hooks, fish-spears and harpoons with one, or sometimes two, rows of barbs. The Magdalenians were skilled hunters able to kill reindeer (as at Pincevent, near Montereau (Seine-at-Marne) as they forded rivers in their annual migration.

LEROI-GOURHAN (1955) has described a complex Magdalenian habitat at Pincevent close to the bank of the Seine, which had relatively comfortable huts or tents with separate parts for eating, sleeping, working and for refuse disposal (middens). With the Magdalenian, sculpture and cave painting reached their highest expression in the polychrome wall painting of the caves of Altamira, near Santander in Spain and, above all, at Lascaux in the Dordogne (Figs. 12.27 and 12.28).

12.23 **Acheulian shelter propped against the wall of a cave, Lazaret, Nice** (reconstruction by H. de Lumley *in* Chaline, Le Quaternaire, publ. Doin).

12.22 **Human Industries.**
1 – Pebble culture; a: chopper; b: chopping tool; c: polyhedron (multidirectional shaping).
2 – Abbevillian (Chellean); a: hand axe; b: Clactonian flake tool.
3 – Acheulian; more symmetrical hand-axe.
4 – Levalloisian; flake struck from a prepared core, with the resulting facetted surfaces.
5 – Tayacian; so-called Quinson point.
6 – Mousterian; a: scraper; b: knife with retouched back.
7 – Perigordian; burin (graving tool).
8 – Aurignacian; a: scraper; b: pointed scraper.
9 – Solutrean; a: "laurel leaf" tanged point; b: bone needle with eye.
10 – Magdalenian; harpoon.
11 – Neolithic; polished stone axe-head.

e) The post-Glacial or Holocene: The last 10,000 years covers four "civilisations"; the hunters of the **Mesolithic**. the farmers of the **Neolithic**; the metal-workers of the Proto-historical period (copper, bronze and iron ages); and the writers of the historical period.

These cultures, especially the last three, were not everywhere simultaneous, however. For example, Sumerian writing appeared in Meso-potamia about 6000 B.P. while western Europe was still occupied by the farmers of the Neo-lithic. At the present day, some of the tribes of

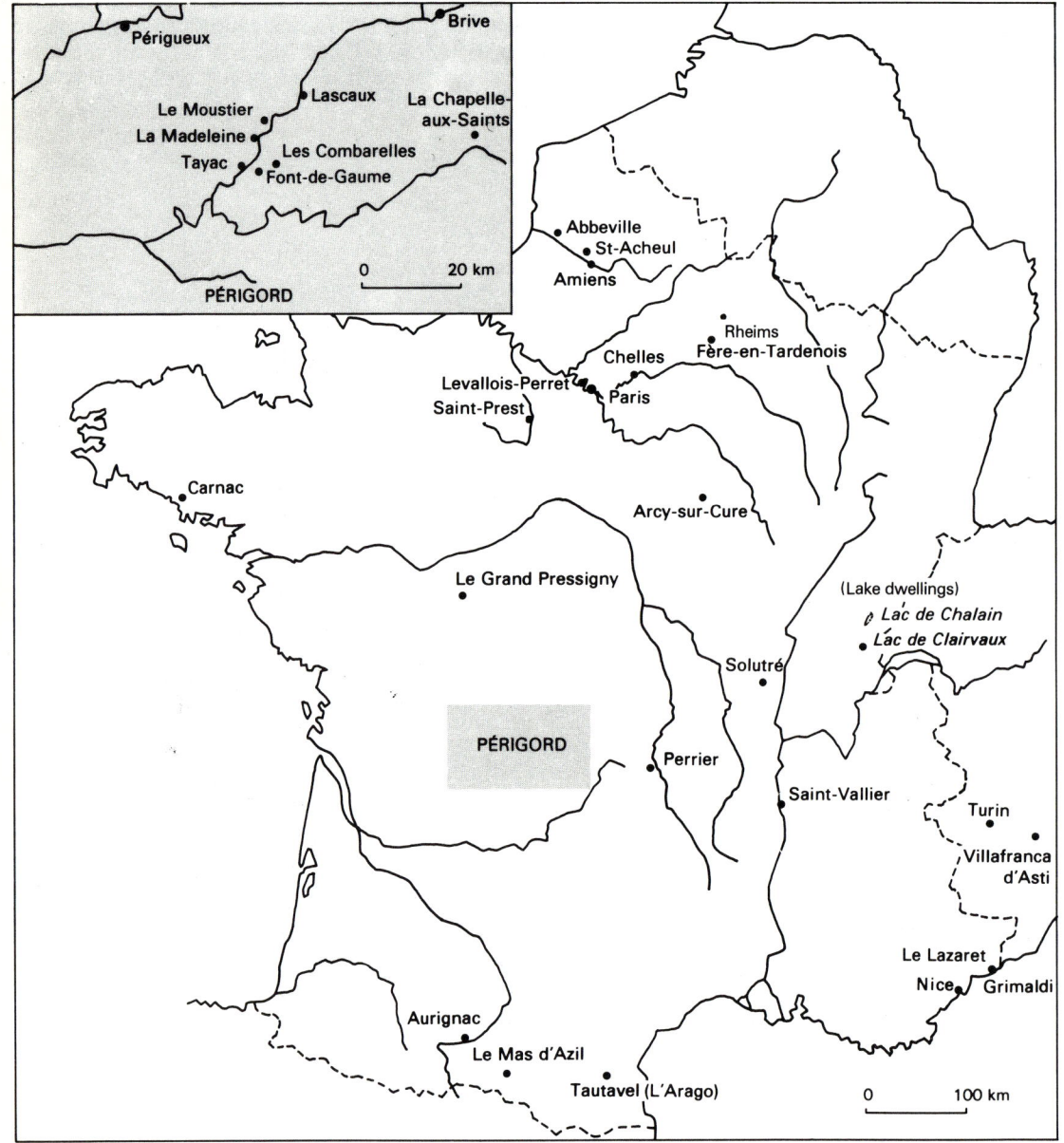

12.24 Prehistoric localities in France and Piedmont (Italy). The Périgord sites (except Lascaux and Chapelle-aux-Saints) are close to the village of Les Eyzies-de-Tayac.

Guyana, Indonesia and Africa might still be regarded as Neolithic.

1) **Mesolithic:** The Mesolithic industries covered about 5000 years from 10,000 to 5000 B.P. Included in this period of time are the Pre-Boreal period about 9000 B.P. in which pine forests spread across Europe; the Boreal period, a dry, warm interval in which the Evergreen Oak (*Quercus ilex*) spread northwards to Normandy and hazel became widespread (about 8000 B.P.); the Pre-Atlantic period which was slightly more humid and finally, the warm and humid Atlantic period during which sea level rose – the Flandrian transgression.

The Mesolithic industries were largely based on stone flakes. In France the early part was the Azilian (from Mas d'Azil, Haute Garonne) with an abundance of small scrapers and knives, made from thin flakes with secondary flaking along one edge. The later part was the Tardenoisian (from Fère-en-Tardenois, Aisne) with microliths, in particular, micro-burins, and tools made from the antlers of deer.

Mesolithic man was both hunter and fisherman. He appears to have successfully hunted the mammoth and may have contributed to its extinction. He also appears to have been a "hunter of fish" since a canoe dated about 8,000 B.P. has been found in a peat bog in Holland. He seems to have led a sedentary, rather than a nomadic, life and his diet included a great quantity of snails, and other molluscs, especially oysters.

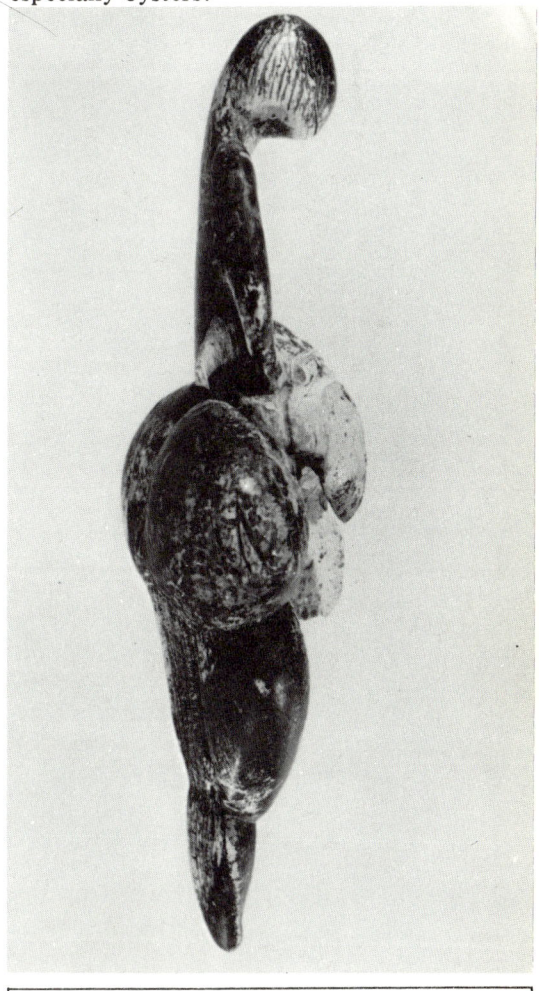

12.26 **Venus of Lespugne** (Haute-Garonne) carved in the tusk of a mammoth (length 15cm) (Musée de l'Homme, Paris).

12.25 **Venus with horn, Laussel** (Dordogne) (Musée de l'Homme, Paris).

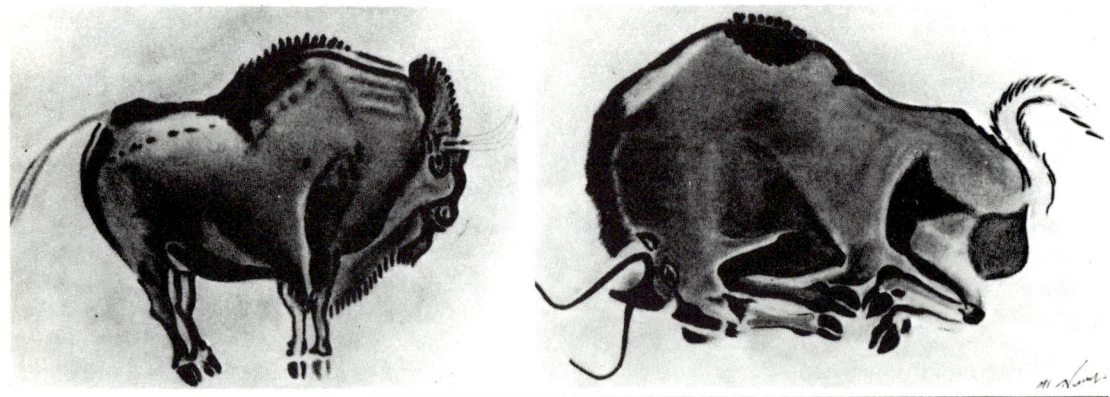

12.27 **Altamira (province of Santander, Spain).** Polychromatic bison; paintings restored by the Abbé Breuil (Photothêque du Musée de l'Homme, Paris).

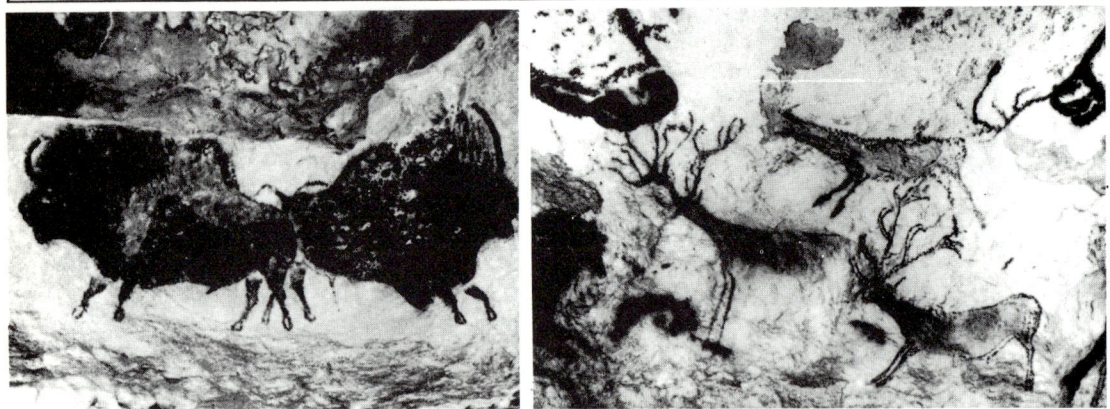

12.28 **Lascaux (Dordogne).** Rock paintings of bison and deer. (Photothêque du Musée de l'Homme, Paris).

12.29 **Dolmen (standing stones) at Carnac (Morbihan)** (photo: Pomerol).

12.30 **Head of a "sorcerer" engraved by shepherds on glacially polished rock wall near Mont-Bego (Maritime Alps)** (photo: Pomerol).

2. Neolithic: Following the transitional Mesolithic, the Neolithic marks the beginning of modern times, and is characterised by the use of polished stone axes. Neolithic man was the first farmer, working the soil with a hoe, cutting down the forests with stone axes, grinding grain with a millstone, making pottery and living in villages. Some of these, the lake villages of Switzerland and the Jura (Fig. 12.24) were constructed on piles at the edge of lakes.

Neolithic men were also the first miners. They dug shafts and galleries for the extraction of fresh flint which was easier to work than that found at the surface. This tradition persisted in France until the middle of the last century in the Cher valley, Saint Aignan, for the manufacture of gun-flints. (A similar industry continued at Brandon in Suffolk until the 1930's – Ed.) In the Sahara, this was the epoch of fishermen living round the lakes formed in the last pluvial period, which corresponded to the "climatic optimum" of 5500 B.P.

3) Protohistorical and Historical periods: The working of copper began about 7000 years ago in Asia Minor, but only 3000 years later did the use of this metal reach western Europe. Bronze working followed that of copper in Asia Minor about 4500 years ago and took about 1000 years to reach Europe, while iron working began about 320 B.P. The iron age in Europe comprised two stages. The first is represented by iron implements from a burial ground at Hallstadt in upper Austria, and the second by the site at La Tène at the eastern end of the Lake of Neuchâtel in Switzerland. This site was occupied from about 2700 to 2000 B.P. by the Gauls. The large amount of wood necessary for iron smelting favoured a site such as this, rather than a site nearer the Mediterranean.

The art of writing which arrived in western Europe at about this time had been invented by the Sumerians more than 3000 years earlier.

It was at the beginning of the age of metals (4000-3000 B.P.) that the great megaliths were built. The *dolmens* were burial chambers and the *menhirs* were perhaps religious structures. Already a farmer, hunter and fisherman, Man domesticated the dog about 3500 B.P.; became a stock-breeder; invented the plough and harnessed the horse. He developed social patterns as evidenced by the towns and monuments of Iran, Egypt and Mesopotamia,

12.31 **Aepyornis.** This ratite (running bird) from the Quaternary of Madagascar (from *aipys*, high, *ornis*, bird) was about 3m high. Its eggs had a volume of about 10 litres, which is six times as large as an ostrich egg and 120 times larger than a hen's egg (Fig. 12.32). Remains of the eggs are especially abundant in the Quaternary bed known as the "Aepyornian". It became extinct only a few centuries ago.

12.32 **The comparative dimensions of the eggs of Aepyornis (left), ostrich (right) and domestic fowl (centre).**

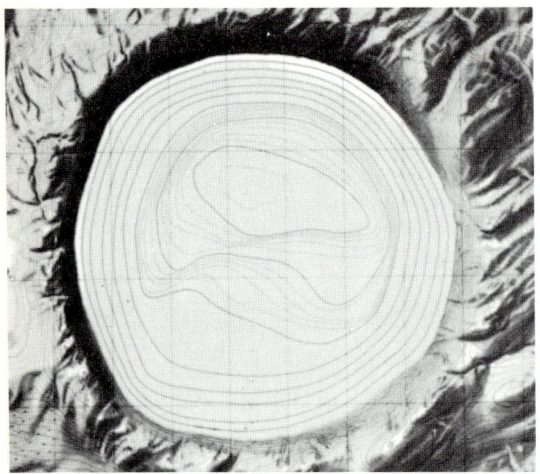

12.33 **A Quaternary meteorite crater** which has largely escaped erosion: the New Quebec crater (Canada). It has a diameter of 3.2km, a depth of 246m and the lip is 163m high (Ministry of Energy, Ottawa).

12.34 **Post-Würm ravine cut in the Chalk of Blaincourt (Oise) at the south eastern extremity of the Pays de Bray anticline.** The cutting of this ravine at the bottom of an asymmetrical periglacial valley, is one of the consequences of the recent tectonism which has affected the Pays de Bray. (See Guide Géologique du bassin de Paris, Publ. Masson) (photo: Pomerol).

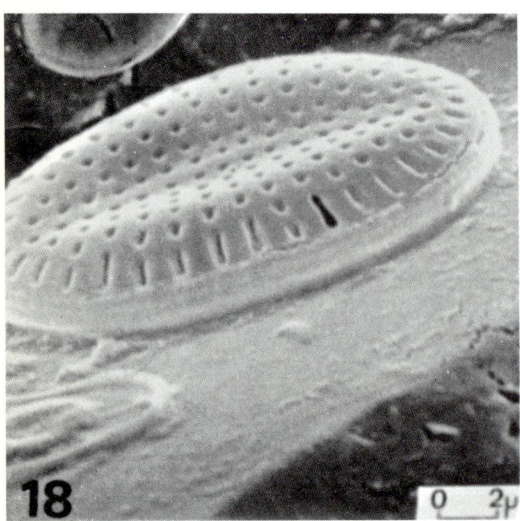

12.36 **Diatoms petrified on grains of quartz.** As has been seen so often throughout these pages, there is a close link between organisms and minerals. These pictures illustrate the mobility of silica between diatom and sand grain in the process of diagenesis. The illustration on the left shows a diatom newly enveloped by a deposit of silica, with (at the bottom left) a diatom almost completely dissolved. The illustration on the right shows the cementation of a broken diatom by the deposition of silica which preserves it (photo: Le Ribault).

but he also wiped out a number of large mammals, such as the wild ox and the bison, and several birds including *Aepyornis* (Figs. 12.31 and 12.32), the dodo and *Dinornis* (Fig. 12.11, p. 232).

Thus, one cannot avoid being struck by the acceleration of technical progress which accompanied the development of Man. It took 3.5 million years to invent fire, but only 5000 years to pass from stone implements to the

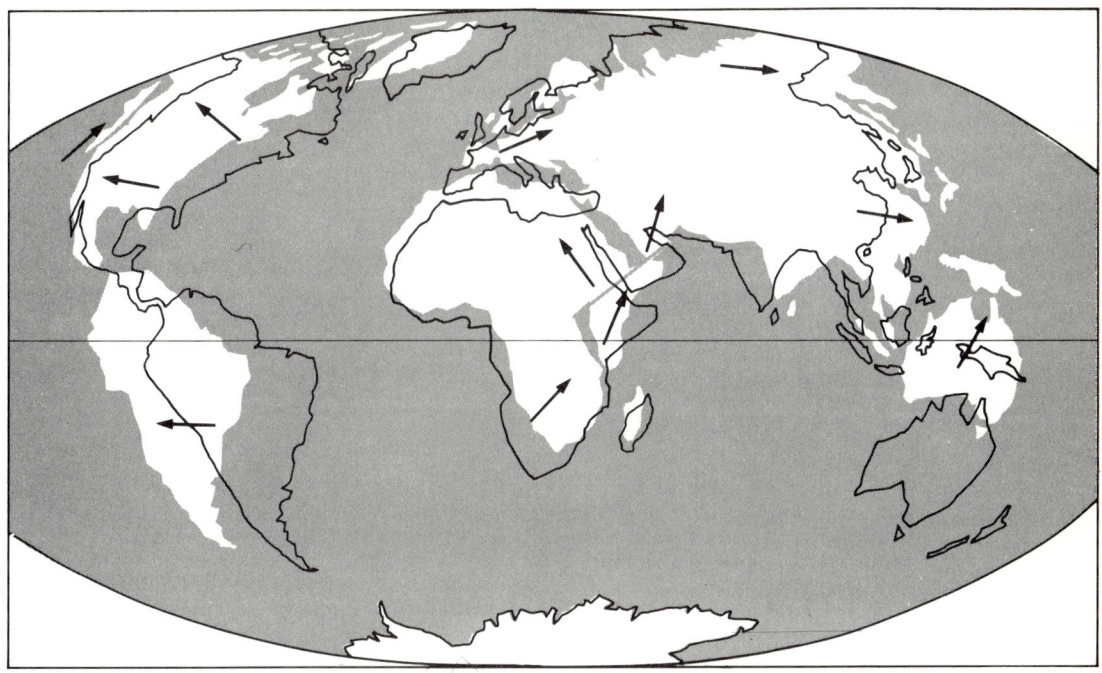

12.35 **The world in 50 million years time.** The position of Antarctica remains the same, but it has turned slightly clockwise. The Atlantic, and in particular the South Atlantic, and the Indian Ocean continue to grow at the expense of the Pacific. Australia has drifted towards the north east and will collide with the Eurasian plate. East Africa has split along the Rift Valley, while the movement of the remainder of Africa towards the north-west has resulted in the reduction in size of the Mediterranean and the closure of the Bay of Biscay. That part of California situated to the west of the San Andreas fault has continued to move towards the north-west as it does today, at about 2cm per year. In 10 million years, Los Angeles will pass the position of San Francisco and in 60 million years will reach Alaska on its way to the Aleutian Islands. . . . (after Dietz and Holden, compare with Fig 1.3, p. 16).

atomic bomb. Three thousand years elapsed between the discovery of iron and the invention of the internal combustion engine, but only 50 years from electricity to the atom and only 50 years from the first aviators to the conquest of the Moon. Sadly, in the field of behaviour, it had to be admitted – and deplored – that having domesticated the animals, Man has sometimes enslaved his fellows and that his thirst for power has led to the death of more than 50 million victims in two great world wars. Thus the physiological evolution of the earlier eras has given way to a psychological evolution infinitely more rapid. There can be no doubt, today, that the evolutionary potential of the earth rests in the mind and hands of Man. What use will he make of it? The evocation of such perspectives is beyond the competence of the geologist, historian of a remote past, who should be cautious in projecting this history into the future . . .

Despite the singular phenomena which are Man and glaciations, it is evident that the planet will continue to evolve in the future as it has done in past epochs. Faults will continue to move (2cm laterally per year for that of San Andreas in California; the same, but vertically this time, for those of Afar in Ethiopia), the continents to deform, the plates to separate. To all the questions expressed or implied in this book, one more may be added. What will the earth be like in 50Ma? (Fig. 12.35).

Man of science and thinker, the stratigrapher espouses the history of his planet, studies its development and in it finds a very beautiful lesson in humility. Seizing during a brief flash of his life the time which escapes from him and the space which dazzles him, he gathers, in homage, a fragment of eternity . . .

BIBLIOGRAPHY

AGASSIZ, L. 1837. Des glaciers, des moraines et des blocs erratiques. *Actes, Soc. Helv. Sci. nat.,* **22,** v-xxxii.

ALIMEN, H. 1936. Étude sur le Stampien du Bassin de Paris. *Mém. Soc. géol. Fr.* (N.S.), **31,** 309 pp. 7pl.

ALVINERIE, J. 1969. *Contribution sédimentologique à la connaissance du Miocène aquitain. Interprétation stratigraphique et paléogéographique.* Thèse Sc., Bordeaux.

——, MAYEUX, C. and RECHINIAC, A. 1964. Esquisse sédimentologique de la coupe de Biarritz. *Colloque sur le Paléogène (Bordeaux, 1962),* **1,** 407-424, *Mém. Bur. Rech. géol. minières.* No. 28.

ANDREIEFF, P. & MARIONNAUD, J. M. 1972. Observations préliminaires sur la limite Éocène – Oligocène dans la série classique du Médoc (Gironde). La position stratigraphique des "grès et calcaire à Anomies". *C. r. Séances Acad. Sci., Paris.* **274-D,** 1637-1640.

ANDRIEUX, J. 1971 *La structure du Rif central.* Thèse, Fac. Sci., Montpellier (1970) and *Mém. Serv. géol. Maroc.* **235.**

ARAMBOURG, C. 1954a. Le gisement de Ternifine. *Arch. Inst. de Paléontologie Humaine* mém., **32,** 37-190.

AUFFRET, J. P. 1973. Découverte du Bartonien en Manche orientale. *C. r. Séances Acad. Sci. Paris,* **276-D,** 1965-1968.

——, BIGNOT, G., BLONDEAU, A. 1975. Géologie du bassin tertiaire de la Manche orientale au large du Pays de Caux. *Phil. Trans. R. Soc.,* **A279,** 169-176.

AUGUSTA & BURIAN. *Encyclopedia Universalis.*

AZZAROLI, A. & CITA, M. B. 1967. *Geologia stratigrafica.* La Goliardica, Milan.

BALDI, T. & SENES, J. 1968. Sur la position du Miocène de la Paratethys centrale dans le cadre du Tertiaire de l'Europe. *Geol. carpat.,* **19,** 95-106.

BANDY, O. L. 1968. Cycles in Neogene Palaeoceanography and Eustatic Changes. *Palaeogeography, Palaeoclimatol., Palaeoecol.,* **5,** 63-75.

BARBOT DE MARNEY, N. 1869. *Esquisse géologique du gouvernement de Kerson.* St. Petersburg (in Russian).

BEA, F. & KIEKEN, M. 1971. Les evaporites dans les formations tertiaires de l'Aquitaine méridionale. *Bull. Cent. Rech. Pau,* S.N.P.A., **5,** (2).

BERTIN, L. 1939 *Géologie et Páleontologie.* Larousse, Paris.

BESSON, L., DERES, F. & PAIRIS, J. L. 1972. Age des "Grès d'Annot" au Nord de la localité-type (Alpes de Haute-Provence). *C. r. Acad. Sc., Paris,* **275** (D), 2603-2606.

BEYRICH, E. 1854. Ueber die Stellung des hessisches Tertiärbildungen. *Mber. K. preuss, Akad. Wiss.* (1854), 640-666.

BIELY, A., RAKUS, M., ROBINSON, P. & SALAJ, J. 1972. Essai de correlation des formations miocènes au sud de la Dorsale tunisienne. *Notes Serv. géol. Tunis,* **38,** 73-92.

BIGNOT, G. 1970. Le Liburnien. Essai de mise au point. *Paléobiologie continentale,* **I,** No. 3, 19-37, Montpellier.

BLANC, A. C. 1942. Variazioni climatiche ed oscillazioni della linea di riva nel Mediterraneo centrale durante l'era glaciale. *Geol. Meers v. Binnengewässer,* Berlin, **5,** 2, 190.

BLONDEAU, A. 1969. Remarques sur *Nummulites germanicus* BORNEMANN. *Nachr. Ak. Wissensch. Göttingen mat.-phys. Kl.,* **14,** 129-135. 1 pl.

BOILLOT, G. 1963. Sur la fosse de la Hague (Manche centrale). *C. r. Séances Acad. Sci., Paris,* **257-D,** 3936-3966.

——, HORN, R. & LEFORT, J. P. 1972. Évolution structurale de la Manche occidentale au Secondaire et au Tertiaire. *Mem. Bur. Rech. géol. minières,* No. 79, 79-86.

BOLLI, H. M. 1966. Zonation of Cretaceous to Pliocene marine sediments based on Planktonic Foraminifera. *Boln, Inf. Asoc. venezol. Geol., Min., Petrol,* **9,** No. 1, 3-32.

BOMBITA, G. 1964. Observations et propositions concernant la nouvelle division de l'Éocène. *Colloque sur le Paléogène (Bordeaux, 1962),* **2,** 941-948 *Mém. Bur. Rech. géol. minières,* No. 28.

BOULANGER, D. 1968. *Révision de la Nummulitique de la Chalosse, du Béarn et du Bas-Adour (Landes et Basses-Pyrénées).* Thèse Sciences, Paris.

BOURCART, J. 1960-2. La Méditerranée et la Révolution du Pliocène. *Livre à la Mémoire du Professeur Paul Fallot,* **1,** 169-174, *Mém. hors-Série: Soc. géol, Fr.*

BOURDIER, F. 1959. L'époque quaternaire dans la region parisienne. *Bull. soc. archéol. et hist. Chelles,* **1-2,** 5-18.

BOUSSAC, J. 1911. Études stratigraphiques et paléontologiques sur le Nummulitique de Biarritz. *Annls. Hébert,* **5,** 96p.

BREUIL, H. 1939. Le gisement de Chelles; ses phénomènes, ses industries. *Quartär,* Bd. **2,** Berlin.

BRONGNIART, A. 1810 *in* CUVIER, G. & BROGNIART, A. Essai sur la Géographie Minéralogique des Environs de Paris, avec une Carte géognostique et des Coupes de Terrain. *Mem. Cl. Sci. Math. Phys. Inst. imp. Fr.* (1810) 278 p.

BROUSSE, R., DELIBRIAS, G., LABEYRIE, J. & RUDEL, A. 1969. Éléments de chronologie des éruptions de la chaine des Puys. *Bull. Soc. géol. Fr.* (7), **11,** 770-793.

BUCKLAND, W. 1823. *Reliquiae diluvianae: or Observations on the organic remains contained in caves, fissures and diluvian gravel, and on other geological phenomena, attesting the action of a universal deluge.* London. J. Murray, 303 pages.

BUROLLET, P. F. 1956. Contribution à l'étude stratigraphique de la Tunisie centrale. *Ann. Mines Géol., Tunisie,* No. 18.

BUSSON, G. 1970. *Le Mésozoique saharien. Essai de synthèse des données des sondages algéro-tunisiens.* C.N.R.S., Paris.

BUSSON, G., FLANDRIN, J. & LAFFITTE, R. 1955. Observations stratigraphiques récentes concernant la paléogéographie de l'Algérie pendant les temps éocènes. *C.R.Ac.Sci., Paris,* **240,** 1445-1447.

CAIRE, A. 1957. Étude géologique de la région des Biban (Algérie). *Publ. Serv. Carte. géol. Algérie n.s. Bull.* No. 16.

CASIER, E. M. 1946. La faune ichthyologique de l'Yprésien de la Belgique. *Mèm. Mus. R. Hist. nat. Belg.* **104,** 3-267, 6 pls.

CASTANY, G. 1951. Étude géologique de l'âtlas tunisien oriental. *Ann. Mines Géol. Tunisie,* No. 8, 633 pp., 243 fig.

—— 1954. Les plages quaternaires à *Cardium* des grands chotts du Sud tunisien. *Quaternaria,* Rome, **I,** 117-130.

CAVELIER, C. 1964. L'Oligocène inférieur du Bassin de Paris. *Colloque sur le Paléogène (Bordeaux 1962),* **1,** 65-73. *Mém. Bur. Rech. géol. minières,* No. 28.

—— 1972. L'age priabonien supérieur de la zone à *Ericsonia subdisticha* en Italie et l'attribution des Latdorf Schichten allemands à l'Éocène supérieur. *Bull. Bur. Rech. géol. minières* (2), sect. IV, 15-24.

—— & LE CALVEZ, Y. 1965. Présence d'*Arenagula kerfornei* (Allix) espèce "biarritzienne", à la partie terminale du Lutétien supérieur de Foulangues (Oise). *Bull. Soc. géol. Fr.,* (7), **7,** 284-286.

CHAVAILLON, J. 1971. Présence éventuelle d'un abri oldowayen dans le gisement de Melka Kontouré (Ethiopie). *C. r. Séances Acad. Sci. Paris,* **273-D,** 623-625.

CHOUBERT, G. 1957-65. Essai de corrélation des formations continentales et marines du Pleistocène au Maroc. (comm. V., Cong. INQUA, Madrid – Barcelona, 1957). *Notes Serv. Géol. Maroc.* **185,** 25, 9-28.

—— & FAURE-MURET, A. 1960-2. Évolution du domaine Atlasique Marocain depuis les temps paléozoiques. *Livre à la mémoire du Professeur Paul Fallot,* **1,** 446-514. *Mémoire hors-série: Soc. géol. Fr.*

CICHA, I. & TEJKAL, J. 1959. Zum Problem des. sog. Oberhelvets in dem Karpatischen Becken. *Vesn. ustred. Ustavu geol. CAV* 34, No. 2.

CITA, M. B. 1972. Pliocene biostratigraphy and chronostratigraphy. *Initial Reports of Deepsea Drilling Project,* XIII, pt. 2, 1343-1379.

——, GELATI, R., PREMOLI-SILVA, I. & ROBBA, A. 1971. Proposal for the definition of superstages for the middle and upper Miocene. *Ve Congrès du Néogène Méditerranéen, Lyon* 1971, **2,** 589-594. *Mém. Bur. Rech. géol. minières,* No. 78.

COGELS, P. & VANDENBROECK, E. 1879. Observations géologiques faites à Anvers à l'occasion des travaux de creusement des nouvelles cales sèche et de prolongement du bassin du Kattendijk. *Ann. Soc. roy. Malac. de Belgique,* **XIV,** 29-79.

COPPENS, Y. 1971. Les faunes de vertébrés du Pliocène et du Pleistocène ancien d'Afrique. *V^{e}Congrès du Néogène Méditerranéen, Lyon* 1971 **1,** 121-129. *Mém. Bur. Rech. géol. minières* No. 78.

CRUSAFONT, M. 1948. El sistema miocénico en la depression espanola de Vallés-Penedés. *Proc. Int. Geol. Congr., 18th Sess.,* **11,** 33-42. London.

CURRY, D. 1962. Age determinations of some rocks from the floor of the English Channel. *Proc. Ussher Soc.,* **1,** 5-6.

—— 1966. Problems of correlation in the Anglo-Paris-Belgian Basin. *Proc. Geol. Assoc., London,* **77,** 437-467.

——, HAMILTON, D. & SMITH, A. J. 1970. Geological and shallow subsurface geophysical investigations in the Western Approaches to the English Channel. *Rep. No. 70/3. Inst. geol. Sci.,* 12pp.

—— —— —— 1971. Geological evolution of the Western Channel and its relation to the nearby continental margin. In DELANY, F. M. (ed.). *Working Party 31 Symposium, Cambridge,* 1970: The geology of the East Atlantic continental margin. 2. Europe *Rep. No. 70/14. Inst. geol. Sci.,* 129-142.

CUVIER, G. 1825. *Recherches sur les ossemens fossiles où l'on rétablit les caractères de plusiers animaux dont les révolutions du globe ont detruit les espèces. 3° edn.* 10 vols, atlas 4°, 2 vols., 260pp. *Dufour & d'Ocagne, Paris.*

D'ALBISSIN, G. 1956. Essai de paléogéographie de l'Ile de France au Sannoisien. *Bull. Soc. géol. Fr.* (6), **6,** 57-69.

DAMOTTE, R. & FEUGUEUR, L. 1963. L'âge du calcaire de Vigny (S-et O) à partir de données paléontologiques nouvelles. *C. r. Séances Acad. Sci. Paris,* **256-D,** 3864-6.

DAVID, L., BALLESIO, R. & MONGEREAU, N. 1967. Première étude sur les Bryozoaires du Pliocène Rhodanien. *Giornale de Geologia,* (2), **35,** 107-115.

DEBELMAS, J. 1970. *Alpes (Savoie et Dauphiné).* Guides Géologiques Régionaux, 213pp, Masson, Paris.

DE GEER, G. 1934. Geology and Chronology. *Geogr. Annaler,* **16,** 1-52.

DE HEINZELIN, J. & GLIBERT, M. 1957. *Lexique Stratigraphique Internationale,* 14a VII, p. 63.

DE LAPPARENT, A. 1883. *Traité de Géologie.* 1st Edition. F. Savy, Paris.

DE LUMLEY, H. 1969. Une cabane acheuléenne dans la grotte du Lazaret (Nice). *Mém. Soc. pre-hist. franc.,* **7,** 234pp.

DEMARCQ, G. 1970. Étude stratigraphique du Miocène rhodanien. *Mém. Bur. Rech. géol. minières* No. 61, 257p., 56 fig., 4 pl., 1 tabl.

DENIZOT, G. 1927. *Les formations continentales de la région orléanaise.* Vendôme, Launay et Fils, 582pp. (Thèse).

DEPÉRET, C. 1892. Note sur la classification et le parallélisme du système miocene. *C. r. Somm. Soc. géol. Fr.* cxlv-clvi.

—— 1895. Observations à propos de la note sur la nomenclature des terrains sédimentaires, par MM. Munier-Chalmas et de Lapparent. *C. r. Somm. Soc. géol. Fr.* (3), **23,** 33-36.

DE ROUVILLE, G. 1853. *Description géologique des environs de Montpelier.* Boehm, Montpellier. 221p.

DE SERRES, M. 1830. De la simultanéité des terrains du sédiment supérieures. *From* La Géographie Physique, vol. 5 of *Encyclopédie Méthodique,* 125 pp.

DESNOYERS, J. 1829. Observations sur un ensemble de dépôts marins plus récens que les terrains tertiaires du bassin de la Seine, et constituant une formation géologique distincte: précédées d'un apercu de la non-simultanéité du bassins tertiaires. *Annales des sciences naturelles* (Paris), **16**, 171-214, 402-491.

DESOR, E. 1846. Sur le terrain danien, nouvel étage de la craie. *Bull. Soc. géol. Fr.* (2), **4**, 179-182.

DEWALQUE, G. J. G. 1868. *Prodrome d'une description géologique de la Belgique,* Librairie polytechnique de Decq, Brussels, 442 p.

DIETZ, R. S. & HOLDEN, J. C. 1970. The Breakup of Pangea, *Scient. Amer.* 30-41.

DOLLÉ, J. 1968. Stratigraphie et micropaleontologie de la partie sud de la Mer du Nord. *Bull. Inf. Géol. Bassin de Paris,* **15**, 13-6.

DOLLFUS, G. 1880. Essai sur l'étendue des terrains tertiaires dans le bassin anglo-parisien, *Bull. Soc. géol. Normandie,* **6**, 584-605.

—— 1900. Le Miocène dans la région de l'Ouest. *Bull. serv. Carte géol. Fr.,* **XI** (73): 100-101.

—— 1907. Classification des couches de l'Éocène supérieur du Nord de Paris. *Bull. Soc. géol. Fr.* (4), **7**, 347-354.

D'ORBIGNY, A. 1852. *Cours élémentaire de paléontologiae et géologie stratigraphiques.* Vol. **2**, Masson, Paris, 848 pp.

DROOGER, C. W. & MARKS, P. 1971. Proposal of four superstages in the Neogene, *Ve Congrès du Néogène Méditerranéen, Lyon,* 1971, *Mém. Bur. Rech. géol. minières,* No. 78.

DUJON, S. C. 1971. Interprétation d'un relief du fond de la mer du Nord comme dû a un dépôt morainique frontal ancien (Warthe). Consequence paléogéographique. *C. r. Somm. Soc. géol. Fr.* 163-165.

DUMONT, A. H. 1839. Rapport sur les travaux de la carte géologique en 1839, avec une carte géologique des environs de Bruxelles. *Bull. Acad. r. Belg. Cl. Sci.,* **6**, 464-485.

—— 1849. Rapport sur la carte géologique du Royaume. *Bull. Acad. r. Belg. Cl. Sci.,* **16**, 351-373.

—— 1851. Note sur la position géologique de l'argile rupélienne et sur la synchronisme des formations tertiaires de la Belgique, de l'Angleterre et du nord de la France. *Bull. Acad. r. Belg. Cl. Sci.,* **18**, 179-195.

DURAND, S. 1960. Le Tertiaire de Bretagne. Étude stratigraphique, sédimentologique et tectonique. *Thèse. Mém. Soc. géol. et min. Bretagne,* **XII.**

DURAND-DELGA, M. 1967. Structure and geology of the Northwest Atlas mountains *in* Guide book to the Geology and history of Tunisia. L. Martin (ed.), *Petr. Expl. Soc. Libya,* 59-81.

—— 1969. Mise au point sur la structure du nord-est de la Berbérie. *Bull. Serv. géol. de l'Algérie,* No. 39, 89-131.

ELLENBERGER, F. 1958. Étude géologique du Pays de la Vanoise. *Mém. Carte géol. Fr.*

ESTÉOULE-CHOUX, J. 1967. Contribution a l'étude des argiles du Massif armoricain. Argiles des altérations et des bassins sédimentaires tertiaires. *Mém. Soc. géol. Bretagne,* **14** (1971), 391 pp.

FABIANI, R. 1912. Nuove osservazioni sul Terziario fra il Brenta e l'Asticco. *Atti. Accad. Sci. veneto-trent.-istriana,* anno 5, **1**, 1-36.

FABRE, A. 1939. Description géologique des terrains tertiaires du Médoc et essai sur la structure tectonique du département de la Gironde. *Thèse, Paris.*

FEUGUEUR, L. 1963. L'Yprésian du Bassin de Paris. Essai de monographie stratigraphique. *Thèse,* 1958, *Mém. Expl. Carte géol. dét. Fr.*

FLINT, R. F. 1971. *Glacial & Quaternary Geology,* New York, N.Y., 892 pp.

FONTES, J.-C. 1968. Le gypse du bassin de Paris. Historique et données récentes. *Colloque sur l'Éocène, Paris* (1968), *Mém. Bur. Rech. géol. minières,* No. 58, 359-386.

FORBES, E. 1846. On the connexion between the distribution of the existing fauna and flora of the British Isles, and the geological changes which have affected their area, especially during the epoch of the Northern Drift. *Mem. Geol. Surv. G.B.,* **1**, 336-432.

FOUCAULT, A. 1972. Sur les rapports entre les chaînes alpines et péripacifiques. *C. r. Séances Acad. Sci. Paris,* **275-D**, 1195-1198.

FUCHS, T. 1894. Tertiärfossilien aus den kohleführenden Miocänablagerungen der Umgebung von Krapina und Radoboj und über die Stellung der sogenannten ''Aquitanischen Stufe''. *Mitt. Jahrb. Kgl. Ungar. Geol. Anst.,* **10**, 161-175.

GABUNIA, L. & RUBENSTEIN, M. 1968. On the correlation of the Cenozoic deposits of Eurasia and North America based on the fossil mammals and absolute age data. *XXI Intern. Geol. Congr.,* **10**, 9-17.

GERMANEAU, J. 1971. Remarques sur la présence d'augites du Massif central entre Seine et Loire. *C. r. Somm. Soc. géol. Fr.* 269-270.

GERVAIS, P. 1867-9. *Zoologie et paléontologie générales. Nouvelles recherches sur les animaux vertébrés vivants et fossiles.* Paris, 263 pp. 41 fig.

GIGNOUX, M. 1913. Les formations marines pliocènes et quaternaires de l'Italie du sud et la Sicilie. *Annls. Univ. Lyon* (n.s.) **1**, fasc. 36, 1-692.

GINSBURG, L. 1971. Les faunes de Mammifères burdigaliens et vindoboniens des Bassins de la Loire et de la Garonne, *Ve Congrès du Néogène Méditerranéen, Lyon* 1971, **1**, 153-167, *Mém. Bur. Rech. géol. minières,* No. 78.

GLACON, J. 1970. La série Mio-Pliocène du Hodna et des régions voisines (Algérie du Nord). *C. r. Séances Acad. Sci. Paris,* **271-D**, 945-948.

—— & ROUVIER, H. 1967. Précisions lithologiques et stratigraphiques sur le ''Numidien'' de Kroumirie (Tunisie septentrionale). *Bull. Soc. géol. Fr.* (7), **9**, 410-417.

GRAMBAST, L. 1962. Flore de l'Oligocène supérieur du Bassin de Paris. *Ann. Paléónt. Fr.,* **68**, 80.

GRAVES, L. 1847. *Essai sur la topographie géognostique du département de l'Oise.* Beauvais.

HARDENBOL, J. 1968. The ''Priabonian'' type section. Colloque sur l'Éocène, Paris, 1968. *Mém. Bur. Rech. géol. minieres,* No. 58, 629-635.

HARTENBERGER, J. L. 1970. Les Mammifères d'Egerkingen et l'histoire des faunes de l'Éocène d'Europe. *Bull. Soc. géol. Fr.* (7), **12**, 886-893.

HAUG, E. 1907. Observations diverses sur la classification de l'Éocène supérieur et moyen. *Bull. Soc. géol. Fr.* (4), **7**, 354.

—— 1907-1911. *Traité de géologie,* 4 vols, 2024 p. 135 pl., Paris.

HEYBROEK, P., HAANSTRA, U. & ERDMAN, D. A. 1967. *Observations on the geology of the North Sea.* Report P.D. No. 9, Bataafse Internationale Petroleum Mij., N.V. The Netherlands.

HOERNES, M. 1836. Die fossilen Mollusken der Tertiär-Beckens von Wien. *Abh. K. K. Geol. Reichsanstalt,* **3**, 1-733.

HOFFSTETTER, R. 1976. Histoire des mammifères et dérive des Continents. La Recherche, No. 64, 124-138.

HOTTINGER, L. & SCHAUB, H. 1960. Zur Stufeneinteilung des Palaeocaens und des Eocaens. Einführung der Stufen Ilerdien und Biarritzien. *Eclog. geol. Helv.,* **53**, 453-480.

HSÜ, K. 1973. The desiccated deep basin model for the Messinian events. *In Messinian events in the Mediterranean.* K. ned. Akad. Weten., 60-7.

HYDE, H. A. & WILLIAMS, D. A. 1944. The Right Word. *Pollen Analysis Circ.,* **8**, 6.

JAUZEIN, A. 1967. Contribution a l'étude géologique des confins de la dorsale tunisienne. *Annls Mines Géol.,* No. 22, Tunis.

—— & ROUVIER, H. 1965. Sur les formations allochtones de Kroumirie (Tunisie septentrionale). *C. r. Somm. Soc. géol. Fr.,* 36-38.

KAY, M. & COLBERT, E. H. 1965. *Stratigraphy and Life History,* 736 p. John Wiley, London.

KIEKEN, M. 1970. Résumé des connaissances acquises au cours des vingt dernières années dans le Hodna, le Titteri et la partie occidentale des Biban (département d'Alger). *Bull. Bur. Rech. géol. minières* (2) sect IV, No. 1, 45-76.

—— 1973. Évolution de l'Aquitaine au cours du Tertiaire. *Bull. Soc. géol. Fr.* (7) **15**, 40-50.

KLINGEBIEL, A. & PUECHMAILLE, C. 1966. Définition et interprétation des formations infra-éocènes du bassin nord-aquitain. *C. r. Séances Acad. Sci. Paris,* **262-D**, 2135-2137.

KROMM, F. 1968. Répartition des facies et position stratigraphique des formations ilerdiennes en Catalogne oriental. *Colloque sur l'Éocène, Paris* 1968, **3**, 209-217, *Mém. Bur. Rech. géol, minières* No. 69.

KUMMEL, B. 1961. *History of the Earth: an introduction to historical geology,* xiv, 610 pp. figs. San Francisco.

LAMANON DE PAUL, R. 1782. Descriptions des divers fossiles trouvés dans les carrières de Montmartre. *Journal de Physique,* **19**, pt. 1, 173-194.

LANKESTER, R. 1914. Description of the Test Specimen of the Rostro-Carinate industry found beneath the Norwich Crag. *Occ. Pap. R. Anthrop. Inst.,* No. 4.

LAFITTE, R. 1942. Plissements post-Pliocène et mouvements quaternaires dans l'Algérie occidentale. *C. r. Séances Acad. Sci. Paris,* **215-D**, 372-374.

LARSONNEUR, C. 1972. Données sur l'évolution paléogéographique post-hercynienne de la Manche.

Colloque sur la géologie de la Manche, 203-214, *Mém. Bur. Rech. géol. minières,* No. 79.

LEAKEY, L. S. B. 1934. *Adam's Ancestors,* London.

—— 1960. The discovery of *Zinjanthropus boisei. Current Anthropology,* **1**, 76-77.

LE PICHON, X., BONNIN, J., FRANCHETEAU, J. & SIBUET, J. C. 1971. Une hypothèse d'évolution tectonique du golfe de Gascogne. *Actes. Symp. Histoire structurale du golfe de Gascogne,* Technip, Paris, VI, 11-1 to 44.

LE PLAY, F. 1842. Explorations des terrains carbonifères du Donetz exécutée de 1837 à 1839 in DEMIDOFF, A. de *Voyage dans la Russe méridionale et le Crimée, par la Hongrie, la Valachie et la Moldavie,* t. **4**, x +516 p.

LEROI-GOURHAN, A. 1955. *Les hommes de la préhistoire: les chasseurs,* Bourrelier, Paris, 128 p.

LEYMERIE, A. 1862. Apercu géognostique des petites Pyrénées et particulièrement de la montagne d'Ausseing. *Bull. Soc. géol. Fr.* (2), **19**, 1091.

LORENZ, C. 1962. Le Stampien et l'Aquitanien ligures. *Bull. Soc. géol. Fr.* (7), **4**, 657-665.

—— 1964. La série aquitanienne de Millesimo (Italie). *Bull. Soc. géol. Fr.* (7), **6**, 192-204.

LOTZE, F. 1964. The distribution of evaporites in space and time. *In* A. E. M. Nairn, Ed., *Problems in Palaeoclimatology,* Interscience, London, pp. 491-509.

LYELL, C. 1833. *Principles of Geology, etc.,* vol. **III**, John Murray, London, xviii + 393 + 109 pp.

—— 1839. On the relative ages of the Tertiary deposits commonly called "Crag" in the counties of Norfolk and Suffolk, *Mag. nat. hist.* **3**, (N.S.) 313-330.

—— 1859. *Manual of elementary geology,* 5th ed., supplement, 36 pp. pp. 11-13.

MAGNE, J. & RAYMOND, D. 1972. Dans le Nord de la Grand-Kabylie (Algérie) le Numidien a un âge compris entre l'Oligocène moyen et le Burdigalien inférieur. *C. r. Séances Acad. Sci, Paris,* **274-D**, 3052-3055.

MANGIN, J. P. 1959-1960. Le Nummulitique Sud-Pyrénéen à l'Ouest de l'Aragon. *Inst. Est. Pirenaicos,* 631 pp., 113 figs.

—— 1965. La nomenclature stratigraphique et les étages du Paléogène. *C. r. Somm. Soc. géol. Fr.,* 169-172.

MARIE, P. 1964. Les facies du Montien (France, Belgique, Hollande), *Colloque sur le Paléogène (Bordeaux, 1962). Mém. Bur. Rech. géol. minières,* No. 28, 1077-1102.

MARTINI, E. 1970. The Upper Eocene Brockenhurst Bed, *Geol. Mag.* **107**, 225-228.

——. 1971. Standard Tertiary and Quaternary calcareous nannoplankton zonation. *In* FARINACCI, A. (ed.), *Proceedings of the II Planktonic Conference,* Roma 1970. Edizioni Tecnoscienza, Rome, 739-785.

MATHERON, P. 1878. Recherches paléontologiques dans le Midi de la France. *Terrains tertiaires,* 12 pp., Marseilles.

MATTAUER, M. 1963. Le style tectonique des chaines telliennes et rifaines. *Geol. Rdschau* Bd. 53, H.1, 296-313.

—— & HENRY, J. 1974. The Pyrenees *in* A. M. Spencer, Ed., *Mesozoic-Cenozoic orogenic belts: Data for orogenic studies. Geol. Soc. Lond. spec. Pub.* No. 4, pp. 3-21.

—— & SEGURET, M. 1971. Les relations entre la chaine des Pyrenees et le golfe de Gascogne, *in Histoire structurale du Golfe de Gascogne.* Technip, Paris, p. IV, 4-1 to IV 4-24.

MAYER-EYMAR, K. 1857. *Essai d'un tableau synchronistique des terrains tertiaires d'Europe.* Zurich.

—— 1858. Versuch einer neuen Klassification der Tertiär-Gebilde Europa's. *Verh. schweiz. naturf. Ges.,* **42**, 165-199.

—— 1867. *Catalogue systematique et descriptif des fossiles des terrains tertiaires qui se trouvent au Musée fédéral de Zurich.* Zurich, Vol. 2, p. 13.

—— 1868. *Tableau synchronistique des terrains tertiaires supérieurs.* Zurich.

—— 1893. Le Ligurien et le Tongrien en Égypte. *Bull. Soc. géol. Fr.* (3) **21**, 7-43.

MILNE-EDWARDS, A. 1867-71. *Recherches Anatomiques et Paléontologiques pour servir à l'Histoire des Oiseaux Fossiles de la France,* **1**, 475 pp., 96 pl., **2**, 632 pp., 104 pl. Masson & Cie, Paris.

MONTENAT, C. 1977. Les bassins néogènes du Levant d'Alicante et du Murcia (Cordillères bétiques orientales – Espagne): stratigraphie, paléogéographie et évolution dynamique. *Docum. Lab. géol. Fac. Sci. Lyon,* No. 69, 340 p., 7 pl.

MOURLON, M. 1887. Sur les dépôts rapportés par Dumont à ses systèmes laekenien et tongrien. *Bull. Acad. r. Belg. Cl. Sci.* (3) **14**, 598-616.

MUNIER-CHALMAS, E. 1891. Equivalent marin du Calcaire de Brie lacustre. *C. r. Somm. Soc. géol. Fr.,* (3) **19**, 110.

—— & DE LAPPARENT, A. 1893. Note sur la nomenclature des terrains sédimentaires. *Bull. Soc. géol. Fr.* (3) **21**, 438-488.

NAUMANN, C. 1866. *Lehrbuch der Geognosie,* **III**, p. 8.

NEMKOV, G. I. 1967. *Nummulitidy Sovyetskogo Soyuza i ikh biostratigraficheskoye znacheniye.* 318 pp., 44 pl., Nauka, Moscow.

NOLF, D. 1973. Stratigraphie des formations du Panisel et Den Hoorn (Eocène belge). *Bull. Soc. belge géol. paléont. hydrol.,* **81**, 75-94.

PAPP, A., CICHA, I., ROGL, F., SENES, J., STEININGER, F. & BALDI, T. 1968. Zur Nomenklatur des Neogens in Österreich. *Verh. geol. Bundesanst.,* (1968), 9-27.

PARETO, L. 1865. Note sur les subdivisions que l'on pourrait établir dans les terrains tertiaires de l'Apennin septentrionale. *Bull. Soc. géol. Fr.* (2) **22**, 209-275.

PENCK, A. 1905. Glacial features in the surface of the Alps. *Journ. Geol.* **13**, 1-19.

PERCONIG, E. 1966. Sull'esistenza del Miocene superiore in facies marina nella Spagna meridionale. *C. r. 3° Congres C.M.N.S., Berne, 1964, Comité sur la Stratigraphie du Néogène méditerranéen.*

—— 1971a. État actual de nos connaissances sur l'étage Andalousien. *Ve Congrès de Néogène Méditerranéen, Lyon, 1971,* **2**, 659-662. *Mém. Bur. Rech. géol. minières,* No. 78.

—— 1971b. Mise au point du stratotype de l'Andalousian. *ibid.* **2**, 663-673.

PETERLONGO, J. M. 1972. *Massif Central.* Guides géologiques régionaux. Masson, Paris.

PINOT, J. P. 1971. La Manche *in* Encyclopedia Universalis, **X**, 409-412.

PLAZIAT, J. C. 1966. *Contribution a l'étude stratigraphique, paléontologique et sédimentologique*

du "Sparnacien" des Corbières septentrionales (Aude). Thèse 3e cycle, Paris, 358 pp., 32 pl.

—— 1973. Principales étapes de la paléogéographie dans la domaine peripyrénéenne, au cours de l'Eocène inférieur et moyen. *C. R. Réunion des Sciences de la Terre,* 1 p., 3 fig.

POLVÊCHE, J. 1956. La terminaison méridionale des nappes sud-telliennes dans la region de Tiaret (Algerie). *Bull. Soc. géol. Fr.* (6) **6**, 643-652.

POMEROL, C. 1964. Découverte de paléosols de type podzol au sommet de l'Auversien (Bartonien inférieur) de Moisselles (Seine-et-Oise). *C. r. Acad. Sci. Paris,* **258**, 974-976.

—— 1965. Les sables de l'Éocène supérieur (étages Ledian et Bartonien) des bassins de Paris et de Bruxelles. *Mém. Expl. Carte géol. dét. Fr.*

—— 1967. Esquisse paléogéographique du Bassin de Paris à l'ère Tertiaire et aux temps Quaternaires. *Rev. Géogr. phys. Géol. dynam.* (2), **9**, f.1 55-86.

POŽARYSKA, K. 1971. La limite Crétacé-Tertiaire en Pologne. *Conférences,* No. 88, 15-30.

PRESTWICH, J. 1852. On the structure of the strata between the London Clay and the Chalk, etc., Part III, The Thanet Sands. *Q. Jl. geol. Soc. Lond.,* **8**, 235-268.

RAOULT, J-F. 1969. Rélations entre la Dorsale kabyle et les flyschs sur la transversale du Djebel Rhedir (Nord du Constantinois, Algérie): phases tangentielles éocènes, paléogéographie. *Bull. Soc. géol. Fr.* (7), **11**, 523-543.

RAYMOND, D. 1970. Formations "telliennes" et flysch littoraux: leurs rapports à l'ouest d'Azeffoun (Port Gueydon), Grande Kabylie, Algérie. *Soc. Hist. Nat. Afr. Nord., Bull.,* **61**, 46-55.

RAYNER, D. H. 1967. *The stratigraphy of the British Isles.* 453 pp. Cambridge University Press, Cambridge.

RENAULT-MISKOVSKY, J. 1967. Contribution à la Paléoclimatologie du Wurmien II en Languedoc méditerranéen (grotte de l'Hortus-Hérault) d'après l'étude des pollens, premiers résultats. *Bull. de l'AFEQ,* **4**, 305.

RENEVIER, E. 1873-4. *Tableau des terrains sédimentaires (in 4°) plus un text explicatif.* Rouge et Dubois, Lausanne, 34 pp. 9 pl.

—— 1896. Chronographe géologique. Texte explicatif suivi d'une répertoire stratigraphique polyglotte. *Bull. Soc. vaudoise Sc. nat.,* **12**.

REY, R. 1966. Malacologie continentale oligocène dans l'Ouest de l'Europe. *Rev. scient. Bourbonnais,* 53-129.

ROTH (TELEGDI), L. 1879. Geologische Skizze der Kroisbach-Ruster Bergzuges und des südlichen Teiles des Leitha-Gebirges. *Földt. Közl.,* **9**, 144.

ROTH, P. H. 1970. Oligocene calcareous nannoplankton stratigraphy. *Eclog. geol. Helv.,* **63**, 779-781.

RUSSELL, D. E. 1964. Les Mammifères paléocènes d'Europe. *Thèse,* Paris. *Mém. Mus. Nat. hist. Nat. (Sér. C), Sc. Terre,* **13**, 324 pp. 16 pl.

RUTOT, A. 1882. Résultats de nouvelles recherches dans l'Éocène supérieur de la Belgique. *Ann. soc. Malac. de Belgique Bull.,* **17**, 168-184.

SCHAUB, H. 1963. Ueber einige Entwicklungsreihen von *Nummulites* und *Assilina* und ihre stratigraphische Bedeutung. pp. 282-297 in *Evolutionary trends in Foraminifera,* 355 pp. Elsevier, Amsterdam.

260

SCHIMPER, W. P. 1874. *Traité de Paléontologie végétale*, Vol. **3**, 896 pp., J. B. Baillière & Fils, Paris.

SEGUENZA, G. 1868. La formation zancléene, ou recherches sur une nouvelle formation tertiaire. *Bull. Soc. géol. Fr.*, (2), **25**, 465.

SEGURET, M. 1970. *Étude tectonique des nappes et des séries décollées de la partie centrale du versant sud des Pyrénées*. Thèse, Montpellier.

SELLI, R. 1960. Il messiniano Mayer-Eymar 1867, proposta di un neostratotipo. *Gior. Geologia*, **28**, 1-33.

—— 1967. The Pliocene-Pleistocene boundary in Italian marine sections and its relationship to continental stratigraphies. *Progr. Oceanogr.*, **4**, 67-82.

SMILEY, T. L. 1958. The geology and dating of Sunset Crater, Flagstaff, Arizona. *N. Mex. geol. Soc.*, Guidebook, 9th Field Conference, p. 186-190.

SORGENFREI, T. & BUCH, A. 1964. Deep tests in Denmark 1935-1959. *Geol. Surv. Denmark, 3 Series*, No. 36, 146 pp.

STACHE, G. 1872. Geologische Reisenotizen aus Istrien. *Verh. k. k. Geol. Reichsanst. Jg.*, **10**, 215-221.

STEHLIN, H. G. 1909. Remarques sur les faunules de Mammifères des couches éocènes et oligocènes du Bassin du Paris. *Bull. Soc. géol. Fr.* (4), **9**, 488-520.

STEININGER, F. & SENES, J. 1968. Eggenburgien. Die Eggenburger Schichtengruppe und ihr Stratotypus-Chronostratigraphie und Neostratotypen. *Miozän der Zentralen Paratethys II*, Bratislava, 1971.

SUESS, E. 1866. Über die Bedeutung der sogenannten "brackischen Stufe" oder "cerithien-schichten". *Sitzber. Akad. Wiss. Wien.*, Bd **54**, Abt 1.

TAMBAREAU, Y. 1972. Thanétien supérieur et Ilerdien inférieur des Petites Pyrénées du Plantaurel et des Chaînons audois. *Thèse sciences nat., Toulouse*, 384 pp. 21 fig., 5 tab., 20 pl.

TAVERNIER, R. & DE HEINZELIN, J. 1962. Introduction au Néogène de la Belgique. Symposium sur la stratigraphie du Néogène nordique. *Mém. No. 6, Soc. belg. Géol. Paléont. Hydrol.*, 7-28.

TEISSEYRE, W. 1907. Stratigraphie des régions pétrolifères de la Roumanie. *Troisième Congr. Inst. du Pétrole*, Guide des excursions, p. 39, Bucarest.

TERS, M. 1961. La Vendée littorale. Étude de géomorphologie. *Thèse*. Rennes. Oberthur edit.

THALER, L. 1966. Les Rongeurs fossiles du Bas-Languedoc dans leurs rapports avec l'histoire des faunes et la stratigraphie du Tertiaire d'Europe. *Mém. Mus. Nat. Hist. Nat.*, **17**, 295.

TOURENQ, J. 1972. L'Augite, indicateur stratigraphique et paléogéographique des épandages détritiques en provenance du Massif Central au Cénozoïque. *C. r. Séances Acad. Sci. Paris*, **275-D**, 9-12.

VAN DER MEERSCH, B. 1966. Nouvelles découvertes de restes humains dans les couches Levalloiso-Mousteriennes du gisement de Qafzeh (Israel). *C. r. Séances Acad. Sci. Paris*, **262-D**, 1434-1436.

VASSEUR, G. 1881. Recherches géologiques sur les terrains de la France occidentale. *Thèse*, Paris.

VIGNEAUX, M., MAGNE, A., VEILLON, M. & MOYES, J. 1954. Aquitanien et Burdigalien. *C. r. Acad. Sci. Paris*, **239**, 818-820.

VILLATTE, J. 1962. *Étude stratigraphique et paléontologique du Montien des Petites Pyrénées et du Plantaurel*. Thèse, Toulouse. 331 pp. Privately printed. Toulouse.

VILLOT, M. 1883. Étude sur le bassin de Fuveau et sur un grand travail à y exécuter. *Annls. Mines* (8), **4**, 1-66.

VINCENT, G. & RUTOT, A. 1878a. Note sur l'absence du Système diestien aux environs de Bruxelles et sur de nouvelles observations relatives au Système laekenien. *Annls. Soc. géol. Belg.*, **5** *(Mémoires)*, 56-66.

WEBER, C. 1973. Le socle antetriasique sous la partie sud du Bassin de Paris d'après les données géophysiques. *Bull. Bur. Rech. géol. minières*, **II**, 219-343.

WHITE, H. J. O. 1921. A short account of the geology of the Isle of Wight. *Mem. geol. Surv. U.K.*, 219 pp.

WINNOCK, E. 1971. Géologie succincte du Bassin d'Aquitaine, *in Histoire structurale du Golfe de Gascogne*. Technip, Paris, pp. IV 1-1 to IV 1-30.

COLLOQUIA ON CENOZOIC STRATIGRAPHY
(post-1960)

Symposium sur la Stratigraphie du Néogène nordique (Gand, 1961). 1 vol., 248 pp., Mémoire de la Soc. Belge de Géol., No. 6.

Colloque sur le Paléogène (Bordeaux, 1962). 2 vol., 1107 pp. *Mém. Bur. Rech. géol. minières*, No. 28.

Comptes rendus de la 3e session du Comité sur la Stratigraphie du Néogène méditerranéen (Berne, 1964). Leiden (Holland) E. J. Brill, publisher.

Compte rendu du Colloque pour l'étude du Néogène nordique (Rennes, 1965), 1 vol., 136 pp. *Mém. Soc. géol. de Bretagne*, **XVI**.

The Quaternary of the United States, 7th Congress of I.N.Q.U.A. (Denver, 1965), 1 vol., 925 pp., Princeton University Press, New Jersey.

Committee on Mediterranean Neogene Stratigraphy, Proceedings of the fourth session (Bologna, 1967), 4 vol., 1600 pp. Giornale di geologia, Annali del museo geologico di Bologna, 1968.

Colloque sur l'Éocène (Paris, 1968) 3 vol., 1400 pp. *Mém.* *Bur. Rech. géol. minières*, Nos. 58, 59 & 60.

Colloquium on Eocene Stratigraphy (Budapest, 1969), 1 vol., 380 pp. Hungarian Geological Institute.

Colloquium on Neogene Stratigraphy (Budapest, 1969) 1 vol., 300 pp. Hungarian Geological Institute.

Études francaises sur le Quaternaire, 8e Congrès de l'I.N.Q.U.A. (Paris, 1969) 1 vol., 275 pp. suppl. au bull. de l'A.F.E.Q.

Études sur le Quaternaire dans le Monde, 8e Congrès de l'I.N.Q.U.A. (Paris, 1969) 2 vol., 1054 pp. suppl. au bull. de l'A.F.E.Q.

Colloque sur la Géologie de la Manche (Paris, 1971) 1 vol., 328 pp. *Mém. Bur. Rech. géol. minière*, No. 79.

Congrès du Néogène méditerranéen (Lyon, 1971) Documents Lab. géol. Univ. Lyon (Le Néogène rhodanien) & *Mém. Bur. Rech. géol. minières*, No. 78.

Le contenu de l'Ilerdian et sa place dans le Paléogène Séance specialisée du 18 novembre, 1975. *Bull. Soc. géol. Fr.* (7) 17, 123-223 (14 papers).

OTHER GENERAL PUBLICATIONS

Lexique Stratigraphique International. C.N.R.S., Paris.
Guides géologiques régionaux. (France only) Masson, Paris.
Stratotypes of Mediterranean Neogene Stages.
Vol 1, 1971, *Giornale de Geologia,* Bologna, II, 37 (G. C. Carloni, P. Marks, R. F. Rutsch & R. Selli Eds.).
Vol. 2, 1975, Veda, Slovak Academy of Sciences, Bratislava (F. F. Steininger & N. L. Nevesskaya Eds.).

A discussion on the geology of the English Channel. Phil. Trans. Roy. Soc. Lond., A279, 1-295, 1975.
A correlation of Tertiary rocks in the British Isles. Geol. Soc. Lond., Special Report No. 12, 72 pp., 1978 (D. Curry *et al.*).
Les étages francais et leurs stratotypes. Mém. Bur. Rech. géol. minieres, No. 109, 1980 (C. Cavelier & J. Roger, Eds.).

LIST OF TABLES AND TABULAR FIGURES

262

LIST OF PALAEOGRAPHIC MAPS

LIST OF SELECTED PLATES

N.B. The photographs of skeletons of Vertebrates are from the Collection of the Natural History Museum of Paris.

GEOGRAPHICAL INDEX

Figures in Roman type refer to text, figures in italics refer to figures and tables.

PALAEONTOLOGICAL INDEX

Figures in Roman type refer to the text, figures in italics refer to figures and tables.

A

Aepyornis *253*
Algae 33, *79*
Alveolina 15, *24, 31, 44*
A. corbarica *24*, 97, 128
A. cucumiformis *24*, 93, 96, 128
A. ellipsoidalis *24*, 128
A. elliptica *70*
A. elongata *24*, 69, 97, 98, 103, 122, 208
A. levis *24*
A. moussoulensis *24*, 128
A. oblonga *24*, 93, 97, 115
A. primaeva *24*, 95, 128
A. trempina *24*, 128
Amblypods *31*, 35
Ammonites 12, *14, 22, 31*
Amphicyon 95, 143, 155
Ampullina *67, 79*
A. crassatina 78, *86*
Anchitherium 156
Ancilla glandiformis *174, 176*, 194
Anomia 93, 94
A. tenuistriata *70*
Anoplotherium 38, 76
Anthracotherium 38, 95, 156, 162
Arca *79*, 155
Archanthropoids 242
Archiacina armorica 77, 103, 209
Arctica islandica *141*, 224, *225*, 238
A. morrisi 61, 115, 119
A. scutellaria 61, 115
Arctocyon 23, 34, 62, 214
Arsinoitherium 38, *39*
Artiodactyls *22*, 23, *31*, 36, 38
Assilina 15, 22, 23, *42*
A. exponens 43, 136, 188
A. spira *41*
Astarte 121
Athleta spinosus *71*
Aturia aturi *142*, 170, 176
Aurochs 234
Australopithecus *240*, *243*, *244*
Avicula defrancei 30, *79*

B

Baculites anceps 207
Baluchitherium *148*
Bats 23
Bayania lactea *71*
Belemnites 12, *14, 22, 31*
Belosepia blainvillei 33, *79*
Beryx lerichei
Betulaceae 56
Birds, *31*, 33, *151*
Bison 234, *252*
Bos 234
Brachiopods *14*
Brontotherium 40
Brotia melanioides 65
Bryozoans *33*
Bulimus cylindricus 105

C

Carcharodon 33, *34*
Cardita bazini *86*
C. jouanneti *139, 141*, 161, 170, 174
C. pectuncularis 61
C. planicosta *70*
C. scaldensis 165
Cardium 160, 165, 176, 200
C. stampense *86*
Carnivores *22*, 23, *31*, 35
Cassidaria nodosa *71*
Cephalopods 30
Ceratoichthys pinnatiformis *131*
Cerithiidae 30, *32*

Cerithium diaboli 105, 131
C. giganteum 69
C. lamellosum *71*
C. tricarinatum *79*
C. tuberculosum *79*
Cetaceans *31*, 35
Chama *14, 31, 70*
Chancelade Man 245
Chapelle-aux-Saints Man 243
Chara 27, *74, 76, 84*
Charophytes *27, 74, 76, 84*
Chlamys (see Pecten) *141, 174*
Clavilithes longaevus *79*
Clypeaster 102, 131, 132, *142*, 161, 174, 184
Coccoliths *50, 51*
Condylarths *22, 31*, 34
Congeria *142, 175,* 176
Corals 33, *70, 79*
Corbicula cuneiformis 65
C. fluminalis 210
Corbis lamellosa *70*
Corbula gallica *70*
Coryphodon 37, 62, 63
Crassatella bellovacensis 63
Crassostrea longirostris 29, 77, 103
Creodonts *31*, 35
Cro-Magnon Man 245
Cucullea crassatina 61, *63,* 119
Cuvillierina eocenica 103, 207
Cycad *14, 31*
Cyclococcolithus leptoporus
Cyclonephelium
Cyrena convexa 28, 76
C. gravesi 67

D

Deer 232, 252
Deinotherium giganteum *142, 144*, 155
Diatoms *150, 254*
Diatryma 34, *151*
Dichobune 38
Dicotyledons *14, 31*
Didacna praetrigonoides *175*
Didus ineptus *232, 233*
Dinoceras 37
Dinoflagellates 54, *55*
Dinornis 232
Discoaster *14, 50, 51, 140*
D. gemmeus *22*
D. multiradiatus *22, 24*, 130
Discocyclina 15, 31, 45, 96, 128
D. seunesi
Discorbis vesicularis 206
Discorinopsis kerfornei 66, 69, 97, 103, 122
Ditrupa strangulata 69, 116
Dodo 232
Dreissenidae 176
Dreissenomya aperta *175*

E

Echinoderms 33, 70
Echinolampas 33, 94, 142
Edentates 35
Elephas 144
E. meridionalis *230, 233*
E. primigenius *233, 235*
Entelodon magnus 76, 95
Eohippus 38, 214
Equus 144, *146*, 234
Ericsonia subdisticha *24*, 25
Ervilia podolica 170, 176
Eulepidina 132, 189, 190
Eupsammia trochiformis 33, 67
Eusmilius *148*
Exelia velifer 132

F

Fabiana 188
F. cassis 97, 122, 208
Fasciculithus tympaniformis *22*
Fontechevade Man 244
Foraminiferids *22, 24, 27,* 40, *47, 48, 87, 140*

G

Gastornis 62
Gastropods *14*
Globigerapsis semiinvoluta 24
Globigerina 14, *48*, 197, 204
G. bulloides *48*, 209, 219
G. daubjergensis *22*, 95, 123, 128, 134
G. eugubina *22*
G. nepenthes *138*, 139, 170, 176, 198
G. pachyderma 165, 168, 219, 224, 238
Globigerinoides 29, 102, 137, *138*, 160, 170, 172, 191, 193, 194
Globorotalia *14, 31*
G. aequa *24*
G. angulata *22*, 47
G. aquiensis 115
G. aragonensis *24*, 97
G. cerroazulensis *24*
G. crassaformis *48*, 140, 144
G. crassula 172
G. danica *22*, 123
G. eugubina 95, 123
G. hirsuta 194
G. inflata 140, 172
G. margaritae 137, *138, 139, 140*, 170, 172
G. menardii *138, 139, 140*, 170, 194
G. pseudobulloides 13, *14, 22*, 132, 207
G. pseudomenardii *22, 24*, 196
G. pusilla pusilla *22*
G. subbotinae *24*, 97, 136
G. trinidadensis *22*, 47
G. truncatulinoides *141*, 224
G. uncinata *22*, 196
G. velascoensis *22, 24*, 47, 128, 130, 196
Globotruncana *14*, 22, *31*, 106, 124, 128, 196
Globotruncanella mayaroensis *22*
Glycymeris angusticostata 78
G. obovata *86*
G. pilosa 165
G. terebratularis 61
Glyptodon *218*, 219
Grasses *14*
Grimaldi Man 245
Gyrogona medicagulina 76

H

Hantkenina aragonensis 24, 132
Heliolithus *22, 51*
Helix ramondi 30, 110, 112, 161
Hercoglossa 30
Hipparion 143, *162*, 169, 194
H. gracile *147*
Hippopotamus 230
Hominids *14, 31, 240*
Homo erectus 242
H. habilis *241*, 242
H. heidelbergensis 242
H. neanderthalensis *243*
H. sapiens 243
Hyaenodon *39*, 95
Hyalinea balthica *141, 184*, 224, 238
Hyracotherium 37, 63, 214

I

Indricotherium *148*
Inoceramus *14*

STRATIGRAPHIC INDEX
Stages, Formations, Tectonic Phases

*Reference should also be made to the Palaeontological and Geographical Indexes for horizons designated by fossil- or place-names.
Figures in Roman type refer to the text; figures in italics refer to figures and tables; figures in bold type refer to definitions.*

A

Abbevillian *226*, **245**
Acheulian *226*, **245**
Almian 136
Alpine *13*, 18
Andalusian 139, *140, 164,* **170**
Antwerp Sands 165
Anversian *164, 165, 166*
Aquitanian 29, 40, 88, *93, 95,* 110, 138, **160**, *164, 190*
Argile à lignites *58,* 64
Argile de Laon *58,* 66
Argile plastique *58,* 63
Argile verte de Romainville *58,* 76
Arikareean *27,* 138
Armagnac Molasse *93,* 161
Arvinchean 13, 14, 105
Asschian **116**
Assian **116**
Astian *140, 172, 226*
Attic 13
Aurignacian *225, 246,* **247**
Auversian *59,* **72**, *75, 77,* 130
Azilian *246,* 251

B

Badenian *164,* **176**
Bagshot Beds 121
Bakhchisaraian **136**
Barstovian **138**
Barton Beds *114,* 122
Bartonian *24, 26, 59,* **69**, *93,* 94, 100, *102, 103,* 109
Begudian *96*
Bembridge Beds 25, *123*
Biarritzian *24, 26, 67, 69,* 94, **98**, 103
Blackheath Beds 121
Bodrakian *135,* **136**
Bolderian *164, 166*
Boom Clay *118*
Bormidian 132
Bracklesham Beds 114, 121
Bridgerian *24,* 214
Brockenhurst Beds 25, 124
Brunhes normal *226,* 237
Bruxellian 68, **115**
Bulitian *24*
Burdigalian 89, 93, 138, 153, **160**, *164*

C

Caillasses *58,* 69
Caillasses d'Orgemont *58*
Calabrian *140,* **223**, *226,* **238**
Calcaire à Anomies *93,* 94
Calcaire à Astéries *93, 96*
Calcaire de Beauce *58,* 78
Calcaire de Blaye 94
Calcaire de Brie *58,* 77
Calcaire de Couquèques 94

Calcaire d'Etampes 78
Calcaire de St. Palais 94
Calcaire grossier *58,* 66
Calcaire grossier de Mons 115
Calcaire pisolitique 58
Cambrian *11*
Campanian *59*
Carboniferous *11*
Carpathian *164,* 176
Chadronian *24,* 27
Chattian *27, 29, 93, 95,* 110, **126**, 132, 161
Chellean *226,* **245**
Clactonian **245**
Claiborne Formation 215
Clarendonian *138*
Clarkforkian *22*
Conglomérat de Cernay 62
Conglomérat de Meudon 63
Coralline Crag 168
Cretaceous *11, 13, 22*
Cromer Forest Bed 168, 230
Cromerian **168**, *226,* 230
Cuisian 23, *24,* 26, *59,* **65**, *93,* 97, 103
Cyrena Marls 110

D

Dacian *164,* **176**
Danian 13, *22, 28,* 89, **123**
Dano-Montian 12, *22,* 23, **57**, *59, 60, 61,* 95, *124*
Delmontian *138*
Den Hoorn formation 115
Deurnian *164,* 165
Devonian *11*
Diestian *164,* 165
Donau *225*
Donzacq Marls *96,* 97
Dragonian *22*
Duchesnian *24,* 27
Dunkerquian **240**

E

Eemian *226,* **239**
Egerian *164,* **174**
Eggenburgian *164,* **174**
Emilian *238*
Eocene *11,* 12, *13,* 21, *22, 24,* **25**, 54, 63, 66, *74, 134, 183, 187, 189,* 197
Epi-Villafranchian *226*

F

Falunian 164
Fish shales 110
Flandrian *226,* **239**
Formazione gessososolfifera 170

G

Garumnian 95, 128

Gauss normal *226*
Gilbert *226*
Girondian **138**, 160
Glaises à Cyrènes *58*
Glauconie grossière 66, 67
Gompholite 108
Grès de Belleu *58*
Grimaldian *226,* **239**
Grimmertingen Sands 116, *118*
Günz **225**, *226*
Gypse *58,* 74, 80

H

Hamstead Beds 122
Headon Beds 25, 122
Heersian 115
Helminthoid Flysch 106, 108
Helvetian *93,* 139, 153, *159,* **162**, *164*
Helvetic 13
Hemmoorer Stufe 167
Hemphillian *138*
Hemingfordian *138*
Holocene *11,* 12, *13,* 20, **223**, *246*
Houthalenian *164,* 165

I

Icenian **168**, 230
Ilerdian *22, 24,* 26, 56, 65, 90, *93,* **128**
Illyric 13
Inkermanian *134*

J

Jackson Formation 216
Jurassic *11*

K

Kachian *134*
Kallo Sands 165
Kansas *226*
Karpatian *164*
Kattendijk Sands 165

L

Landenian *22,* 61, **115**
Langhian 139, *164,* **170**
Laramide *13,* 14, 213, 219
Lattorfian 25, *26,* 110, *114,* **124**
Ledian 68, 69, **116**, *118*
Lehrte greensands 124
Leitha Limestone 174
Levalloisian **245**, *246*
Liburnische Stufe 129
Lignites du Soissonnais 64
Londinian 26
London Clay 121